Computational Fluid Dynamics: Concepts and Applications

Computational Fluid Dynamics: Concepts and Applications

Edited by **Maria Forest**

NY RESEARCH
PRESS

New York

Published by NY Research Press,
23 West, 55th Street, Suite 816,
New York, NY 10019, USA
www.nyresearchpress.com

Computational Fluid Dynamics: Concepts and Applications
Edited by Maria Forest

International Standard Book Number: 978-1-63238-088-3 (Hardback)

Printed in China.

Contents

Preface

This book discusses essential elements of computational fluid dynamics like cooling system design, flow modeling in sedimentation, hypersonic aerodynamics, air conditioning system, greenhouse environment, photochemical reaction engineering, hydraulic component design, fluid-structure interaction and atmospheric reentry problem. It will be accessible to advanced scientists and research engineers who are attempting to sharpen their computational fluid dynamics skills.

This book is a result of research of several months to collate the most relevant data in the field.

When I was approached with the idea of this book and the proposal to edit it, I was overwhelmed. It gave me an opportunity to reach out to all those who share a common interest with me in this field. I had 3 main parameters for editing this text:

1. Accuracy - The data and information provided in this book should be up-to-date and valuable to the readers.
2. Structure - The data must be presented in a structured format for easy understanding and better grasping of the readers.
3. Universal Approach - This book not only targets students but also experts and innovators in the field, thus my aim was to present topics which are of use to all.

Thus, it took me a couple of months to finish the editing of this book.

I would like to make a special mention of my publisher who considered me worthy of this opportunity and also supported me throughout the editing process. I would also like to thank the editing team at the back-end who extended their help whenever required.

Editor

1

Computational Fluid Dynamics

Victor Udoewa[1] and Vinod Kumar[2]
[1]*George Washington University,*
USAID Development Engineer,
AAAS Science & Technology Policy Fellow, AAAS, 2009-2011,
[2]*Mechanical Engineering,*
University of Texas at El Paso,
USA

1. Introduction

Computational Fluid Dynamics (CFD) is the emerging field of fluid mechanics in which fluid flow problems are solved and analyzed using computational methods and numerical algorithms. In fluid mechanics, there are generally three routes of work in the field, three ways to conduct experiments. The first category is theoretical, or analytical, fluid mechanics. Theoretical fluid mechanics includes theorizing, manipulating and solving equations with pen and paper. The Navier-Stokes equation governing incompressible fluid flow is an example of theoretical fluid mechanics. Secondly, many engineers and physicist work in the area of experimental fluid mechanics. Experimental fluid mechanics involves conducting actual physical experiments and studying the flow and the effect of various disturbances, shapes, and stimuli on the flow. Examples include waves generated by pools, air flow studies in actual wind tunnels, flow through physical pipes, etc. Lastly, a growing number of engineers, mathematicians, computer scientists, and physicists work in the area of computational fluid dynamics (CFD). In CFD, you may still run an experiment of waves across water, an airplane in a wind tunnel, or flow through pipes, but now it is done through the computer Instead of actual, physical, 3D objects. A computer model is created, and computer programmers code the equations representing the physical laws that govern the flow of the molecules of fluid. Then the flow results (such as velocity and pressure) are output into files that can be visualized through pictures or animation so that you see the result just as you do with physical experiments.

In cases where an analytical, or theoretical, solution exists, CFD simulations and the mathematical models, which are coded in the computer program, are corroborated by comparison to the exact solutions. This comparative check is called validation. CFD is not yet to a point where solutions to problems are used without corroboration by existing, known, analytical or exact solutions when available. Validation is not to be confused with verification, however; validation is a check to make sure that the implemented, coded model accurately represents the conceptual, mathematical description and the solution intended to be modeled.

Still, there are many times when there is no analytical solution. In these cases, one often uses a computational approach. In such cases without a known solution, CFD is used to approximate a solution. Most often, CFD is used when a computational solution is faster, cheaper, or more convenient. Convenience may be due to time or safety or another reason.

If we wanted to create a database of information about 3D flow around a cylinder for different types of fluids at varying speeds, CFD is quite helpful. It would take quite a long time to change the fluid in our flow container and clean the container for every type of fluid we desired. It would also take some time to change the speed of the flow. In this case, it's much faster to simulate it computationally. Then, anytime we wanted to change the speed of the flow, we simply would change one number in a computer input file. Or if we wanted to change viscosity and density for the fluid (representing a different fluid) we would just change the corresponding values in a computer input file. In this case, CFD is faster.

Now imagine if you were doing space defense work for a government, and you were asked to do fluid dynamics simulation of the combustion dynamics during an explosion or when a space shuttle launched (1). It would take immeasurable amounts of money to do test launches over and over as you studied the combustion dynamics of space shuttle thruster ignition. And it would take large amounts of money to test explosive devices, especially considering the damage they cause. In these cases, CFD is, again, quite helpful. The only costs in CFD are the time of creating a computer model, choosing the right mathematical model, coding it, and the power and computer time required to solve the equations. But it is plain to see that CFD is cheaper.

What if you were hired to help design material for the outfits of swimming athletes? Your company gave you the job of studying sharks and their agile ability to swim and maneuver through the water. So you start by trying to study the fluid dynamics around the shark skin (2). How convenient is it to locate sharks and place them in some type of testing container where you have probes and measuring devices located? How convenient is it to place probes on the body of the shark itself? How safe is it to work with the sharks in that manner? No, it's better to create a computer model of a shark and get the information for the shape, design, feel, and density of its skin and to use this information to run simulations. It is clear that CFD is more convenient in this situation. Sometimes a CFD simulation can be all three—safer, cheaper, and more convenient.

Imagine a situation in which two paratroopers, jumping from both side doors of a military cargo aircraft, always crash into each other down below (3). In order to analyze the fluid dynamics of the problem to see what air flow forces are affecting the paratrooper paths, you would need to perform test jumps with paratroopers. However, that is potentially injurious and not safe. You would also have to rent the plane, pay for the rental by hour, hire the test pilot, and pay for all the equipment for the jump. That is expensive. Lastly, the organization of the use of the military aircraft and personnel and equipment takes many months, and it can take from 6 to 12 months to plan the test. In this case a CFD experiment is more convenient: faster, cheaper, and safer.

Usually solving CFD problems involves three stages. First there is the pre-processing stage. In this stage, the geometric boundaries of the problem are defined. In 3D, a volume is created (in 2D, an area) over which the equation will be solved. This volume is broken into smaller units or cells creating a mesh (though there are meshless methods for computing CFD problems). This may be uniform or non-uniform. Along with constitutive equations, the particular equations are chosen for the problem in order to properly physically model the flow. These equations may be manipulated depending on the mathematical method being used. Boundary conditions are prescribed along the boundary. For time dependent problems, initial conditions are prescribed.

Secondly, the problem is solved numerically. At the least, we usually solve for velocity and pressure, but the list of unknowns can be longer depending on the mathematical model (equation) chosen for the physical situation. Other unknowns may include temperature, energy, and density. The numerical solution is usually computed iteratively for steady-state solutions. For time-dependent problems, a step in time is taken, and the equation is numerically solved again, eventually producing a solution for every time step.

The final step in the CFD process is post-processing. In this stage, the solution is analyzed usually with the help of visualization and possibly animation for dynamic, or time-dependent, problems. It is in this stage that CFD results are usually compared to any previous experimental results or known analytical, or theoretical, solutions. This comparison is usually called validation. Today, confidence in CFD is growing, but we have not arrived at the point of trusting CFD solutions without validation. Even if a particular model is validated, we still corroborate the results of a simulation with experimental or analytical results.

Because of the hybrid nature of CFD, advances in CFD are usually made in three areas: computational and applied mathematics, mechanical/chemical engineering, and computer science/electrical engineering. Some researchers work on new theoretical, mathematical models creating new discretization methods (ways to discretize the problem in order to numerically solve it over the discrete units or cells), or turbulence models. They might publish in applied mathematics or computational mathematics journals. Others work on computer architecture (such as different types of supercomputers or computer clusters), coding techniques such as parallel programming, or speeding up the computational processes through faster mesh generation and mesh reordering. They might publish their results in electrical engineering or computer science journals. Lastly others might use CFD to concentrate on new insights in the engineering aspect of the problem such as the mechanics of bird flight or sharkskin-inspired speedo design for less water resistance, or resistance to blood flow inherent to certain veins. They might publish their results in engineering journals next to experimental or theoretical engineering results. Their focus is on the application more so than the math or computer science.

2. Pre-processing

Pre-processing refers to the work that must be done prior to the actual computational experiment or simulation. This work can be reduced to four general areas: geometry definition, volume division, model choice and definition, and boundary condition definition. For the purposes of this article, we will ascribe the work of coding and the choice of computational implementation to the processing stage called simulation. Sometimes researchers refer to mesh generation as pre-processing in general because a mesh is generated when the volume is divided.

2.1 Geometry definition

The first step is to define the computational domain of the problem. The purpose of this definition is to confine the problem to a finite space and limit the computation. It is true that a plane flying in the air has some residual effects on air flow patterns 1,000 miles away, but because of the negligible nature of those effects, we are relatively safe in looking at the effect of the plane on air flow within a reasonable vicinity of the plane thereby limiting our

computational work and making the problem finite. It would be quite a task to compute the effect of the airplane on the air flow at a certain height around the entire earth.
To limit the flow in this way usually requires defining a domain inside which we will compute the flow, outside of which we will not compute the flow. Usually the geometry of the domain is chosen to be a box of some sort, usually a rectangular prism in 3D or a rectangle in 2D. However, any closed shape may be chosen as long as the shape closes off an inside computational domain from an outside space in which computations will not take place.
A closed space does not imply, however, an empty domain box. For example imagine that we are simulating intravenous blood flow (4) (5) around a cancerous growth. The vascular domain is modeled by a 3D cylindrical prism (our domain box), but we still have an object inside. In this case, our object is a semi-spherical cancerous growth on the surface of a wall of the vein. Traditionally, our domain must be totally closed, so the surface of the domain goes from the wall of the vein, joins the surface of the cancerous growth, and continues on the other side rejoining the wall of the vein creating a closed 3D space that does not go under the tumor but continues over the surface of it. Likewise, if we were calculating flow around a sphere (6), the domain box would be the outer half of the domain surface. The inner half of the surface would be the sphere inside the domain box. Just as in the cancer example, we are not calculating the flow inside the sphere, just as we were not calculating the blood flow in the tumor. But the tumor and sphere form part of the boundaries of the domain helping to close off the computational space in which we are interested in the velocity and pressure of the fluid. Remember, domain boxes may contain objects inside which no flow is calculated, but whose surface forms part of the surface of the domain helping to limit the computational space and better define where the fluid flows.

2.2 Volume division

The second step is volume division or mesh generation. Why we must divide the volume is not obvious until one remembers that it is easier to solve a flow problem over a smaller area or volume than a larger one. So dividing the volume into smaller units transforms the large problem over the entire domain into a large number of smaller problems over smaller sub-domains. However, the real reason we divide the volume is because we seek to find, for example, the velocity and pressure of the fluid at various points throughout the domain volume or area. In order to do this, we fill the inner domain volume or area with nodes—points at which we will calculate, in this instance, velocity and pressure. Once we have filled the inner volume or area with nodes, we connect the nodes with edges (and sides in 3D) creating smaller sub-volume or sub-area elements. For example, if our computational box is a rectangular prism and we fill it with nodes, we can connect the nodes to create quadrilaterals or tetrahedrals. If the domain is a rectangle and we fill the rectangle with nodes, we can connect all the nodes to create small rectangles or triangles. This network of rectangles/triangles and quadrilaterals/tetrahedrals creates a mesh of nodes; a mesh has been generated.
Remembering that the purpose of volume or area division is to create more manageable sub-volumes or sub-areas, it behooves us to evenly space out the distribution of nodes. If we do not, we may find that there are large spaces (volumes or areas) with no nodes. This is problematic because it means some of the sub-volume or sub-area elements will still be large; though our goal is to make them small.

Our second goal is to make each sub-element evenly shaped. In 2D, evenness in both directions means that 2D rectangles tend towards squares or that our triangles are equilateral. In 3D, if evenness is desired in all 3 directions, our quadrilaterals tend toward cubes and the tetrahedrals tend to be equilateral. Oblong and unevenly shaped sub-elements also create spaces (areas or volumes) with fewer nodes than parts of the domain with evenly shaped sub-elements. In these cases, our second goal serves a similar purpose as our first goal: to divide the domain into smaller, more manageable sub-domains with evenly distributed nodes.

An important aspect of mesh generation is choosing the appropriate size, or refinement, of the mesh sub-elements, such that important aspects of the flow are properly resolved. The general rule is that no fluid particle should advance through multiple sub-elements, cells, or units in one time step. Therefore our third goal is to increase the refinement (or the number of nodes) in areas or sub-volumes of increased fluid velocity or vorticity or any interesting fluid flow phenomena that you would like to capture computationally. This will allow us to visualize it later.

So far, the realm of geometry definition and mesh generation fall into the computer science side of CFD. When generating the mesh by defining a geometry and dividing that geometric area or volume, one must decide if one will use uniform sub-elements or non-uniform sub-elements. Definitely in parts of the domain with special flow requiring increased refinement, the elements in those parts will not match the refinement of elements elsewhere. But in the general flow one can still choose a uniform, structured mesh or a non-uniform, unstructured mesh. The ability to create non-uniform meshes is important in CFD because of the physical nature of fluids to occupy and fill any void left unoccupied. When dealing with complex geometries and small nooks, crannies, and crevices of an automobile or a model of a city block, it helps to have unstructured, non-uniform meshes that allow for the modeler to create the best shapes to fit the 2D or 3D space (7).

Likewise, the division of the volume or area in mesh generation requires the modeler to choose between quadrilaterals and rectangular prisms or triangles and tetrahedrals (there are other choices of shapes, as well, such as wedges and pyramids in 3D). Generally, quadrilaterals and rectangular prisms have a more accurate solution than triangles and triangular prisms, but there are ways to increase the accuracy of the latter. Because of the non-uniform and sharp geometries found in fluid problems by nature of fluids, triangles and triangular prisms work better geometrically for CFD applications. Normally, CFD researchers will utilize triangles and triangular prisms and then increase the number of interpolation points inside these elements so that no accuracy is lost. Interpolation points are points inside an element at which the solution is calculated. From these interpolation points, we can approximate the solution at any location inside an element.

All of the choices in dividing the volume and discretizing the mesh have the potential to introduce errors. Such errors, due to bad distribution of nodes or parts of the domain where the refinement is too low, are called discretization errors. These are errors that would disappear if we appropriately divided the volume or area or appropriately discretized the mesh.

2.3 Physical model definition

The first two steps dealt with the computer science side of CFD and there are many CFD engineers who work on geometric mesh discretization and mesh partitioning methods. Step three deals with the computational and applied mathematical side of CFD—choosing the appropriate mathematical model. When preparing to model a certain fluid flow situation, one must decide which equation accurately describes the fluid flow one wishes to simulate. If no equation currently exists, the CFD engineer must do work in the theoretical side of CFD and formulate a new equation or a more specific equation for his or her specific fluid flow situation. If equations do exist, the CFD engineer must simply choose the correct equation for the fluid flow. This is not a trivial step as sometimes the same situation may require different mathematical models at different velocity regimes or different temperature regimes, for instance. So the specific parameters of the flow must be looked at in detail—velocity, viscosity, density, pressure, etc.—so that the correct equation is chosen.

A good example from fluid mechanics is the Navier-Stokes equation which is the basic or fundamental equation for fluid dynamics. If you remove viscosity from the equation, the Navier-Stokes equations become the Euler equations. Since all fluids have some amount of viscosity this approximation is important in flows in which the viscosity is negligible (8) such as sonic flows. A plane flying at sonic speeds will have air sliding past it, relatively, as if it had no viscosity. So the use of the Navier-Stokes equations also depends on the velocity of the flow, or more accurately the Reynolds number which governs the ratio of the kinematic forces to the inertial forces. You can still go further: if you remove vorticity from the Euler equations, you arrive at the full potential equations. The point in this illustration is that choosing the correct mathematical model is important, sometimes difficult, and always specific to the flow situation.

The choice of the mathematical model affects the unknown values you will compute. Some CFD simulations are really computational fluidothermodynamics because temperature and energy are calculated as well (9). For compressible flows, density is an unknown value and we would seek to solve for this value of density in the simulation. So the choice of mathematical modeling affects what unknowns we will compute. More accurately, the unknowns we want to compute in a given situation (along with other details about the flow situation) may help positively affect our choice of a mathematical model or the need to formulate a new one.

Remember that errors can be introduced at this step as well. If a inappropriate or poorly approximating governing equation is chosen, this affects the final solution. If a governing solution is chosen or formulated for which no analytical solution or experimental solution exists, we lose the opportunity for validation to reduce errors. Any simplification in the model or any untrue assumptions the mathematical model uses introduces errors as well. All of these types of errors can be classified as physical approximation errors because they deal with the physical, mathematical model (not the geometric model).

2.4 Boundary condition definition

After a mathematical model is chosen to model the physical phenomenon, usually boundary conditions must be chosen. This is the fourth step of the pre-processing stage, and this step falls on the applied math side of CFD. Usually we deal with boundary-value problems which require values to be assigned along the boundary of the domain of the problem in order to solve the problem throughout the 2D or 3D space.

For instance, in some problems, one may specify the value of the unknown on the boundary. Imagine prescribing the value of the velocity of the fluid on the boundary. Such a boundary condition is called a Dirichlet or a direct boundary condition because you are setting the

value of the unknown. If one specifies the value of a derivative of the unknown, the boundary condition is called a Neumann or natural boundary condition.
For example in hydrodynamic flow around a submerged rock with moss, a CFD researcher would usually place the boundary condition of free-stream velocity on the entrance side and exit side of the domain box assuming that the entrance side is sufficiently upstream from the rock so as to still be undisturbed, and the exit side is sufficiently downstream that the flow conditions have returned to free-stream conditions. The prescription of free-stream velocity would be Dirichlet or direct boundary conditions. The same researcher might assume stress (a derivative of velocity) to be zero in the direction perpendicular to the side surfaces of the domain box. When she prescribes stress in that direction she is setting Neumann or natural boundary conditions in the direction perpendicular to the side surfaces of the domain box.
Such boundaries are spatial boundaries. For time-dependent problems, there are temporal boundaries in a sense. Time dependent problems require an initial condition, prescribed values for the unknowns set at the temporal start of the simulation. In the same hydrodynamic example, let us say we want to simulate the flow when a rock, half the size of the stationary, mossy rock, was thrown into the river passing next to the mossy rock and hitting the riverbed. To start the simulation we need to have the steady flow of the river around the mossy, stationary rock without the 2nd rock thrown in. Once we have computed this flow, we can use the values of velocity and pressure from this flow as initial conditions for a simulation of a moving 2nd rock that is falling to the bottom of the river. From there the simulation will march in time and use the flow results from the previous time step as initial conditions for the next time step.

3. Simulation

The second phase of CFD work is the actual simulation or the "processing" work once the pre-processing work is completed. However, there are still some pre-simulation decisions to be made. CFD work is done through computers which not only decreases the time it takes to perform calculations, but also increases the amount of calculations that can be done in a given time period. As computing power has increased over the years, CFD has been used to solve larger and larger problems.
Large problems, however, were traditionally reserved for supercomputers. Supercomputers are large computers made up of multiple computers or CPUs. A desktop or laptop computer could only handle so many calculations due to hard drive limitations on different types of computer memory. As computers in general become more advanced, not only has the memory capacity of supercomputers increased, but so has the memory of desktops and laptops increased. This has created a cycle where problems solved by supercomputers today are solved by desktop computers and laptop computers tomorrow. And the problems solved by desktop and laptop computers today were only solved by supercomputers yesterday. For example, historically, a simulation of flow past an automobile was done on supercomputers (10) (11). Today one can create a model of an automobile and run a flow simulation of air flow past the automobile with one desktop or laptop machine. This example is one of many indicative of this ever-improving cycle.
Besides memory the other limitation on computing ability in today's world is clock speed. CFD workers are dependent upon computer scientist researchers to continue to increase the clock speed of microprocessors. In general, the faster computers become, the faster is the

speed CFD scientists and engineers can compute solutions to problems. And as the speed of computations increases, the time to do computations decreases, and CFD scientists and engineers can compute larger problems (as long as they have the memory capacity for the calculations and storage for the solution). Currently, as this article goes to press, the fastest machine in the world is the K computer which computes at 10 Petaflops (12). Flops are floating point operations per second, and the prefix peta means 10^15. Therefore, the K computer can compute 10^16 floating point operations per second.

Supercomputers come in varying shapes and sizes. A small supercomputer may have 36-100 CPUs. A larger supercomputer could have 50,000 different nodes, or CPUs. For example, the K computer has 68,544 CPUs, each with 8 cores (octo-core) for a total of 548,352 cores (12). As well, today we have small supercomputing clusters, in which different CPUs are linked together to act as a supercomputer. Often one will find Linux clusters arranged in this way. Each node of a cluster can actually contain multiple processors itself acting as a single computer. A computer or node with 2 processors is called a dual core machine or node. A computer or node with 4 processors is called a quad-core machine or node. Processors in a multicore machine or cluster can use memory in different ways. Some use a shared memory architecture. In this case, all processors can access all the memory because it is completely shared between all processors. Some may have distributed memory where each processor or node has access only to its own memory. Finally there are hybrid machines like the K machine. Each of the 68,544 CPUs has its own distributed memory. But inside each CPUs memory is a system of 8 cores that share memory.

Historically supercomputers used vector-based architecture, but this created a niche market since codes for such machines could not simply be run on non-vector based machines like a desktop computer. Today laptop computers are very similar to supercomputers because many supercomputers use bus-based architecture which is a modified architecture allowing a desktop computer to run more than one processor like a quad core Linux machine (13).

Because most CFD work is done on clusters or supercomputers, CFD programmers often learn parallel programming, a type of computer programming with instructions or directives for communication and transmission of information between processors/cores or nodes on a supercomputer or cluster. Parallel programming is especially important because the purpose of supercomputing is to divide the large problem into smaller pieces given to each node to process. However, in order to solve the larger problem, the nodes must communicate especially and specifically about border regions of the partitioned mesh.

Once the mesh unit and resolution are chosen in the pre-processing stage, the mesh is partitioned and a piece of the mesh is given to each processor or node. However, sometimes researchers will re-order the mesh for large problems. Inefficient mesh partitioning contributes increasing costs of calculations and time for larger and larger problems. To facilitate calculations, mesh re-ordering schemes seek to minimize communication. Communication is minimized most when each processor or node manages a contiguous portion of the mesh. In this case, each processor only shares geometric nodes on the borders of mesh portions with processors that work on neighboring portions of the mesh. Imagine the opposite situation in which mesh elements are randomly distributed. A processor might have to communicate with 8 other processors if 8 of the bordering elements lie on 8 distinct processors! In CFD work, communication can take more than 50% of the computational time depending on the specific problem and its size. So it is very important to minimize this as much as possible, to leave more computational work for actual mathematical computations.

Examples of mesh partitioning and re-ordering methods include MATLAB's MESHPART (14), METIS (15), and PARMETIS (16) (17). Ordering the mesh so that each CPU has access to cells or units that are connected to each other is important. We also re-order and partition to maintain proper load balance so that no CPU has more work than any other.
All of these choices—mesh unit geometry, mesh resolution, partitioning, and order can affect the ability of each processor to solve the resulting algebraic system of equations. Therefore, researchers work on the parallelism of such computer codes. There are many parallel programming language directives such as OpenMP (18), and MPI (19) and languages such as Manticore (20) and NESL (21).
After the mesh partitioning and re-ordering scheme is chosen, the next step a CFD engineer takes is choosing a discretization method. The mesh partitioning lies mostly in the realm of computer science, but this next step of discretization lies in the area of computational and applied mathematics. The mathematical model and governing equation has been chosen, but the CFD scientists or engineer must choose how to discretize the solution of this model over the entire domain.

3.1 Numerical discretization methods

In computational and applied mathematics, there are different numerical discretization methods (22). Three popular methods are finite difference, finite volume, and finite volume methods. Each of these classes of methods contains many variations usually specific to an application area. Other methods include boundary element methods, higher-resolution methods, and meshless methods like spectral methods.

3.2 Finite difference, finite volume, and finite element methods

The finite difference method is probably the oldest of the main three (23). It lends itself quite well to orderly and structured geometries. It is not used as commonly as the finite element method or finite volume method, probably due to the geometric limitations on applications. Still, there are modern finite difference codes that employ overlapping grids and embedded boundaries allowing the use of the finite difference method for difficult or irregular geometries. However, it is the easiest method to code and is often taught first in courses that teach numerical discretization methods.
The finite volume method is a method in which the governing partial differential equation is solved over smaller finite control volumes (24). Since the governing equations are cast in a conservative manner over each control volume, the fluxes across the volumes are conserved
In terms of tests, applications, validation, and literature, the most robust of all methods is the finite element method (25). The finite element method is a type of residual method in which a residual equation is weighted and integrated over the domain. Since the domain is broken into many elements, this integration actually takes places over each element in the mesh. The finite element method requires more memory than the finite volume method but is also more stable than the finite volume method.
There are other methods as well. Boundary elements methods include methods in which the boundary is meshed into separate sub-elements (26) (27). In 3D, a boundary element method domain would be a 2D surface. In 2D, a boundary element method domain would be represented as a 1D surface or edge. There are also immersed boundary methods to deal with situation in which elastic structures interact with fluid flows (28).
There are numerous other discretization methods. Each of the above discretization methods can be used with functions of varying order. Every increase in the order of the functions

used in the discretization method (from linear to quadratic, quadratic to cubic, from cubic to quartic, etc.) carries increased computational costs since higher order functions have more terms and more coefficients. In fact, higher order functions require more sample points or interpolation points to properly resolve them. Accuracy comes at a price. Still, such accuracy is sometimes warranted in cases where there are sharp gradients or shocks in velocity, pressure, density, or temperature, for instance. In these cases, some of the previous discretization schemes fail and introduce what we call "spurious oscillations" which are not actually physical but a byproduct of computational approximation. In these cases, CFD engineers may choose to use higher-order discretization methods or shock capturing methods such as Total Variation Diminishing (TVD) schemes (29), Essentially Non-Oscillating (ENO) schemes (30), and the Piecewise Parabolic Method (PPM) (31).

Spurious oscillations are an example of discretization errors mentioned earlier. They are also called numerical errors because they are not physical. Specifically, such numerical error resulting in spurious oscillations is often called dispersive error or dispersion. Truncation error is the type of numerical error resulting from the difference between the partial differential equation and the finite equation that we actually code.

Additionally, with CFD we can experience a third error—computer error. For example, for one calculation or one floating point operation, computer roundoff is usually negligible. However, when doing repeated calculations over and over in simulations dealing with billions of nodes at which we solve for multiple unknowns at each node, computer roundoff error can build.

3.3 Turbulence models

Another area of CFD modeling is turbulence modeling. Turbulence modeling is a difficult yet still potentially fruitful area of CFD research because of the computational difficulties it poses. Turbulence is a complicated phenomenon that occurs over a wide range of time scales and length scales. This is where the trouble lies making it impossible to fully resolve turbulent phenomena when using most approaches. Because of the wide ranges of time periods over which turbulent periodicity manifests and the wide range of length scales on which turbulence acts, most CFD engineers and scientist choose to resolve a certain length scale and time scale range of turbulence and model the rest. In this case resolving turbulence means that we actually compute and calculate turbulent quantities like vorticity and velocities and pressures in turbulent regions. Modeling turbulence means that we add an expression into our equation, the effect of which is approximately to create the effect of fully resolved turbulence on our unknown values such as velocity and pressure.

As one increases the range of time scales and length scales over which the turbulence is resolved, one must increase the refinement of the simulation both in length (refinement of the mesh) and in time (temporal refinement—the size of the time steps). This increased refinement increases the computational costs (the number of equations to be solved and the time it takes to compute the entire simulation). If one tends toward the other end of the model-resolve spectrum models all turbulence of all length and time scales, the computational costs decrease but one loses accuracy in the simulation. CFD researchers usually tend to resolve a range of turbulent length and time scales and model the rest. Usually the range is related to the range of interest of the flow simulation. For instance, if someone is simulating gas dynamics in the inner-ballistics of a particular weapon, it would be best to resolve turbulence on the length range of the chamber inside the weapon and model anything much larger and much smaller. Likewise, any turbulence that occurs over a longer period than the time it takes to fire the weapon would have smaller effects on the simulation; it would be best to model turbulence on longer time scales and much shorter time scales.

The most expensive turbulence scheme is Direct Numerical Simulation (DNS) (32). In DNS, all length scales of the turbulence are resolved and little or no modeling is done. It is not used in cases of extremely complex geometries which can create prohibitive expense in resolving the turbulence especially in special geometrically complex portions of the domain.

Less expensive than DNS, the Large Eddy Simulation (LES) resolves turbulence on large length scales as the name suggests (33). A model is used to represent sub-grid scale effects of turbulence. The computational cost of turbulence at small length scales is reduced through modeling.

Reynolds-averaged Navier-Stokes (RANS) is the oldest approach to turbulence modeling and it is cheaper than LES. It involves solving a version of the transport equations with new Reynolds stresses introduced. This addition brings a 2nd order tensor of unknowns to be solved as well. Examples of RANS methods are K-ε methods (34), Mixing Length Model (35), and the Zero Equation model (35).

The Detached-eddy simulation (DES) is a version of the RANS model in which portions of the grid use RANS turbulence modeling and portions of the grid (or mesh) use an LES model (36). Since RANS is usually cheaper to implement than LES, DES is usually more expensive than using RANS throughout the entire domain or grid and usually cheaper than using LES throughout the entire grid. If the turbulent length scales fit within the grid dimensions or the particular portion of the grid is near a boundary or wall, a RANS model is used. However, when the turbulence length scale exceeds the maximum dimension of the grid, DES switches to an LES model. Care must be taken when creating a mesh over which DES will be used to model turbulence due to the switching between RANS and LES. Therefore thought must be given to proper refinement to minimize computation while maximizing accuracy (especially refinement near walls). DES itself does not utilize zonal functions; there is still one smooth function used across the entire domain regardless of the use of RANS or LES in certain regions.

There are many more turbulence models including the coherent vortex simulation which separates the flow field turbulence into a coherent part and a background noise part, somewhat similar to LES (37). Many of the contributions today are coming through different versions of RANS models.

3.4 Linear algebraic equation system

Once the mathematical equation has been chosen in the pre-processing stage, the mesh has been partitioned and distributed, and the numerical discretization is chosen and coded, the last part of the computer program is to solve the resulting algebraic system for the unknowns. Remember in CFD we are calculating the unknowns (velocity and pressure, for instance) at all nodes in the domain. The algebraic system usually looks like $Ax = b$. Usually with many steady problems, Ax is a linear equation system. Therefore we can just invert the matrix A to find the vector x of unknown values. The problem is that for large computations, for instance a computation involving 500 million nodes, inverting the matrix A takes too long. Remember that the number of nodes does not necessarily equal the number of unknowns. If you're calculating 5 unknowns at every node, than you must multiply the number of nodes by the number of unknowns per node to get the value of actual number of unknowns and the length of the unknown vector x. So for such large problems, we solve the equation system iteratively. These systems are solved iteratively with iterations such as Newton or the Picard iteration.

In many cases with unsteady flows, Ax represents a system of ordinary or partial differential equations. In this case, CFD researchers must choose either an implicit method or an explicit method to deal with the time integration. An explicit method calculates the solution at a later time based on the current solution. Another way to rephrase that is that it calculates the current solution based on an earlier solution in time. An implicit method calculates the solution at a later time based on both the solution at a later time and the current solution. Rephrasing that, an implicit method calculates the current solution based on both the current solution and the solution at an earlier time. Once a method is chosen and the time derivatives have been expanded using an implicit or explicit method, Ax is usually a nonlinear algebraic system.

As stated earlier, for large problems, it may be too time-consuming or too costly (computationally) to compute this directly by inverting A. Usually, then, the linear system is solved iteratively as well (38).

One must then choose which type of iterative solver to employ to find the unknown vector. Depending on the characteristics of this equation system, an appropriate solver is chosen and employed to solve the system. The major implication of all computational research in this area is that there is no one perfect iterative solver for every situation. Rather there are solvers that are better for certain situations and worse for others. When operating and computing with no information about the equation system, there are, however, general solvers that are quite robust at solving many types of problems though may not be the fastest to facilitate the specific problem one may be working on at the moment.

A popular class of iterative solvers for linear systems includes Krylov subspace methods of which GMRES appears to be the most general and robust (38) (39). Its robustness makes it a great choice for a general solver especially when a researcher has no specific information about the equation system she is asked to solve. GMRES minimizes residuals over successively larger subspaces in an effort to find a solution.

Still, in recent years, the Multigrid method has shown better optimal performance than GMRES for the same problems, motivating many to use this method in place of GMRES. The Multigrid method minimizes all frequency components of the residual equally providing an advantage over conventional solvers that tend to minimize high-frequency components of the residual over low-frequency components of the residual. Operating on different scales, the Multigrid method completes in a mesh-independent number of iterations (40) (41).

CFD experts also employ methods to simplify the ability of a solver to solve an algebraic equation system by pre-conditioning the matrix. The point of preconditioning is to transform the matrix closer to the identity matrix which is the easiest matrix to invert. Though we rarely invert matrices for such large problems, matrices that are invertible or closer to being invertible are also easier to solve iteratively.

Preconditioning is an art, involving the correct choice of preconditioner according to the type of linear system (just as in choosing the correct solver or iterative method). Examples include lumped preconditioners, diagonal preconditioners, incomplete LU, Conjugate-Gradient, Cholesky or block preconditioners such as block LU or Schwarz (38) (42) (43). The

best preconditioners are those that take advantage of the natural structure of the algebraic system in order to facilitate the transformation of A to a matrix closer to the identity matrix, thereby facilitating the solving of the system.
The condition number roughly measures at what rate the solution x will change with respect to a change in b. Low conditions numbers imply more ease in inverting and solving the system. Characteristics such as condition number and even definite-ness of systems become important in choosing a preconditioner.

3.5 Libraries

Many CFD programs require thousands of lines of code for one specific simulation. To avoid writing a different program for every application, researchers often write general programs and software packages that can be used for a variety of situations. This generality greatly increases the lines of code even more.
To help, there are many CFD and computational solid dynamics (CSD) libraries and applied mathematics libraries for the solving of such differential equations. They include code libraries such as OFELI (44), GETFEM (45), OOFEM (46), deal.ii (47), fdtl (48), RSL (49), MOUSE (50), OpenFOAM (51), etc. Sometimes a CFD engineer may use a package like ANSYS Fluent (52) to do a simulation, and sometimes a CFD researcher may find that the software does not give him the freedom to do what he would like to do (usually for very specific research applications). In these cases, he can write his own code or piece code together using these libraries. The most general Finite Element Library is deal.ii which is written in C++. It contains great support but currently does not support triangles or tetrahedral.

3.6 Moving problems

Moving problems are another class of CFD challenges. One can either use a body-fitted mesh approach (an approach where the mesh fits around the solid body of interest), an overlapping mesh method, or a meshless method. Body-fitted mesh methods must actual move the mesh elements or cells. Overlapping methods have the option of moving the mesh but capture the movement of interest in the overlapping regions. Meshless methods avoid the need to move the mesh as moving the mesh has the ability to introduce error. Usually when a mesh is moving the aspect ratio of the mesh elements must be checked. If the ratio goes beyond some limit, the mesh must be reordered and remade, and the solution from the old mesh must be projected to the new mesh in order to continue the simulation. Each of those steps has the ability to introduce error. Therefore it is best to limit mesh distortion or concentrate it in elements that have the ability to absorb the deformation without reaching the aspect ratio limit for the computation (7) (53).
There are also numerous methods that avoid meshes. Due to the problems and errors that poor refinement and mesh distortion introduce, some people avoid mesh methods altogether. Some meshless methods (54) include spectral methods (55), the vortex method (meshless method for turbulent flows), and particle dynamics methods (56).

4. Post-processing

This is the final stage when the computations are complete. This involves taking the output files full of velocity, pressure, density, and similar. information at each node for each time step and displaying the information visually, hopefully with color. Many CFD engineers also animate the results so as to better show what happens in time or to follow a fluid particle or a streamline.

4.1 Engineering

Up to this point, all of the work we have spoken about deals with the computer science/electrical engineering part or the computational and applied mathematics part of CFD. The mechanical and chemical engineering part of CFD comes in the post-processing and analysis.

The interpretation of the visualized results is also another point at which error can be introduced. CFD scientists and engineers must therefore take care in not overextending the analysis and simultaneously not missing important implications or conclusions that can be drawn from it. Engineers also make sure to corroborate the results with physical experiments and theoretical analysis. Additionally, the choice of problem and the motivation can come from the engineering side of CFD at the very beginning before any pre-processing work. Finally, the engineering analysis of the phenomenon directs the feedback loop of the research work. At this point the CFD worker must decide what values to change, whether it should be run again, what parameters should be maintained at current values, etc. He must decide which parameters' effect should be tested. He must determine if the results make sense and how to properly communicate those results to other engineers, scientists, scientists outside of the field, lay persons, and policy makers.

CFD work has taken exciting directions and comes to bear in many ways in society. When the fluid is water, CFD workers study hydrodynamics. When the fluid is air, they study aerodynamics. When it is air systems, they study meteorology and climate change. When it is the expanding universe, astronomy and cosmology; blood, medicine; oil and fossil flows, geology and petroleum engineering. CFD workers work with zoologists studying the mechanics of bird flight, paleontologists studying fossils, and chemists studying mixing rates of various gases in chemical reactions and in cycles like the nitrogen cycle.

Because of the importance and presence of fluids everywhere, the interaction of fluids with structures throughout the real world, and the application of mechanics and chemical engineering everywhere, CFD remains important. Moreover, the computational tools CFD engineers use can be utilized to solve and help other computational fields. Since the advent and continual innovation of the computer, all scientific fields have become computational—computational biology, computational chemistry, computational physics, etc. The same computational tools used in CFD can often be applied in other areas (predicting the weather involves a linear algebraic system for instance). This allows CFD engineers and scientists to move in and out of fields and bring their engineering analytical skills and critical reasoning to bear in other situations. CFD engineers and scientists are even used to make video games and animations look more physically realistic instead of simply artistic (57).

5. Closing

There are a host of other interesting areas in CFD such as two-phase or multi-phase flow (58), vorticity confinement techniques (similar to shock capturing methods) (59), probability density function methods (60), and fluid-structure interaction (FSI) (61). Other chapters in this book address those, so we will not talk specifically about them here. It is simply important to understand that as computing capability increases as computer scientists increase the clock speed of the microchip all the time, our ability to solve real-life problems increases. Problems involving two liquids or two phases occur all the time, and there are many instances of fluids interacting with deforming solids—in fact that is the general reality. So combining CFD with computational solid dynamics is an important partnership that lends itself well to solving the major challenges facing us in the 21st century.

6. References

[1] Phing, Anselm Ho Yen. *Simulating Combustion Flow in a Rocket Chamber.* Lulea: Lulea University of Technology, 2008.

[2] *A Study of Sharkskin and its Drag Reducing Mechanism.* Freidmann, Elfriede, Portl, Julia and Richter, Thomas. Berlin: Springer-Verlag, 2010.

[3] *Three-Dimensional Aerodynamic Simulations of Jumping Paratroopers and Falling Cargo Payloads.* Udoewa, Victor. 5, Reston, Virginia: AIAA, 2009, Vol. 46. 0021-8669.

[4] *Blood Flow in Arteries.* Ku, David N. 1, Palo Alto, CA: Annual Reviews, 1997, Vol. 29.

[5] *Outflow boundary conditions for three-dimensional finite element modeling of blood flow and pressure in arteries.* Vignon-Clementel, Irene E., et al., et al. 29-32, Stanford : Elsevier, 2006, Vol. 195.

[6] *Numerical investigation of transitional and weak turbulent flow past a sphere.* Tomboulides, Ananias G. and Orszag, Steven A. Cambridge: Cambridge University Press, 2000, Vol. 416.

[7] *Mesh Generation and Update Techniques for 3D Aerodynamic Simulations.* Udoewa, Victor. 7, s.l.: Wiley, 2009, Vol. 29.

[8] *Slip formulation for numerical simulations of jumping paratroopers.* Udoewa, Victor. 7, Johannesburg: Academic Journals, 2009, Vol. 2.

[9] *Models and Finite Element Techniques for Blood Flow Simulation.* Behr, Marek. s.l.: International Journal for Computational Fluid Dynamics, 2006, Vol. 20.

[10] Sawley, Mark L. Numerical Simulation of the Flow around a Formula 1 Racing Car. *EPFL Supercomputing Review.* [Online] November 1, 1997. [Cited: July 13, 2011.] http://ditwww.epfl.ch/SIC/SA/publications/SCR97/scr9-page11.html.

[11] *Computational study of flow around a simplified car body.* Guilmineau, Emmanuel. 6-7, Ottowa, Canada: Elevier, 2008, Vol. 96.

[12] Top 500. Top 500 Supercomputing Sites. *Top500.org.* [Online] Top 500, June 1, 2011. [Cited: July 13, 2011.] www.top500.org/lists/2011/06.

[13] Salisbury, David F. News Release. *Stanford News Service.* [Online] November 19, 1997. [Cited: July 13, 2011.] http://news.stanford.edu/pr/97/971119supercomp.html.

[14] MESHPART a Matlab Mesh Partitioning and Graph Separator Toolbox. *Cerfacs.* [Online] Cerfacs, February 8, 2002. [Cited: July 13, 2011.] http://www.cerfacs.fr/algor/Softs/MESHPART/.

[15] Karypis, George and Kumar, Vipin. *METIS - Unstructured Graph Partitioning and Sparse Matrix Ordering System, Version 2.0.* Minneapolis: s.n., 1995.

[16] Karypis, George, Schloegel, Kirk and Kumar, Vipin. *PARMETIS - Parallel Graph Partitioning and Sparse Matrix Ordering.* Minneapolis: University of Minnesota, 2003.

[17] *Applying Parmetis To Structured Remeshing For Industrial CFD Applications.* Laflamme, S., et al., et al. 1, s.l.: Spring: International Journal of High Performance Computing Applications, 2003, Vol. 17.

[18] OpenMP. OpenMP News. *OpenMP.* [Online] OpenMP, July 9, 2011. [Cited: July 13, 2011.] http://openmp.org/wp/.

[19] Pacheco, Peter S. *Parallel Programming with MPI.* San Francisco: Morgan Kaufmann Publishers, Inc., 1997.
[20] *Manticore: A heterogeneous parallel language.* Fluet, Matthew, et al. Nice, France: DAMP 2007, ACM 2007, 2007.
[21] NESL: A Parallel Programming Language. *Computer Science at Carnegie Mellon University.* [Online] Scandal Lab, August 1, 2005. [Cited: July 13, 2011.] http://www.cs.cmu.edu/~scandal/nesl.html.
[22] Kaw, Autar and Kalu, E. Eric. *Numerical Methods with Applications.* s.l.: Lulu Self-published, 2008.
[23] Oliver, Rubenkonig. *The Finite Difference Method (FDM) - An introduction.* Freiburg: Albert Ludwigs University of Freiburg, 2006.
[24] Leveque, Randall. *Finite Volume Methods for Hyperbolic Problems.* Cambridge: Cambridge University Press, 2002.
[25] Huebner, K. H., Thornton, E. A. and Byron, T. D. *The Finite Element Method for Engineers.* s.l.: Wiley Interscience, 1995.
[26] Ang, W. T. *A Beginner's Course in Boundary Element Methods.* s.l.: Universal Publishers, 2007. 978-1581129748.
[27] Beer, Gernot, Smith, Ian and Duenser, Christian. *The Boundary Element Method with Programming: For Engineers and Scientists.* s.l.: Springer, 2008. ISBN 978-3211715741.
[28] *The immersed boundary method.* Peskin, C. S. s.l.: Acta Numerica, 2002, Vol. 11.
[29] Wesseling, P. *Principles of Computational Fluid Dynamics.* s.l.: Springer-Verlag, 2001.
[30] *Uniformly High Order Accurate Essentially Non-Oscillatory Schemes III.* Harten, A., et al., et al. s.l.: Journal of Computational Physics, 1987, Vol. 71.
[31] *The Piecewise parabolic Method (PPM) for Gasdynamical Simulations.* Colella, P. and Woodward, P. s.l.: Journal of Computational Physics, 1984, Vol. 54.
[32] Pope, S. B. *Turbulent Flows.* Cambridge : Cambridge University Press, 2000. ISBN 978-0521598866.
[33] Garnier, E., Adams, N. and Sagaut, P. *Large eddy simulation for compressible flows.* s.l.: Springer, 2009. 978-90-481-2818-1.
[34] *The Numerical Computation of Turbulent Flows.* Launder, B. E. and Spalding, D. B. 2, s.l.: Computer Methods in Applied Mechanics and Engineering, 1974, Vol. 3.
[35] Wilcox, David C. *Turbulence Modeling for CFD (3 ed.).* s.l.: DCW Industries, Inc., 2006. 978-1928729082.
[36] *Comments on the feasibility of LES for wing and on a hybrid RANS/LES approach.* Spalart, P. R. Arlington, TX: 1st ASOSR CONERFENCE on DNS/LES, 1997.
[37] *Coherent Vortex Simulation (CVS), A Semi-Deterministic Turbulence Model Using Wavelets.* Farge, Marie and Schneider, Kai. 4, s.l.: Flow Turbulence and Combustion, 2001, Vol. 66.
[38] Saad, Y. *Iterative Methods for Sparse Linear Systems, 2nd edition.* s.l.: SIAM, 2003. ISBN 978-0-89871-534-7.
[39] *GMRES: A generalized minimal residual algorithm for solving nonsymmetric linear systems.* Saad, Y. and Schultz, M. H. s.l.: SIAM Journal of Scientific and Statistical Computing, 1986, Vol. 7.

[40] Wienands, Roman and Joppich, Wolfgang. *Practical Fourier analysis for multigrid methods.* s.l.: CRC Press, 2005. 1584884924.

[41] Trottenberg, U., Oosterlee, C. W. and Schüller, A. *Multigrid.* s.l.: Academic Press, 2001. ISBN 012701070X.

[42] Trefethen, Lloyd N. and Demmel, James W. *Numerical Linear Algebra.* s.l.: SIAM, 1997. 0898713617.

[43] Golub, Gene H. and Van Loan, Charles F. *Matrix Computations 3rd Edition.* s.l.: John Hopkins University Press, 1996. 0801854148.

[44] OFELI. OFELI An Object Oriented Finite Element Library. *OFELI.* [Online] OFELI, June 1, 2011. [Cited: July 13, 2011.] http://www.ofeli.net/.

[45] Renard, Yves; Pommier, Julien. GETFEM++. *GNA.* [Online] GETFEM, June 1, 2010. [Cited: July 13, 2011.] http://download.gna.org/getfem/html/homepage/.

[46] OOFEM. OOFEM - free object oriented finite element solver. *OOFEM.* [Online] OOFEM, March 29, 2011. [Cited: July 13, 2011.] http://www.oofem.org/en/oofem.html.

[47] Dealii. Dealii Homepage. *Dealii.* [Online] Dealii, January 9, 2011. [Cited: July 13, 2011.] http://www.dealii.org/.

[48] Google. FDTL - Finite Difference Template Library. *FDTL.* [Online] Google, June 2, 2011. [Cited: July 13, 2011.] http://code.google.com/p/fdtl/.

[49] Michalakes, John. A Runtime System Library for Parallel Finite Difference Models with Nesting. *Argonne National Laboratory.* [Online] May 18, 1995. [Cited: July 13, 2011.] http://www.mcs.anl.gov/~michalak/RSL/.

[50] MOUSE team. MOUSE. *USASK.* [Online] University of Duisburg, January 7, 2010. [Cited: July 13, 2011.] http://homepage.usask.ca/~ijm451/finite/fe_resources/node563.html.

[51] OpenFOAM. OpenFOAM - The Open Source CFD Toolbox. *OpenFOAM.* [Online] OpenFOAM, June 2, 2011. [Cited: July 13, 2011.] http://www.openfoam.com/.

[52] Ansys. Features of Ansys Fluent. *Ansys.* [Online] Ansys, June 3, 2011. [Cited: July 13, 2011.] http://www.ansys.com/Products/Simulation+Technology/Fluid+Dynamics/ANSYS+FLUENT/Features.

[53] Huang, Weizhang and Russell, Robert D. *Adaptive Moving Mesh Methods.* s.l.: Springer, 2011. 978-1-4419-7915-5.

[54] Liu, G. R. *Mesh Free Methods, 2nd ed.* s.l.: CRC Press, 2009. 978-1-4200-8209-9.

[55] Hesthaven, J., Gottlieb, S. and Gottlieb, D. *Spectral methods for time-dependent problems.* Cambridge: Cambridge University Press, 2007.

[56] Belytschko, T. and Chen, J.S. *Meshfree and particle methods.* s.l.: John Wiley and Sons, Ltd., 2007. 0-470-84800-6.

[57] Gourlay, Michael J. "*Fluid Simulation for Video Games*". s.l.: Intel Software Network, 2009.

[58] *Two phase flow in complex systems.* Levy, Salomon. s.l.: Wiley, 1999.

[59] *Numerical Simulation of Vortical Flows Using Vorticity Confinement Coupled with Unstructured Adaptive Grid Refinement.* Murayama, M. and Kato, T. 1, s.l.: Computational Fluid Dynamics Journal, 2001, Vol. 10.

[60] Fox, Rodney. *Computational methods for turbulent reacting flows.* Cambridge: Cambridge University Press, 2003. 978-0-521-65049-6.

[61] *FOIST: Fluid–object interaction subcomputation technique.* Udoewa, V. 9, s.l.: Wiley: Communications in Numerical Methods in Engineering, 2009, Vol. 25.

2

A Computational Fluid Dynamics Model of Flow and Settling in Sedimentation Tanks

Ali Hadi Ghawi[1] and Jozef Kriš[2]
[1]*Department of Civil Engineering, Faculty of Civil Engineering, AL-Qadisyia University*
[2]*Department of Sanitary and Environmental Engineering, Faculty of Civil Engineering, Slovak University of Technology, Bratislava*
[1]*Iraq*
[2]*Slovakia*

1. Introduction

Sedimentation is perhaps the oldest and most common water treatment process. The principle of allowing turbid water to settle before it is drunk can be traced back to ancient times. In modern times a proper understanding of sedimentation tank behavior is essential for proper tank design and operation. Generally, sedimentation tanks are characterized by interesting hydrodynamic phenomena, such as density waterfalls, bottom currents and surface return currents, and are also sensitive to temperature fluctuations and wind effects.

On the surface, a sedimentation tank appears to be a simple phase separating device, but down under an intricate balance of forces is present. Many factors clearly affect the capacity and performance of a sedimentation tank: surface and solids loading rates, tank type, solids removal mechanism, inlet design, weir placement and loading rate etc. To account for them, present-day designs are typically oversizing the settling tanks. In that way, designers hope to cope with the poor design that is responsible for undesired and unpredictable system disturbances, which may be of hydraulic, biological or physico-chemical origin.

To improve the design of process equipment while avoiding tedious and time consuming experiments Computational Fluid Dynamics (CFD) calculations have been employed during the last decades. Fluid flow patterns inside process equipment may be predicted by solving the partial differential equations that describe the conservation of mass and momentum. The geometry of sedimentation tanks makes analytical solutions of these equations impossible, so usually numerical solutions are implemented using Computational Fluid Dynamics packages. The advent of fast computers has improved the accessibility of CFD, which appears as an effective tool with great potential. Regarding sedimentation tanks, CFD may be used first for optimizing the design and retrofitting to improve effluent quality and underflow solids concentration. Second, it may increase the basic understanding of internal processes and their interactions. This knowledge can again be used for process optimization. The latter concerns the cost-effectiveness of a validated CFD model where simulation results can be seen as numerical experiments and partly replace expensive field experiments **(Huggins et al. 2005).**

Generally, many researchers have used CFD simulations to describe water flow and solids removal in settling tanks for sewage water treatment. However, works in CFD modelling of sedimentation tanks for potable water treatment, rectangular sedimentation tanks, and iron removal by sedimentation tank in surface and groundwater treatment plants have not been found in the literature. Moreover, the physical characteristics of the flocs may not be such significant parameters in the flow field of sedimentation tanks for potable water, due to the much lower solids concentrations and greater particle size distributions than those encountered in wastewater treatment.

Design of sedimentation tanks for water and wastewater treatment processes are often based on the surface overflow rate of the tank. This design variable is predicated on the assumption of uniform unidirectional flow through the tank. **Dick (1982),** though, showed that many full-scale sedimentation tanks do not follow ideal flow behavior because suspended solids removal in a sedimentation tank was often not a function of the overflow rate. Because of uncertainties in the hydrodynamics of sedimentation tanks, designers typically use safety factors to account for this nonideal flow behavior **(Abdel-Gawad and McCorquodale, 1984).**

It can be concluded from the discussion that the current ways in which STs are designed and modified could and should be improved. Providing a tool that might lead to sedimentation tank optimization, as well as understanding, quantifying and visualizing the major processes dominating the tank performance, are the main goals of this research.

2. Scope and objectives

This research focuses on the development of a CFD Model that can be used as an aid in the design, operation and modification of sedimentation tanks **(Ghawi, 2008).** This model represents in a 2D scheme the major physical processes occurring in STs. However, effect of scrapers and inlet are also included, hence the CFD Model definition. Obviously, such a model can be a powerful tool; it might lead to rectangular sedimentation tanks optimization, developing cost-effective solutions for new sedimentation projects and helping existent sedimentation tanks to reach new-more demanding standards with less expensive modifications. An important benefit is that the model may increase the understanding of the internal processes in sedimentation tanks and their interactions. A major goal is to present a model that can be available to the professionals involved in operation, modification and design of sedimentation tanks.. The ultimate goal of the project is to develop a new CFD methodology for the analysis of the sediment transport for multiple particle sizes in full-scale sedimentation tanks of surface and groundwater potable water treatment plants with high iron concentration. The CFD package FLUENT 6.3.26 was used for the case study of the effect of adding several tank modifications including flocculation baffle, energy dissipation baffles, perforated baffles and relocated effluent launders, were recommend based on their field investigation on the efficiency of solids removal. An overview of the outline of the project is given in Figure 1.

The specific objectives of this research include:

- Improve the operation and performance of horizontal sedimentation tank in Iraq which have been identified as operating poorly, by predicting the existing flow, coagulant dose to remove iron and flocculent concentration distribution of the sedimentation tank by means of CFD techniques.
- Develop a mathematical model for sedimentation tanks in 2D;

- Introduce a flocculation submodel in the general ST model,
- Introduce a temperature submodel in the general ST model.
- Design CFD model for simulation of sedimentation tanks, i.e. grids and numerical descriptions.
- Develop a model calibration procedure, including the calibration of the settling properties, and validate the models with experimental data.
- Evaluate the suitability of CFD as a technique for design and research of rectangular sedimentation tanks for drinking water treatment plants and iron removal.
- Use CFD to investigate the effects of design parameters and operational parameters.

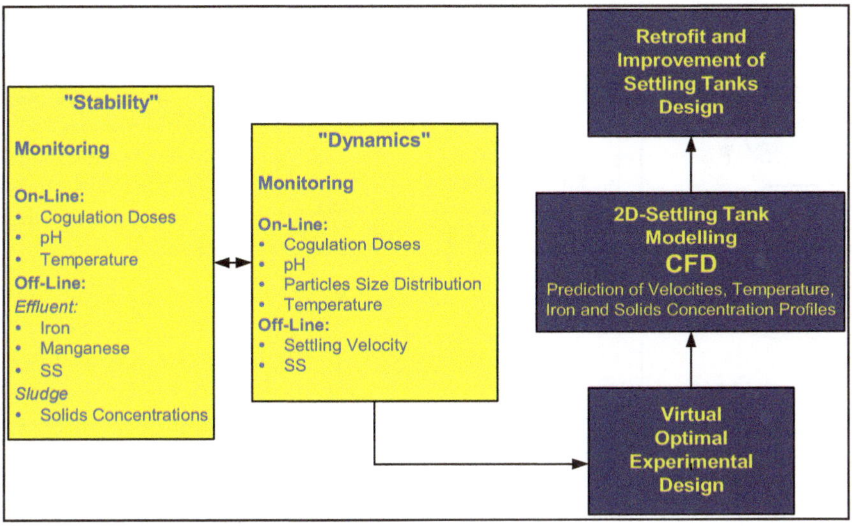

Fig. 1. Overview of the settling tank project

Finally, a CFD model was developed to simulate the full scale rectangular sedimentation tanks at the AL-DEWANYIA purification works in Iraq. The CFD simulations of the AL-DEWANYIA tanks were done by setting up standard cases for each, i.e. a configuration and operating conditions that represented the physical tanks as they were built, and then varying different aspects of the configuration or operating conditions one or two at a time to determine the effect. discrete particles in dilute suspension was simulated, as it is the applicable type for the operating conditions in rectangular sedimentation tanks for potable water treatment.

3. Modelling the settling tank

Figure 2 shows the set-up of the settling tank CFD model which developed in this work. The code predicts fluid flow by numerically solving the partial differential equations, which describe the conservation of mass and momentum. A grid is placed over the flow region of interest and by applying the conservation of mass and momentum over each cell of the grid sequentially discrete equations are derived. In the case of turbulent flows, the conservation equations are solved to obtain time-averaged information. Since the time-averaged equations contain additional terms, which represent the transport of mass and momentum by turbulence, turbulence models that are based on a combination of empiricism and theoretical considerations are introduced to calculate these quantities from details of the mean flow.

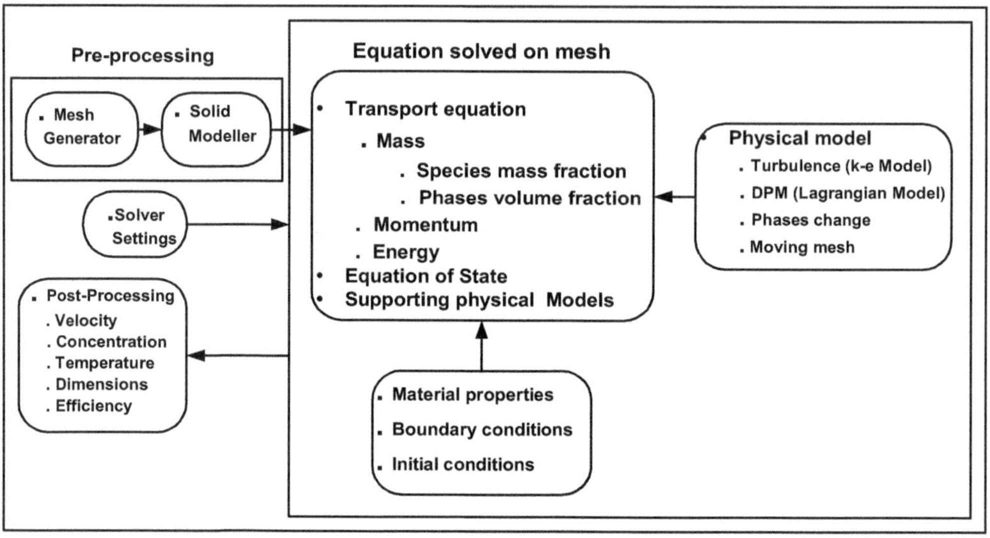

Fig. 2. CFD model

4. Numerical techniques used in Fluent

This section will shortly deal with the methods applied in **(Ghawi, 2008).** The Fluent software utilises the finite volume method to solve the governing integral equations for the conservation of mass and momentum, and (when appropriate) for scalars such as turbulence and solids concentration. In the work (Ghawi, 2008), the so-called segregated solver was applied; its solution procedure is schematically given in Figure 3. Using this approach, the governing equations are solved sequentially, i.e. segregated from one another. Because the governing equations are non-linear (and coupled), several iterations of the solution loop must be performed before a converged solution is obtained.

Concerning the spatial discretisation, the segregated solution algorithm was selected. The k-ε turbulence model was used to account for turbulence, since this model is meant to describe better low Reynolds numbers flows such as the one inside our sedimentation tank. The used discretisation schemes were the simple for the pressure, the PISO for the pressure-velocity coupling and the second order upwind for the momentum, the turbulence energy and the specific dissipation. **Adams and Rodi 1990** pointed out that for real settling tanks the walls can be considered as being smooth due the prevailing low velocities and the correspondingly large viscous layer. Consequently, the standard wall functions as proposed by **Launder and Spalding 1974** were used. The water free surface was modeled as a fixed surface; this plane of symmetry was characterized by zero normal gradients for all variables

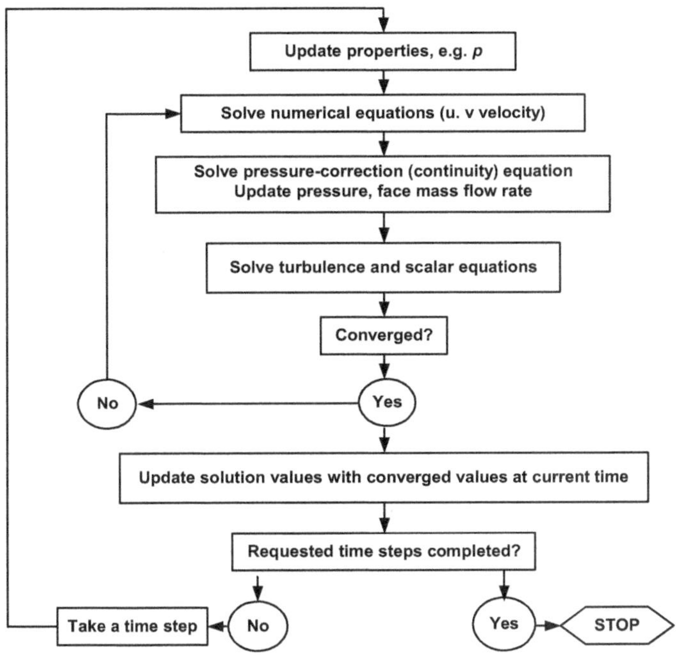

Fig. 3. Solution procedure

5. Experimental techniques for model calibration and validation

The process of developing (incl. calibration), verifying, and validating a CFD code requires the use of experimental, theoretical and computational sciences. This process is a closed loop as presented in Figure 4.

The above clearly indicates that good experimental data are indispensable for settling tank model validation; their quality largely depends on the applied experimental technique.

For the purposes of testing the numerical model presented in this thesis on a full scale tank, the data set gathered laboratories, was selected. Here, a comprehensive experimental study of a working settling tank at AL-DEWANYIA in Iraq were carried out. Velocity and concentration profiles were gathered at 7 stations along the length and 3 stations across the width of the tank for a variety of inlet conditions and inlet and outlet geometries. Volumetric flow rates through the inlets and outlets were measured for each test condition studied. Details of the tank geometry and the experimental conditions for which 3D numerical simulations have been made are given in next sections. The following topics are dealt with which measured in the sites:

Flow rate, (2) Settling velocity, (3) Solids concentration (Turbidity), Iron, and Manganese, (4) Particle size distribution, (5) Velocity of liquid, and (6) Temperature

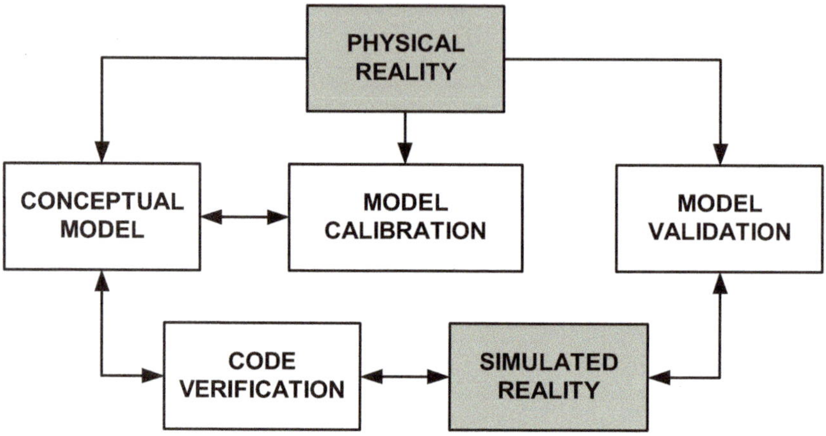

Fig. 4. Process of developing CFD code.

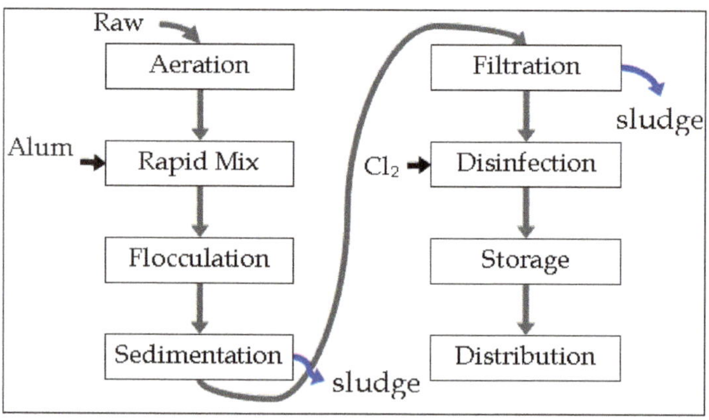

Fig. 5. Layout of AL-DEWANYIA WTP

6. Model development, applications and results

6.1 Introduction

The full-scale horizontal settling tanks at the drinking treatment plant of AL-DEWANYIA were opted for. Most settling tanks of Iraq Water exhibit a horizontal settling tank. This research was focused on this type of settling tanks.

Figure 5 represents treatment of water obtained from a deep well in AL-DEWANYIA WTP a The AL-DEWANYIA WTP were built to remove turbidity and organic material.

6.2 Simulation

To limit computational power requirements, the rectangular settling tank was modeled in 2D. The major assumption in the development of the model is that the flow field is the same for all positions; therefore, a 2D geometry can be used to properly simulate the general features of the hydrodynamic processes in the tank. As a first step, a mesh was generated across the sedimentation tank. As a result, the solutions from the grid of 137,814 quadrilateral elements were considered to be grid independent.

For simulation purposes, the range of the suspended solids was divided into thirteen distinct classes of particles based on the discretization of the measured size distribution. The number of classes was selected in order to combine the solution accuracy with short computing time. Two other numbers, 6 and 15, were tested. While the predictions obtained using 6 classes of particles were found to be different from those resulting from the 13 classes, the difference between the predictions made by the 13 and the 15 classes were insignificant. Therefore, a number of 13 classes were selected as a suitable one. Within each class the particle diameter is assumed to be constant (Table 1). As it can be seen in Table 1, the range of particle size is narrower for classes that are expected to have lower settling rates.

Class	Range of particle size (μm)	Mean particle size (μm)	Mass fraction
1	10-30	20	0.025
2	30-70	50	0.027
3	70-90	80	0.039
4	90-150	120	0.066
5	150-190	170	0.095
6	190-210	200	0.115
7	210-290	250	0.126
8	290-410	350	0.124
9	410-490	450	0.113
10	490-610	550	0.101
11	610-690	650	0.077
12	690-810	750	0.057
13	810-890	850	0.040

Table 1. Classes of particles used to account for the total suspended solids in the STs in AL-DEWANYIA STs.

6.3 The influence of particle structure

The settling velocity of an impermeable spherical particle can be predicted from Stokes' law. However, the aggregates in the water not only are porous but it is well known that they have quite irregular shapes with spatial varying porosity. The flow chart of this computations sequence is presented in Figure 6.

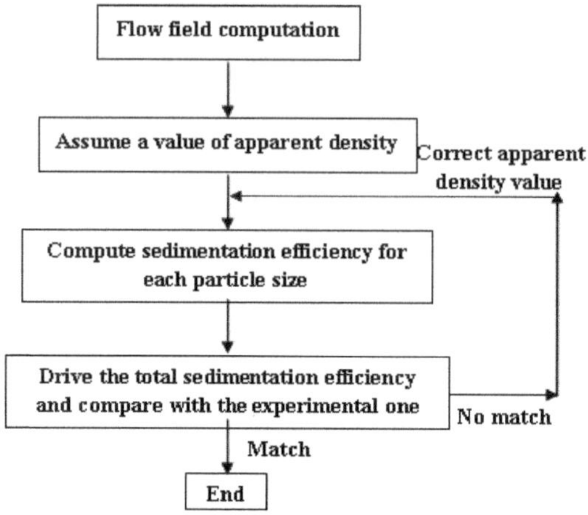

Fig. 6. Flow chart of computation sequence.

6.4 Simulation of existing sedimentation tanks

The AL-DEWANYIA water treatment plant uses lime, and $(NH_4)_2SO_4$ and $Fe_2(SO_4)_3$ to flocculate the and solid concentrations, respectively before entering the sedimentation tanks. There are 4 rectangular tanks at the AL-DEWANYIA WTP .

The Physical and hydraulic data during study periods, and settling tank data for two WTPs are shown in Table 2 .

Geometry	Value
Tank length	30.0 m
Tank width	4.50 m
Hopper depth	2.50 m
Bottom slop	0.00
Weir length	4.50 m
Weir width	0.70 m
Weir depth	0.50 m
loading	Value
SOR	2.7 m/h
Inlet concentration	30-80 mg/l
Density of water	1000 kg/m^3
Density of particulate	1066 kg/m^3
Tank parameter	Value
Average flow rate	60-80 l/s
Sludge pumping rate	5-15 l/s
Inflow temperature average	4°C -11°C , and 20°C -27°C
Inflow suspended solids	25-80 mg/l
Detention time	2.5-3.6 hr
C_{min}	0.17 mg/l
μ	0.002 N.s/m^2

Table 2. Physical and hydraulic data during study periods, and settling tank data.

6.4.1 AL-DEWANYIA WTP

Figure 7 shows the velocity profiles of the existing tanks for a flow rate of 80 l/s and an inlet concentration of 50 mg/l (~75 NTU). High velocities are present at the inlet (0.065 m/s). The flow is further accelerated towards the bottom of the hopper due to the density differences as well as the wedge shape of the hopper. The strong bottom current is balanced by a surface return current inside the hopper. The velocities near the effluent weir are very low.

The solids concentration profile is shown in Figure 8. Note the high concentration downstream of the sludge hopper. The sludge that is supposed to settle in the hopper is washed out of the hopper into the flat section of the tank. Over time a significant amount of sludge accumulates. According to both the field observations and the modeling of the existing process, each of the following reasons (or combination of them) may cause the ST problems, i.e. the flocculant solids blowing out:

1. The location of the existing weir (distributed in a range of 1 meter at the very downstream end of the ST) cause very strong upward currents, which could be one of the major reasons that the flocculant solids were blowing out around the effluent area.
2. The strong upward flow is not only related to the small area the effluent flow passes through but also to the rebound effect between the ST bottom density current and the downstream wall. The "rebound" phenomenon has been observed and reported by many operators as well as field investigators, especially in ST with small amounts of sludge inventory. A reasonable amount of sludge inventory can help dissipate the kinetic energy of the bottom density current.
3. In the existing operation, the bottom density current must be fairly strong due to the lack of proper baffling and the shortage of sludge inventory in the tank.

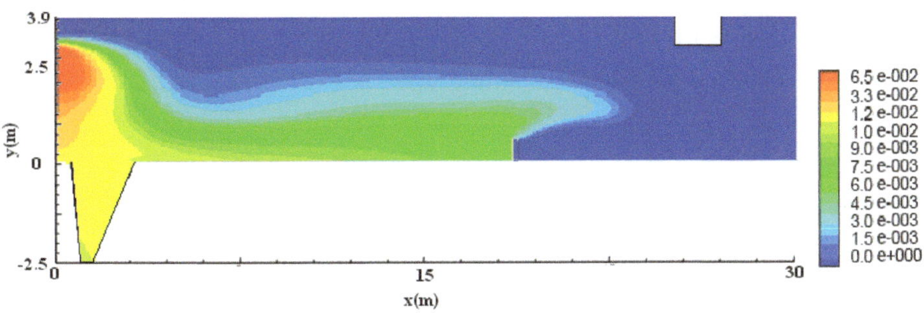

Fig. 7. Velocity contours of existing tank (m/s)

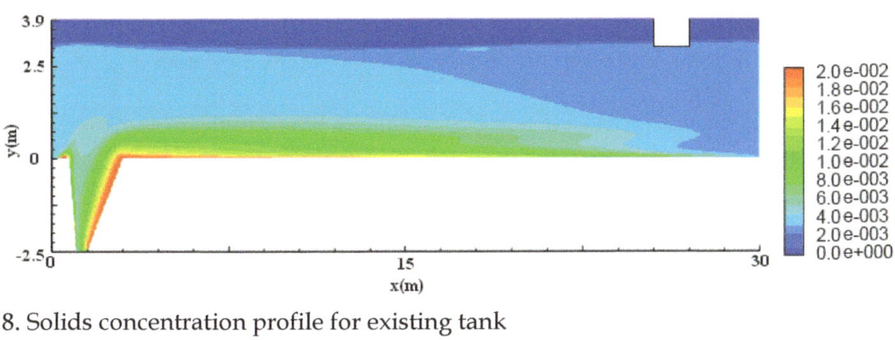

Fig. 8. Solids concentration profile for existing tank

7. Simple improvements to the existing sedimentation tank in WTP

Field data collected from the laboratories during the last 3 years was used to analyze the tank behavior and to enhance the performance of the settling tanks at the AL-DEWANYIA WT Plant. Several tank modifications including flocculation baffle, energy dissipation baffles, perforated baffles and inboard effluent launders, were recommend based on their field investigation

The relationship between the effluent SS and the hydraulic loading is summarised in Table 3 for the existing STs and with different modification combinations. The predicted Effluent SS (ESS) in Table 3 and Figure 9 indicates that the average ESS can be significantly reduced by improving the tank hydraulic efficiency. The comparison of model predictions with the subsequent field data indicates that the significantly improvement of STs performance was obtained by using the minor modifications based on the 2-D computer modeling.

	Q= 50 l/s Influent conc.= 40 mg/l	Q= 70 l/s Influent conc.= 40 mg/l	Q= 80 l/s Influent conc.= 50 mg/l	Q= 80 l/s Influent conc.= 75 mg/l
	Predicted average effluent concentration			
Existing tank	20	30	40	50
Modification 1	12	11	30	22
Modification 2	6	8	12	13
Modification 1 and 2	4	6	7	9
(1) Perforated baffle distance from inlet = 16m; gap above bed = 0.5 m; height above bed = 1.8 m; porosity = 55%				
(2) Length of launder = 12 m.				

Table 3. Performance data for modelled settling tank

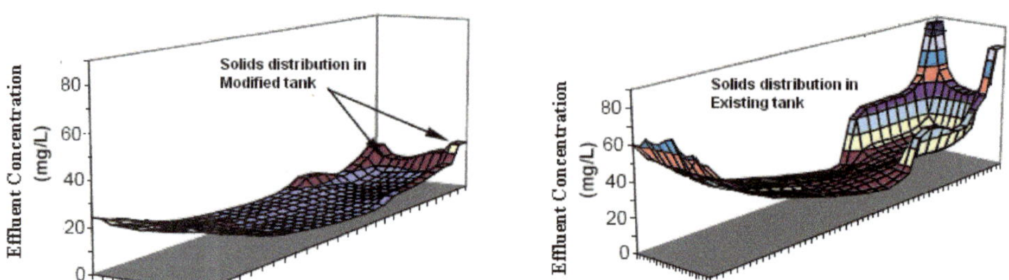

Fig. 9. Comparison of solids distributions on surface layer between existing and modified tanks

8. Modelling the scraper mechanism

The gravitational (and laminar) flow along the bottom, which may go up to 8-15 mm/s near the sump, is blocked for 40 minutes of scraper passage. This is clearly seen in Figure 10. The scraper blade thus constrains the bottom flow discharge by counteracting the gravitational force. Near the floor the velocity increases with height in the shear flow region, but is obviously limited by the scraper's velocity.

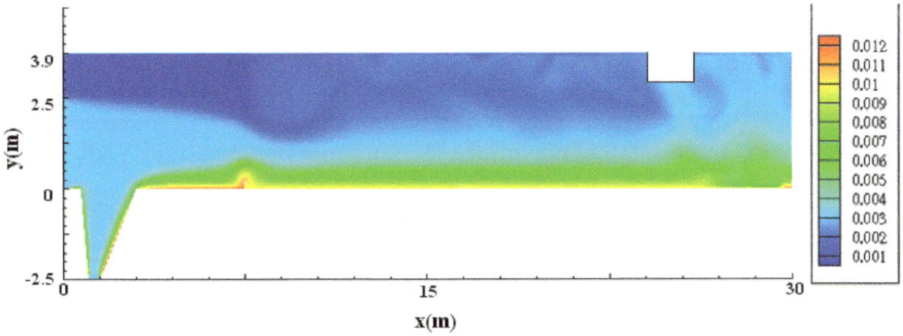

Fig. 10. Effect of scraper on solids concentration profiles

9. Design procedures and guidelines

The design procedures are necessarily based on many assumptions, not normally stated as shortcomings and limitations during the design process. To demonstrate the implications of these assumptions and the way in which these assumption deviate from real tanks. **(Ghawi, 2008)** tried to improve design procedure as show in Table 4.

Improved design procedure	
Step	description
Step 1	Measurement of settling velocity and sludge density
Step 2	Set up of computational grid
Step 3	Simulate tank
Step 4	Evaluate results and check for evidence of the following: - short circuiting - high velocities zones - high overflow concentration - poor sludge removal
Step 5	If none of the above is present, tank size can be reduced to reduce capital cost. If problems are evident, adjust the design by adjusting the: - inlet - position of sludge withdrawal - position of overflow launders Also consider using perforated, porous and deflecting baffles
Step 6	Repeat until a satisfactory tank geometry is obtained and check final geometry for various process changes such as density, concentration and inflow rate.
Step 7	Asses the influence of the settling velocity and sludge density input parameters and repeat steps 3 - 6 if necessary.

Table 4. Proposed CFD enhanced design procedure.

10. Temperature effect

Settling velocity correction factor

In order to define a correction factor for the settling velocities based on temperature difference, the temperature effect on the zone settling velocity has to be determined.
Figure 11 displays graphically the value of the relationship Vs_{T2} / Vs_{T1} and μ_{T2} / μ_{T1} for the data presented in Table 5 at temperatures T_s (summer temperature) and T_w (winter temperature).

CFD Calculated at summer temperature				
SS mg/l	**Settling velocity V m/h**	**Inlet temperature oC**	**Outlet temperature oC**	**Dynamic viscosity μ kg/m.s**
60	1.5	27.5	27.5	8.5e-04
50	1.7	27.5	27.5	8.6e-04
25	1.83	26	26	8.7e-04
15	2.52	25.4	25.4	8.8e-04
CFD Calculated at cooled temperature				
60	0.95	8	9.2	1.3e-03
50	1.05	6.6	6.8	1.35e-03
25	1.9	7.8	8.8	1.29e-03
15	2.7	7	8.9	1.30e-03

Table 5. Settling velocity and dynamic viscosities for summer and winter temperature.

From Figure 11 can be observed that the numerical values of the ratios Vs_{T2} / Vs_{T1} and μ_{T2} / μ_{T1} are very close, suggesting that an easy correction in the zone settling velocity for different temperatures can be made with a correction factor based on the dynamic viscosity of the water at the two temperatures. Figure 12 shows an extended data set indicating the relationships between the ratios Vs_{T2} / Vs_{T1} and μ_{T2} / μ_{T1}.
Fitting a straight line to the data point presented in Figure 13 can find a correction factor for the settling velocities based on temperature

$$V_{ST_2} = V_{ST_1} \left(\frac{10^{\left[\frac{247.8}{T1+133.15}\right]}}{10^{\left[\frac{247.8}{T2+133.15}\right]}} \right) \tag{1}$$

Equation 1 can be applied to correct the settling velocities for difference in temperatures in whichever of the four types of sedimentation, i.e., unflocculated discrete settling, and flocculated discrete settling. Even though equation 1 can be used for a sensitivity analysis on the performance of the model for different seasons, e.g. summer and winter, there is no evidence that the settling properties can be accurately extrapolated from one season to another.

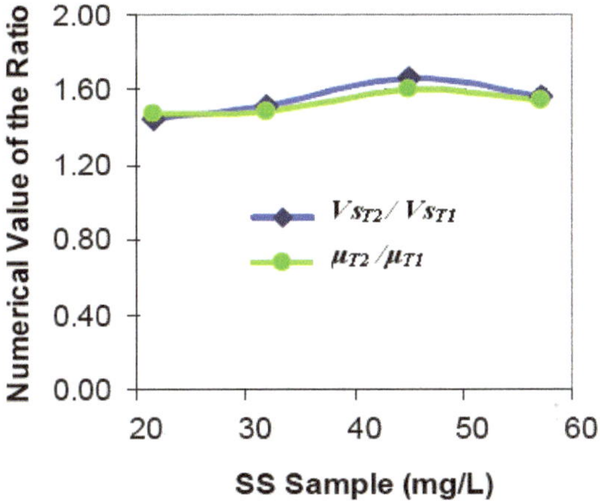

Fig. 11. Ratios of Vs_{T2} / Vs_{T1} and μ_{T2} / μ_{T1} for Different suspended solid (SS) concentrations.

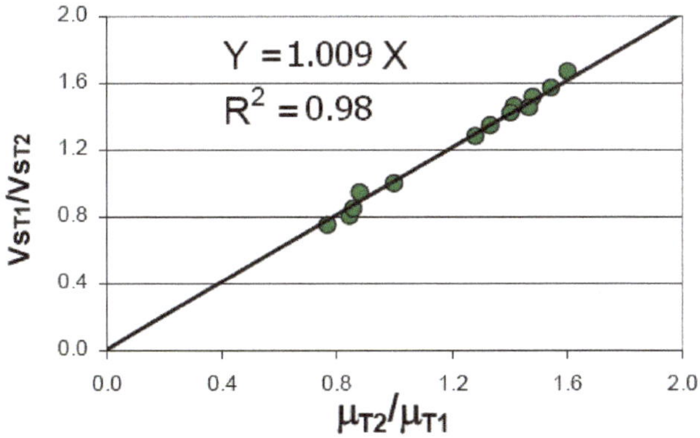

Fig. 12. Effect of Temperature on Settling Velocity.

11. Validation of the model

The validation process involves comparing the model response to actual measured data. The model was validated using measured data from the AL-DEWANYIA WTPs.

After the development of the hydrodynamic model, and turbulence model, the ST model was tested. The ESS predicted by the model was tested during seven days (from a 10 day period) showing a very good agreement with the field data. Figure 13 presents a comparison between the experimentally measured and the simulated values of the floc concentration in the effluent of the existing tanks in AL-DEWANYIA. Apparently, there is a good agreement between measured and predicted values.

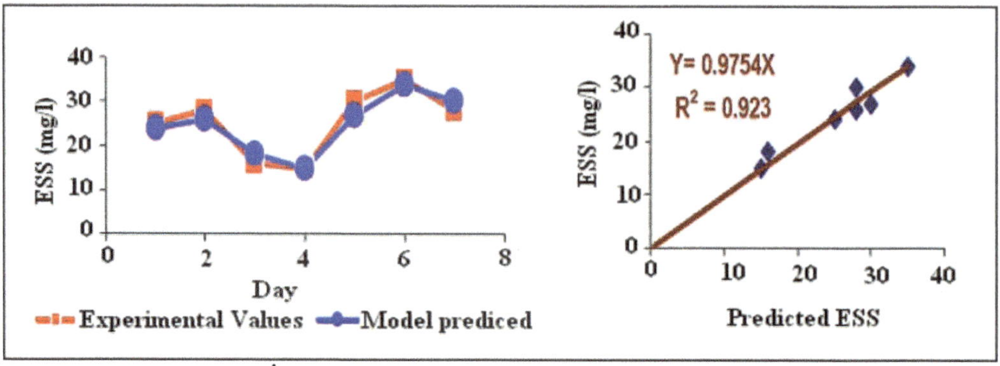

Fig. 13. Validation of the ESS Simulated by the Model

The average values of effluent concentration of improvement are presented in Table 6. The tank operation conditions in the data collection periods (February-April, 2007, (June-August, 2006)) and the CFD model predictions are very close as shown in Table 6. The comparison of model predictions with the subsequent field data indicates that the significantly improvement of tank performance was obtained by using the minor modifications based on the 2-D computer modeling.

	Operation conditions	Effluent concentration (mg/l) and improvement		
	Ave. concentration mg/l	No Modifications	baffle Modifications	Baffle and launder Modifications
Field Data June-August 2006	50	28	-	-
Model Predictions	47	27	6 (+78%)	5 (+82%)

Table 6. Comparison of model predictions with field data

12. Conclusions

The introduction of this study made clear that many factors influence the performance of settling tanks. They may be categorised as physico-chemical and hydraulic influences. To account for them in terms of process operation and design, mathematical models may be utilised. In this respect, Computational Fluid Dynamics (CFD) enables the investigation of internal processes, such as local velocities and solids concentrations, to identify process in efficiencies and resolve them. Although these complex models demand for considerable computational power, they may become an option for the study of process operation and control as computer speed increases. Nowadays, they mostly find applications in the world of settling tank design.

The main purpose of this investigation was to develop a CFD ST model capable of simulating the major processes that control the performance of settling tanks, this goal was achieved. The accomplished objectives of this research include: the development of a compound settling model that includes the representation of the settling velocity for the suspended solids usually encountered in this type of tank (horizontal sedimentation tanks) the inclusion of iron removal effects, a flocculation sub-model, and a temperature sub-model.

These types of sub-models have not been previously incorporated in CFD ST models. The model was rigorously tested and validated. The validation process confirms the utilities and accuracy of the model. An important benefit of this research is that it has contributed to a better understanding of the processes in STs. The results presented in this research clarify important points that have been debated by previous researchers.

This research may also open the discussion for future research and different ways for improving the performance of existing and new STs. In summary, this research has led to more complete understanding of the processes affecting the performance of settling tanks, and provides a useful tool for the optimization of these corn stone units in water treatment. The major conclusions, general and specifics, obtained from this research are:

1. CFD modeling was successfully used to evaluate the performance of settling tank.
2. The usually unknown and difficult to be measured particle density is found by matching the theoretical to the easily measured experimental total settling efficiency. The proposed strategy is computationally much more efficient than the corresponding strategies used for the simulation of wastewater treatment.
3. Solid removal efficiency can be estimated by calculating solids concentration at effluent.
4. High solid removal efficiency was achieved for all cases tested.
5. Baffling inlet arrangement succeeded in controlling kinetic energy decay.
6. Improved energy dissipation due to an improved inlet configuration.
7. Reduced density currents due to an improved inlet configuration.
8. Improved sludge removal due to the inlet configuration.
9. Troubleshoot existing STs and related process operations.
10. The effluent quality can be improved by more than 60% for any cases.
11. Evaluated ST design under the specified process conditions.
12. Develop reliable retrofit alternatives with the best cost-effectiveness.
13. The changes in temperature on STs play an important role on the performance of STs.
14. Scrape is important in the settling process and play a big role in changing the flow field.
15. In this work we improved the STs guidelines design procedure.
16. The fairly good agreement between model predictions and field data.

In general the study demonstrated that CFD could be used in reviewing settling tank design or performance and that the results give valuable insight into how the tanks are working. It can be inferred that CFD could be use to evaluate settling tank designs where the tanks are not functioning properly.

13. Acknowledgement

This chapter is supported by the Scientfic Grant Agency of Ministry of Education-VEGA Projekt No. 1/1143/11 with at the Department of Sanitary and Environmental Engineering, Faculty of Civil Engineering, Slovak University of Technology in Bratislava.

14. References

[1] Huggins D.L., Piedrahita R.H., & Rumsey T., (2005). Use of computational fluid dynamics (CFD) for aquaculture raceway design to increase settling effectiveness, Aquacult. Eng. 33, 167–180.

[2] Dick, R. I. (1982) Sedimentation Since Camp, J Society of Civil Engrs, 68, 199-235.

[3] Abdel-Gawad, S. M. & McCorquodale, J. A. (1984b). Hydrodynamics of Circular primary clarifiers. Canadian J. Civil Engr. ,11, 299-307.

[4] Adams, E.W. & Rodi, W. (1990). Modelling flow and mixing in sedimentation tanks, Journal of Hydraulic Engineering, 116(7):895-913, Jul.

[5] Launder, M. & Spalding, D. (1974). The numerical computation of turbulent flows, Computer methods in applied mechanics and engineering, (3), 269-289.

[6] Ghawi, A.H., (2008). A numerical model of flow and settling in sedimentation tanks in potable water treatment plants, Edícia vedeckých prác (Science Works Edition) , Workbook no. 64, STU Bratislava, pp. 150, ISBN 978-80-227-2964-2

3

Fluid Dynamics Analysis of a Space Vehicle Entering the Mars Atmosphere

Antonio Viviani[1] and Giuseppe Pezzella[2]
[1]*Seconda Università degli Studi di Napoli, Aversa*
[2]*Centro Italiano Ricerche Aerospaziali, Capua*
Italy

1. Introduction

This paper reports the results of design analyses of two Manned Braking Systems (MBS) entering the Mars atmosphere, with the aim of supporting design studies of a planetary entry system. Two lifting body configurations with rounded edge delta-like cross section have been analyzed. The preliminary aerodynamic and aerothermodynamic analyses have considered flight conditions compatible with a manned mission entering the Mars atmosphere. However, neither the mission architecture needed to reach Mars from Earth or neighbour Earth space, nor surface exploration have been addressed.

All the design analyses have been performed at several levels. Indeed, vehicle aerodynamic assessment has been extensively addressed through simplified design approach as, e.g., hypersonic panel methods (HPM); then, a number of fully three-dimensional computational fluid dynamics (CFD) simulations, both with Euler and Navier-Stokes approximations, of the hypersonic flowfield past the entry vehicle have been performed.

The results herein provided have been obtained for a Mars entry scenario compliant with an approach to the red planet both by direct planetary entry and entry after aerobraking (Polishchuk et al., 2006) (Viviani and Pezzella, 2009). These results may be used to provide numerical data for understanding requirements for the human exploration of Mars.

2. Computational flowfield analysis

CFD analyses have been performed to assess the aerothermal environment that the MBS experiments during descent, thus evaluating several surface loading distributions (e.g., pressure and heat flux). To this aim, several fully three-dimensional numerical computations, both for perfect and chemically reacting gas approximation, have been performed.

The vehicle configurations, under investigation in this work, are shown in Fig. 1 These lifting bodies feature an aerodynamic configuration with a compact body about 8 [m] long with a rounded edge delta-like cross section (Hanley et al., 1964). A very preliminary internal layout for a crew of four astronauts is also reported in Fig. 1.

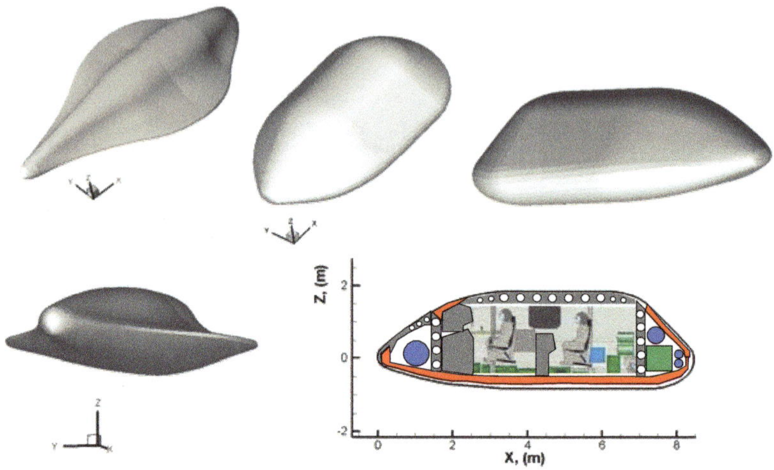

Fig. 1. Vehicle configurations with quotes.

2.1 Freestream conditions

The flight scenario considered so far is summarized in Table 1. They refer to entry conditions compatible with a vehicle entering the Mars atmosphere both from a hyperbolic orbit (HO), e.g., direct planetary entry, and an elliptic orbit (EO) e.g., planetary entry after aerobraking (Gupta et al., 1996).

		Mach [-]	AoA [deg]	Altitude [km]
Perfect Gas		10	10	10.0
		15	40	60.0
		22	40	60.0
	EO PH	22	40	44.2
	HO PH	26	40	52.1
Reacting Gas		10	10	10.0
		10	20	10.0
		10	30	10.0
		10	40	10.0
EO PH	NCW	22	40	44.2
	FCW	22	40	44.2
HO PH	NCW	26	40	52.1
	FCW	26	40	52.1

EO PH (Elliptic Orbit Peak Heating)
HO PH (Hyperbolic Orbit Peak Heating)
NCW (Non-Catalytic Wall)
FCW (Fully Catalytic Wall)

Table 1. CFD freestream conditions

Therefore, thirteen CFD numerical simulations have been performed. As one can see, CFD computations (both Euler and Navier-Stokes) have been performed, both in trajectory-based and space-based design approaches (Hanley et al., 1964). Several Mach numbers and different angles of attack (α) have been investigated and compared each other. The Fluent code together with user defined functions (UDFs), developed in order to simulate mixtures of gas in thermo-chemical non-equilibrium, have been used for CFD computations with a non-equilibrium chemical model suitable for Martian atmosphere (Gupta et al., 1996) (Mack et al., 2008) (Kustova et al., 2009).

For the reacting gas computations, the Martian atmosphere has been considered as a mixture of 95.7% carbon-dioxide, 1.6% Argon and 2.7% nitrogen. The flow has been modelled as a reacting gas mixture of 9 species (Ar, CO_2, N_2, O_2, CO, NO, N, O, C) involved in the chemical reactions of Table 2 (Park et al., 1994) (Anderson, 1989). The reaction mechanism and the related chemical kinetics, taken into account in the present work, are summarized in Table 2, where M is the reacting partner (third body) that can be any of the nine reacting species of the gas mixture.

Non-equilibrium computations have been performed since one of the most challenging problem facing the design of atmospheric entry vehicle is the phenomenon of "real gas behaviour". At hypersonic speeds, the shock wave produced ahead of the vehicle suddenly elevates the gas temperature in the shock layer. So the gas thermal energy may be comparable with the energy associated with a whole range of gas chemical processes such as: molecular vibrational excitation; dissociation of atmospheric molecules into their atomic forms; formation of other chemical species through recombination reactions; and ionisation of both molecular and atomic species (Park et al., 1994).

Reaction	Third Body M	A_r $[cm^3 mol^{-1} s^{-1}]$	β_r	T_d [K]
$CO_2+M \rightarrow CO+O+M$	CO_2,CO,N_2,O_2,NO	6.9×10^{21}	-1.5	63275
	Ar	6.9×10^{20}		
	C,N,O	1.4×10^{22}		
$CO+M \rightarrow C+O+M$	CO_2,CO,N_2,O_2,NO	2.3×10^{20}	-1.0	129000
	Ar	2.3×10^{19}		
	C,N,O	3.4×10^{20}		
$N_2+M \rightarrow N+N+M$	CO_2,CO,N_2,O_2,NO	7.0×10^{21}	-1.6	113200
	Ar	7.0×10^{21}		
	C,N,O	3.0×10^{22}		
$O_2+M \rightarrow O+O+M$	CO_2,CO,N_2,O_2,NO	2.0×10^{21}	-1.5	59750
	Ar	3.0×10^{21}		
	C,N,O	3.0×10^{22}		
$NO+M \rightarrow N+O+M$	CO_2,C,N,O,NO	1.1×10^{17}	0.0	75500
	Ar	5.0×10^{15}		
	CO,N_2,O_2	5.0×10^{15}		
$C_2+M \rightarrow C+C+M$	*All*	2.0×10^{21}	-1.5	59750
$NCO+M \rightarrow CO+N+M$	*All*	6.3×10^{16}	-0.5	24000
$NO+O \rightarrow N+O_2$		8.4×10^{12}	0.0	19450
$N_2+O \rightarrow NO+N$		6.4×10^{17}	-1.0	38370
$CO+O \rightarrow C+O_2$		3.9×10^{13}	-0.18	69200
$CO_2+O \rightarrow CO+O_2$		2.1×10^{13}	0.00	27800

Table 2. Reactions mechanism and rate parameters (Park et al., 1994).

Therefore, the gas mixture has to be considered in thermal and chemical non-equilibrium. Finally, the CFD analysis of the MBS have been preceded by a code validation phase performed considering the available numerical and experimental data for the Mars Pathfinder probe at entry peak heating conditions, as summarized in (Viviani et al., 2010) (Gnoffo et al., 1998) (Gnoffo et al., 1996) (Mitcheltree et al., 1995).

2.2 Numerical results

The aerodynamic analysis of MBS is shown in term of lift (C_L), drag (C_D) and pitching moment (C_{My}) coefficients which are calculated according to Eq. (1) and Eq. (2), respectively.

$$C_i = \frac{F_i}{\frac{1}{2}\rho_\infty v_\infty^2 S_{ref}} \qquad i = L, D \tag{1}$$

$$C_{Mj} = \frac{M_j}{\frac{1}{2}\rho_\infty v_\infty^2 L_{ref} S_{ref}} \qquad j = Y \tag{2}$$

The reference parameters L_{ref} (e.g., longitudinal reference length) and S_{ref} (e.g., reference surface) are the vehicle length (e.g., 8 m) and planform area (e.g., 32 m^2). The pitching moment is computed from the vehicle nose (i.e. 0, 0, 0). Engineering based aerodynamic analysis has been extensively performed by using a 3D Panel Methods code developed by CIRA, namely HPM (Viviani and Pezzella, 2009). This tool, at high supersonic and hypersonic speeds, is able to accomplish the aerodynamic and aerothermodynamic analyses of a complex re-entry vehicle configuration by using simplified approaches as local surface inclination methods and approximate boundary-layer methods, respectively. The SIM typical of hypersonics are based on Newtonian, Modified Newtonian, and Prandtl-Mayer theories (Anderson, 1989). Typical surface meshes of the MBS, used for the engineering level computations, are shown in Fig. 2.

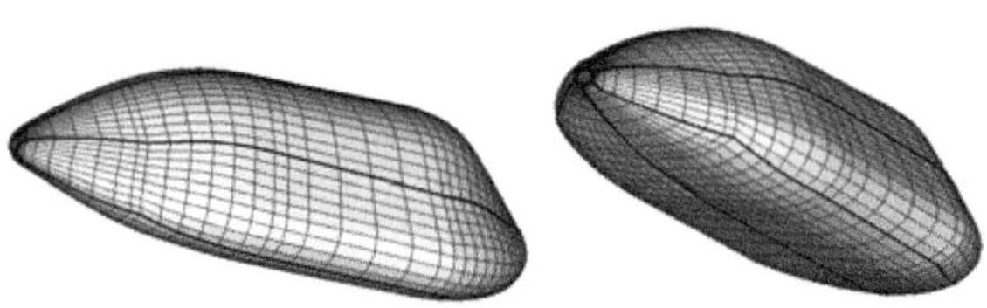

Fig. 2. Example of panel mesh for engineering-based aerodynamic analysis.

MBS aerodynamic results, provided by engineering-based analysis, cover α ranging from 0 to 50 deg. Present CFD computations for the MBS have been carried out on a 3-D multiblock structured grid close to that shown in Fig. 3. The grid consists of about 20 blocks and 900.000 cells (half body). Both computational domains are tailored for the free-stream conditions of Table 1. The distribution of surface grid points has been dictated by the level of resolution desired in various areas of the vehicle such as the stagnation region and the base fillet one, according to the computational scopes. Fig. 3 shows also a close-up view of the 3-D mesh on the vehicle surface and pitch plane. Grid refinement in strong gradient regions of flowfield has been made through a solution adaptive approach.

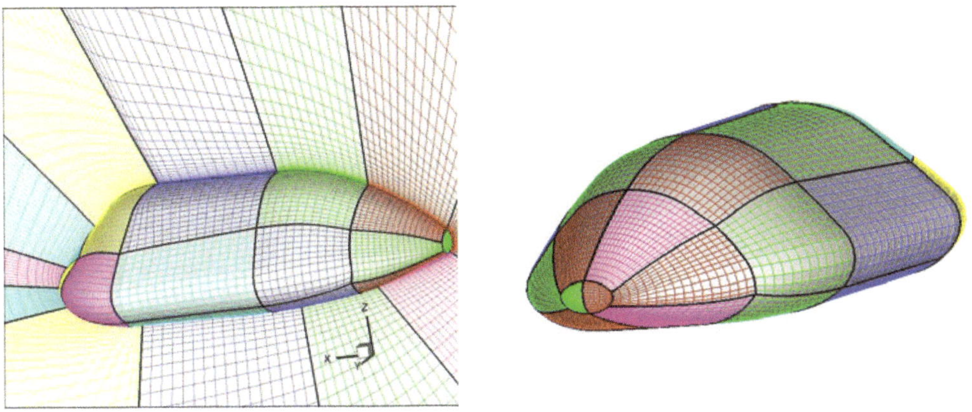

Fig. 3. Example of computational mesh domains for Euler CFD simulations

The preliminary results of CFD simulations performed so far are summarized hereinafter. For example, Fig. 6 shows the static temperature contours on the vehicle symmetry plane and static pressure contours on vehicle surface at M_∞=20 and α=20 deg, considering the Mars atmosphere as a reacting gas mixture. As shown, the MBS bow shock structure around the descent vehicle can be appreciated as well.

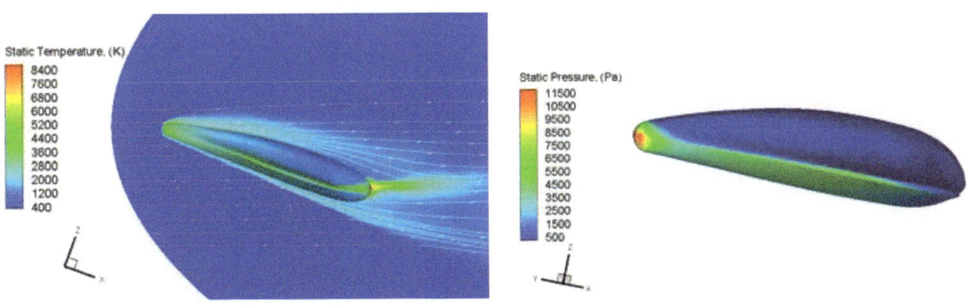

Fig. 4. Results for M_∞=20 and α=20 deg. (Left) static temperature field on vehicle symmetry plane; (right) static pressure contour on vehicle surface for non-equilibrium reacting gas

At the same flight conditions, Fig. 5 reports on chemical dissociation of the flow in the shock layer considering the contours of CO_2 mass fraction on MBS pitch plane. As a consequence, flow dissociation determines a large density ratio ε across the strong bow shock compared with a flow of the same gas where no dissociation takes place (Viviani and Pezzella, 2009) (Anderson, 1989). This results in a thinner shock layer around the entry vehicle (e.g., lower stand-off distance).

Under conditions where dissociation exists, the aerodynamics of vehicle depends primarily on shock density ratio. In fact, the change of aerodynamic characteristics is the result of change in surface pressure acting on the vehicle forebody (Gnoffo et al., 1998) (Viviani et al., 2010).

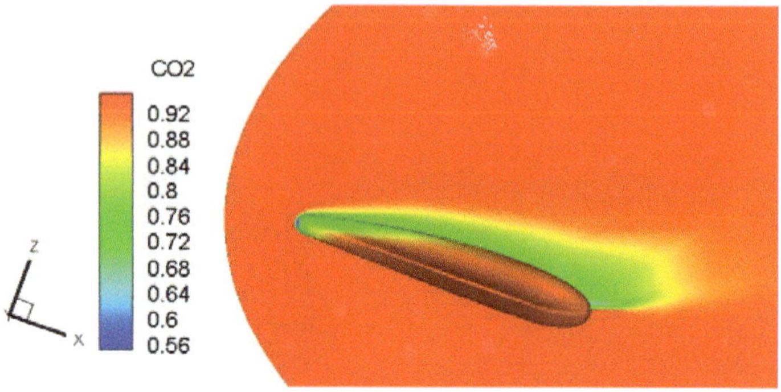

Fig. 5. Results for M_∞=20 and α=20 deg. Contours of CO_2 mass fraction

Moreover, Fig. 6 shows CFD results for M_∞=21 and α=40 deg. The left side reports pressure coefficient contours (C_p) on vehicle surface and on two cross sections; whereas on the right C_p contours on vehicle surface and Mach number contours on three cross sections have been shown.

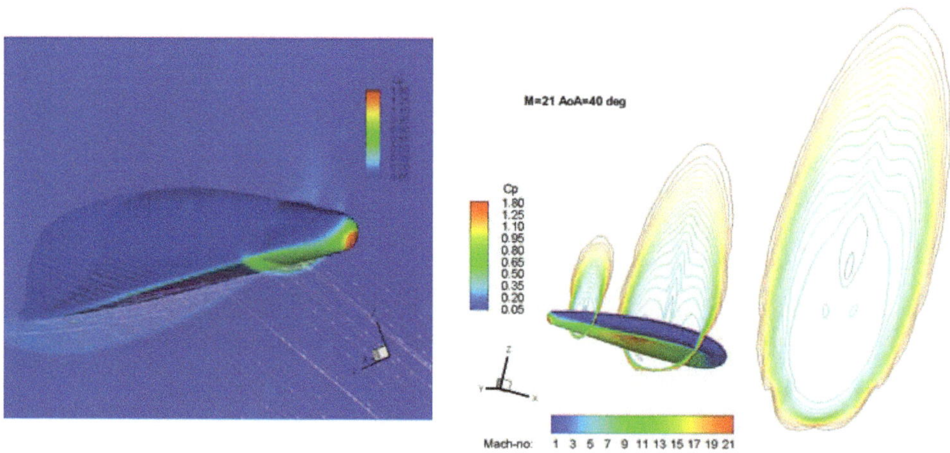

Fig. 6. Results for M_∞=21 and α=40 deg. (Left) pressure coefficient contours (C_p) on vehicle surface and on two cross sections; (right) C_p contours on vehicle surface and Mach number contours on three cross sections

The curves of lift and drag coefficients are shown in Fig. 7. Those curves collect MBS aerodynamic coefficients compared with available numerical data both for perfect gas and reacting gas approximations, reported in order to highlight accuracy of both numerical and engineering-based results (Viviani et al., 2010).

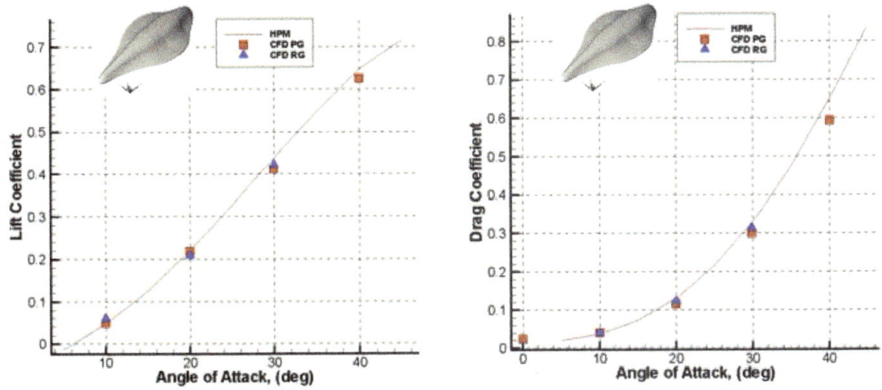

Fig. 7. C_L and C_D versus α. Comparison between panel methods and CFD results for perfect and reacting gas approximations.

As one can see, engineering and numerical data compare very well, thus confirming that engineering-based estimations represent reliable preliminary aerodynamics of a Mars entry vehicle. Moreover, real gas effects increase the aerodynamic drag coefficient whereas the lift is only slightly influenced.

As far as CFD results for the second configuration are concerned, Fig. 8 shows the Mach number contour field that takes place around the vehicle when it is flying at the peak heating conditions of entry by EO (e.g., M_∞=22, α=40 deg, and H=44.20 km).

In particular, the left side of Fig. 8 shows the Mach contour field on the vehicle pitch plane while at the right side of Fig. 8 gives an idea of the bow shock shape that envelopes the vehicle since the Mach field is reported on three different flowfield cross sections.

As shown, a thin shock layer envelopes the entry vehicle windside with a strong expansion that characterizes the flow at the end of the vehicle.

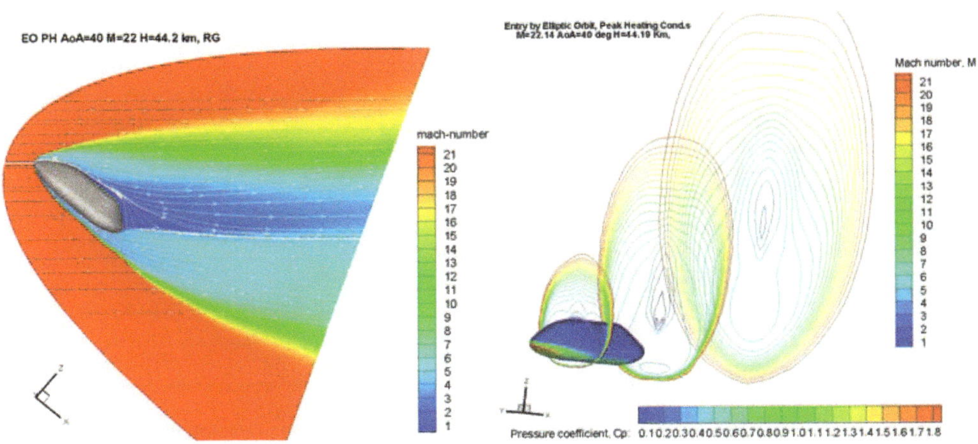

Fig. 8. Mach contours on the vehicle pitch plane and three flowfield cross sections at the EOPH conditions.

The CO mass fraction field around the vehicle for the same freestream conditions is shown in Fig. 9 where some streamtraces colored by Mach number are also reported.

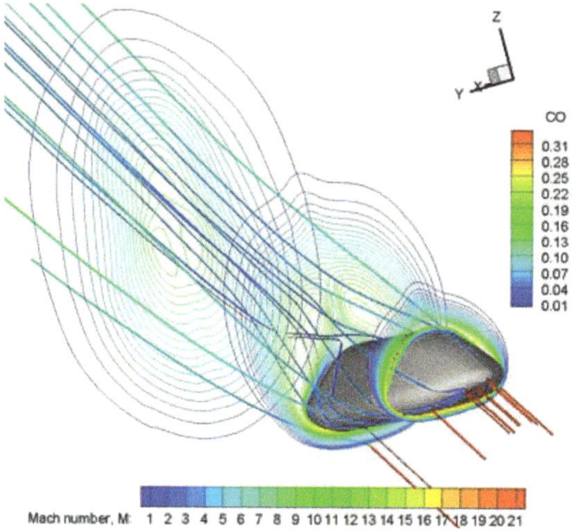

Fig. 9. CO mass fraction at the EOPH conditions on three cross sections with streamtraces colored by Mach number

As shown the CO concentration reaches its maximum value close to the body. Fig. 10 shows the temperature comparison among non-equilibrium flow (right side of pilot) and perfect gas computation, evaluated at three flowfield cross sections (x=1.5 [m], 5.5 [m] and 9.5 [m]). It is clearly evident how real gas phenomena affect the vehicle shock layer, thus confirming all the conclusions highlighted before.

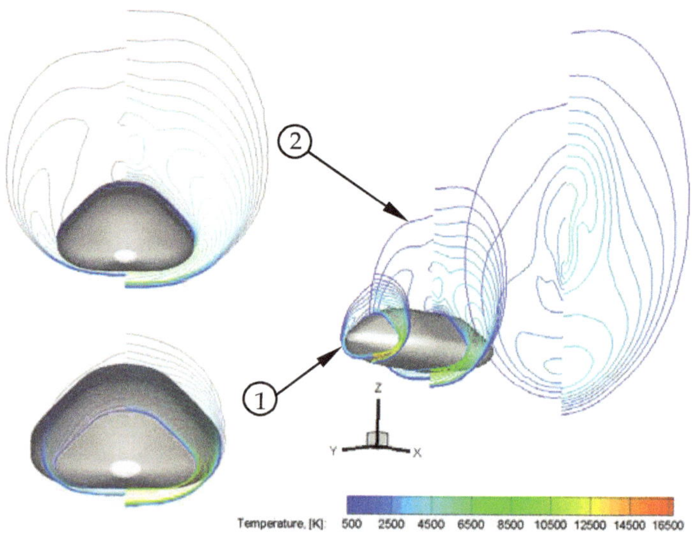

Fig. 10. Temperature comparison between non-equilibrium flow (right side of pilot) and perfect gas computation at x=1.5 [m], 5.5 [m] and 9.5 [m] flowfield cross sections

Finally, the curves of lift, drag, and pitching moment coefficients are shown in Fig. 11. Real gas effects increase both drag and pitching moment coefficients, whereas the lift is only slightly influenced. Vehicle aerodynamics is also summarized in the table of Fig. 11.

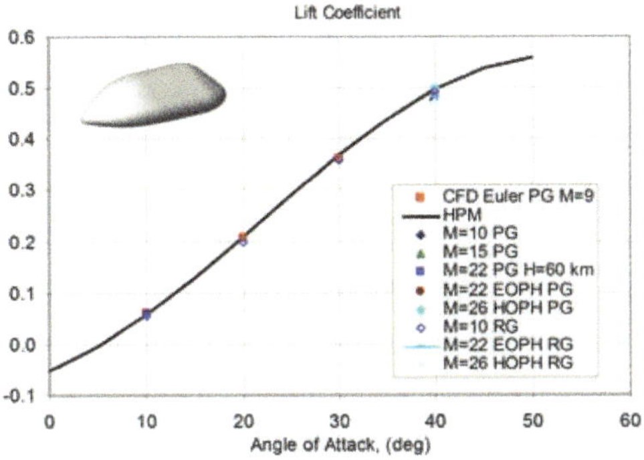

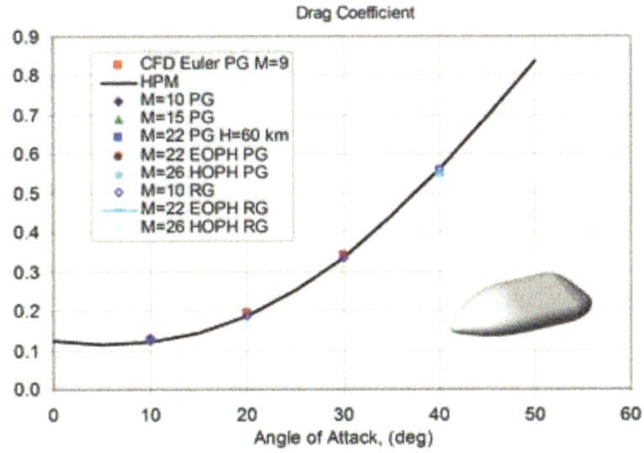

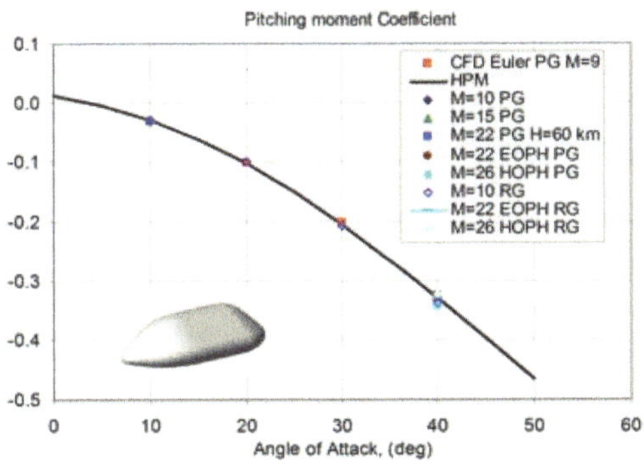

		Mach [-]	AoA [deg]	Altitude [km]	CL [-]	CD [-]	CM* [-]
Perfect Gas		10	10	10.0	0.06150	0.13060	-0.03170
		15	40	60.0	0.47830	0.55360	-0.32940
		22	40	60.0	0.47930	0.55540	-0.33010
	EO PH	22	40	44.2	0.48300	0.55500	-0.33200
	HO PH	26	40	52.1	0.49920	0.55930	-0.34330
Reacting Gas		10	10	10.0	0.05710	0.12450	-0.02950
		10	20	10.0	0.19940	0.18730	-0.09930
		10	30	10.0	0.35970	0.33420	-0.20650
		10	40	10.0	0.49340	0.56100	-0.33770
EO PH	NCW	22	40	44.2	0.48368	0.55014	-0.32980
	FCW	22	40	44.2	0.48369	0.55018	-0.32990
HO PH	NCW	26	40	52.1	0.47122	0.53627	-0.32220
	FCW	26	40	52.1	0.47875	0.54536	-0.32560

EO PH (Elliptic Orbit Peak Heating)
HO PH (Hyperbolic Orbit Peak Heating)
NCW (Non-Catalytic Wall)
FCW (Fully Catalytic Wall)
* pole @ vehicle nose

Fig. 11. Lift, Drag and pitching moment coefficients versus α. Comparison between panel methods and CFD results for perfect and non-equilibrium gas computations.

3. Conclusion

The paper deals with the flowfield analysis of two braking systems for manned exploration mission to Mars.

A number of fully 3D Navier-Stokes and Euler CFD computations of the hypersonic flowfield past two lifting body vehicles have been performed for several freestream conditions of a proposed Mars entry loading environment. These evaluations have been aimed at carrying out only a preliminary design of the MBS configuration, in compliance with the Phase-A design level.

The range between Mach 2 and Mach 26 has been analyzed, to provide both aerodynamic databases according to both the space-based and trajectory-based design approaches. Numerical results show that real gas effects increase both the aerodynamic drag and pitching moment coefficient, whereas the lift is only slightly influenced.

4. References

Anderson, J. D., *Hypersonic and High Temperature Gas Dynamics*, McGraw-Hill Book Company, New York, 1989.

Gnoffo, P., Weilmuenster, K., Braun, R., Cruz, C., *Influence of Sonic-Line Location on Mars Pathfinder Probe Aerothermodynamics*, Journal of Spacecraft and Rockets, vol. 33 No 2, March-April 1996.

Gnoffo, P., Braun, R., Weilmuenster, K., Mitcheltree, R., Engelung, W., Powell, R., *Prediction and Validation of Mars Pathfinder Hypersonic Aerodynamic Data Base*, 7th AIAA/ASME Joint Thermophysics and Heat Transfer Conference. June 15-18, Albuquerque, NM (USA). 1998.

Gupta, R., Lee, K., Scott, C., *Aerothermal Study of Mars Pathfinder Aeroshell*, Journal of Spacecraft and Rockets, vol.33 No 1, Jan.-Feb. 1996.

Hanley, G. M., Lyon, F. J., *The Feasibility of Spacecraft Deceleration by Aerodynamic Braking at the Planet Mars*, Proc. of the 1th AIAA Annual Meeting. Washington, D.C. June 29-July 2, 1964. AIAA-64-479.

Hannemann, V., Mack, A., *Chemical Non Equilibrium Model of the Martian Atmosphere*, Proc. of the 6th European Symposium on Aerothermodynamics for Space Vehicles. Versailles, France. 3-6 Nov. 2008 – ESA SP-659, January 2009.

Kustova, E.V., Nagnibeda, E.A., Shevelev, Y. D., Syzranova, N.G., *Comparison of Non-Equilibrium Supersonic CO_2 Flows with Real Gas Effects near a Blunt Body*, Proc. of the 6th European Symposium on Aerothermodynamics for Space Vehicles. Versailles, France. 3-6 Nov. 2008 – ESA SP-659, January 2009.

Mack, A., *CFD Validation for CO_2 Reentry Applications*, 2nd Inter. ARA Days. Arcachon, France, 21-23 October 2008. AA-3-2008-37.

Mitcheltree, R. A., Gnoffo, P. A., *Wake Flow about the Mars Pathfinder Entry Vehicle*, Journal of Spacecraft and Rockets, Vol. 32 No 5, Sept.- Oct. 1995.

Polishchuk, G., Pichkhadze, K., Vorontsov, K., Pavela, V., *Proposal on application of Russian technical facilities for International Mars Research Program for 2009–2015*, Acta Astronautica, Vol. 59, pp.113-118. 2006.

Park, C., Howe, J. T., Jaffe, R. L., Candler, G. V., *Review of Chemical-kinetic Problems of Future NASA Missions, II: Mars Entries*, Journal of Thermophysics and Heat Transfer, vol. 8, No 1, pp. 9-23, Jan-Mar 1994.

Viviani, A., Pezzella, G., *Aerodynamic Analysis of a Capsule Vehicle for a Manned Exploration Mission to Mars*. 16th AIAA/DLR/DGLR International Space Planes and Hypersonic Systems and Technologies Conference. Bremen Oct. 2009. AIAA-2009-7386.

Viviani, A., Pezzella, G., Golia, C., *Aerodynamic and thermal design of a space vehicle entering the Mars atmosphere*. The Fifth International Conference on Thermal Engineering Theory and Applications. May 10-14, 2010 Marrakesh. Morocco.

Viviani, A., Pezzella, G., Golia, C., *Aerothermodynamic analysis of a space vehicle for manned exploration missions to Mars*. 27th International Congress of the Aeronautical Sciences. ICAS 2010. Nice, France, 19-24 September 2010.

4

Advances in Computational Fluid Dynamics Applied to the Greenhouse Environment

Jorge Flores-Velázquez[1], Guillermo De la Torre-Gea[2], Enrique Rico-García[2], Irineo L. López-Cruz[1] and Abraham Rojano-Aguilar[1]
[1]Department of Agricultural Engineering, University of Chapingo, Chapingo
[2]Department of Biosystems, School of Engineering
Queretaro State University Querétaro
México

1. Introduction

In order to perform an accurate study of the climate inside a greenhouse, it is necessary to use models based on heat and mass transfer (Rico-García *et al.*, 2008). Mathematical models based on mass and energy balances assume a homogeneous greenhouse environment. These models generate a set of nonlinear ordinary differential equations without an analytical solution. However, a more detailed monitoring of the environment inside the greenhouses reflects a two-dimensional and three-dimensional variability of climatic variables. Recently, this problem has been tackled using the fundamental equations of fluid dynamics. The set of numerical methods applied in order to solve those equations are called Computational Fluid Dynamics (CFD). Computational Fluid Dynamics (CFD) provides a numerical solution from an energy balance of a controlled volume, which in comparison with other methods and expensive technologies allows an efficient study of the climate inside the greenhouse. CFD techniques consider the values of the independent variables as primary unknowns in a finite number of places inside the domain, and then a set of algebraic equations are derived from the fundamental equations applied to the domain and can be solved by pre-establish algorithms.

In spite of, the greenhouse is a very complex bio-system, in which there are several physical, chemical and biological interacting process and phenomena, during the last decade, due to the development of computer simulation tools and the increase in computational processing power, it is possible to develop numerical models for the greenhouse environment such as more accurate models for transport phenomena and energy exchange inside the greenhouse. As a consequence, these studies have led improvements in the design of greenhouses (Norton et al., 2007).

According to Boulard *et al.* (2002), CFD is a branch of fluid mechanics that uses numerical methods and algorithms to solve and analyze problems involving fluids flow. Therefore, it is possible with the use of computers to perform millions of calculations to simulate the interaction of liquids and gases with surfaces defined by the boundary conditions. In recent studies the modeling of air flow, CFD has deepened to test their effectiveness in relationships of climatic factors (Bournet and Boulard, 2010). Computational parametric studies on greenhouse structures can help to the identification of design factors that affect greenhouse ventilation under specific climatic conditions (Romero-Gómez *et al*, 2008; Romero-Gómez *et al.*, 2010).

In the last years, many studies have used CFD to investigate the climatic conditions inside greenhouses. CFD has been able to increase the degree of realism by taking into account insect-proof screens and simulation of the crop effect, considering it as a porous medium, among others in 3D models. The results have been able to improve our understanding of the phenomenon of greenhouse ventilation. Therefore, this chapter discusses significant recent studies to understand how the use of CFD has evolved.

2. Fundamental CFD equations and methodology

2.1 Finite element models using CFD

Computational fluid dynamics is based on the governing fluid dynamics equations (continuity, momentum and energy). The set of equations obtained directly from the volume or fixed element in space is known as "conservative form" Euler type. The equations obtained directly from the volume or movement with the fluid element are called "non-conservative form" Lagrange type (Anderson, 1995).

2.2 Substantial derivative

The substantial derivative physically is the exchange rate of any substance that moves with a fluid element. It consists of two parts, where the first part is called the local derivative, which means the rate of change over time in a fixed point. The second part is called the convective derivative, which physically is the exchange rate due to movement of the fluid from one point to another in the field of fluid, where the fluid properties are spatially different. The resulting material can be applied to any field variable fluid, for example: velocity(u), pressure (p) or temperature (T) (Anderson, 1995).

$$\nabla \equiv i\frac{\partial}{\partial x} + j\frac{\partial}{\partial y} + k\frac{\partial}{\partial z} \tag{1}$$

$$V \equiv (u, v, w)$$

$$\frac{\partial}{\partial t}\iiint_V \rho dV + \iint_S \rho V \cdot dS = 0 \tag{2}$$

$$\rho\frac{Du}{Dt} = -\frac{\partial p}{\partial x} + \frac{\partial t_{xx}}{\partial x} + \frac{\partial t_{yx}}{\partial x} + \frac{\partial t_{zx}}{\partial x} + \rho f_x$$

$$\rho\frac{Dv}{Dt} = -\frac{\partial p}{\partial x} + \frac{\partial t_{xy}}{\partial x} + \frac{\partial t_{yy}}{\partial x} + \frac{\partial t_{zy}}{\partial x} + \rho f_y$$

$$\rho\frac{Dw}{Dt} = -\frac{\partial p}{\partial x} + \frac{\partial t_{xz}}{\partial x} + \frac{\partial t_{yz}}{\partial x} + \frac{\partial t_{zz}}{\partial x} + \rho f_z \tag{3}$$

$$\rho\frac{Dw}{Dt}\left(e+\frac{V^2}{2}\right)=\rho q+\frac{\partial}{\partial x}\left(k+\frac{\partial T}{\partial x}\right)+\frac{\partial}{\partial y}\left(k+\frac{\partial T}{\partial y}\right)+\frac{\partial}{\partial z}\left(k+\frac{\partial T}{\partial z}\right)-\frac{\partial(up)}{\partial x}-\frac{\partial(vp)}{\partial y}-\frac{\partial(wp)}{\partial z}$$

$$+\frac{\partial(ut_{xx})}{\partial x}+\frac{\partial(ut_{yx})}{\partial y}+\frac{\partial(ut_{zx})}{\partial z}+\frac{\partial(vt_{xy})}{\partial x}+\frac{\partial(vt_{yy})}{\partial y}+\frac{\partial(vt_{zy})}{\partial z}$$

$$+\frac{\partial(wt_{xz})}{\partial x}+\frac{\partial(wt_{yz})}{\partial y}+\frac{\partial(wt_{zz})}{\partial z}+pf\cdot V \tag{4}$$

1. **Continuity equation**
2. **Momentum equation (a non-conservative)**
3. **Components in x, y and z**
4. **Energy equation (a non-conservative)**

The fundamental CFD equations form a coupled system of nonlinear partial differential equations. So far no analytical solution has been found. It is commonly assumed that the fluid is an ideal gas where the intermolecular forces can be neglected. For an ideal gas, the state equation is:

$$p=\rho RT \tag{5}$$

Where R is the specific gas constant. For a calorically ideal gas we have:

$$e=CvT, \tag{6}$$

Where Cv is the specific heat at constant volume (Rodríguez, 2006, Norton et al, 2007).

2.3 CFD procedure applied to the greenhouse environment

Domain setting and grid generation are the first steps in the CFD modeling process. This implies the choice of a computational domain that is large enough to correctly assess the main mechanisms that occur in the system and to avoid interference with artificial Boundaries (Bournet and Boulard, 2010). The CFD modeling process encompasses three stages: preprocessing, solution and post-processing.

The preprocessing is the most time-consuming activity because of mesh generation that is the basis for an accurate simulation, does require heavy calculations. Therefore, keeping an adequate strategy should allow reliability in the calculations and physically consistent results. The meshing process is based on a serie of activities that can be summarized as follows:

a. Geometry definition (Figure 1)
b. Geometry decomposition
c. Computational mesh generation (Figure 2), refining, giggling, quality
d. Functional and quality meshing
e. Definition of the boundary conditions and
f. Export mesh

One of the most important problems currently in CFD modeling of the greenhouse environment is the time-consuming for a simulation to converge, due to the high number of cells that come from the process of meshing. Thus, for many years was chosen to model only in two dimensions (2D) (Flores-Velázquez, 2010). However, air distribution inner greenhouse has a third-dimensional (3D) pattern, for that, nowadays 3D CFD models are more commonly developed. According to Bournet and Boulard (2010) the calculation process is an iterative procedure that requires the definition of convergence criteria, and the user of the CFD codes must to decide on an appropriate level of convergence, of the solution. Generally, 10^{-4} of the value of a variable at all of the nodes is used. Faster convergence may also be reached by optimizing the grid shape or by assuming the Boussinesq model as density dependent rather than by setting up the problem on the basis of the ideal gas theory. However, this choice may not be applicable in the case of large thermal gradients and may also overlook the influence of temperature on air viscosity.

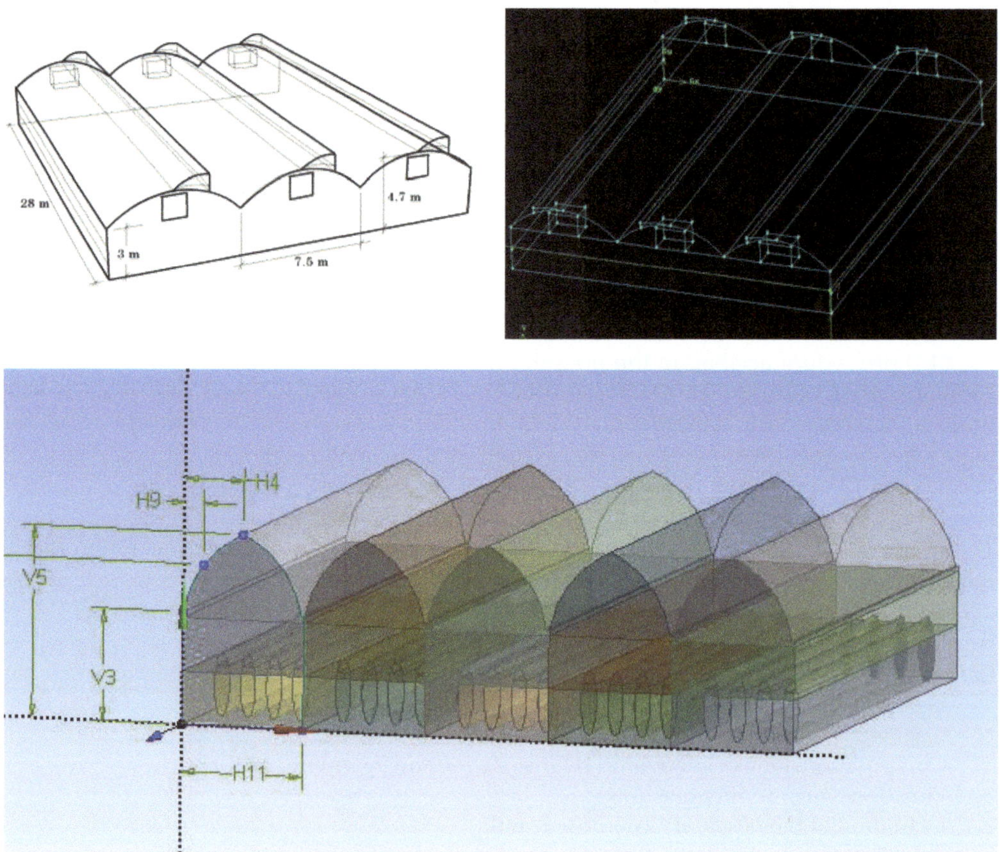

Fig. 1. Geometry generation for a greenhouse.

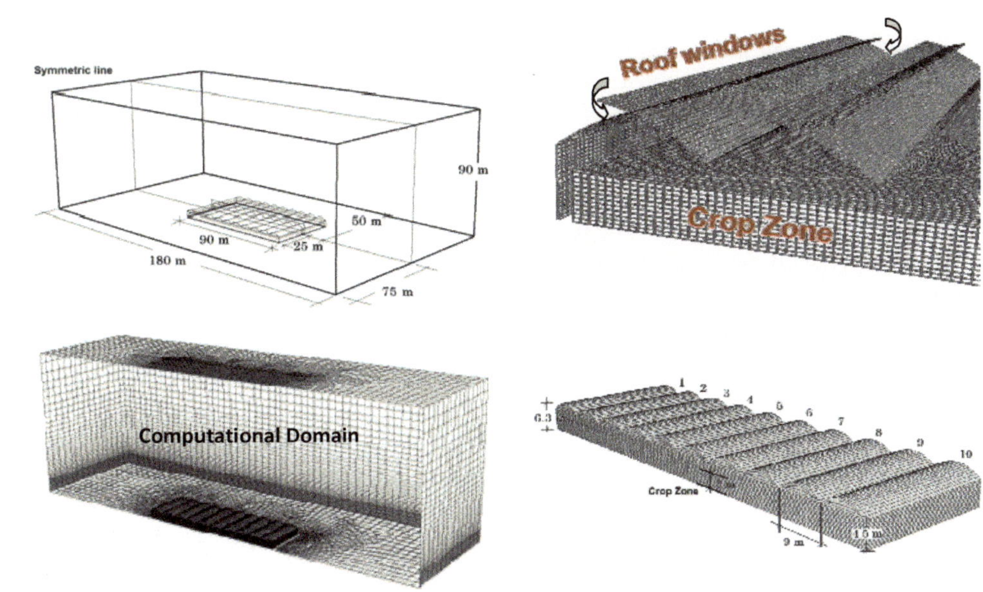

Fig. 2. Computational mesh generation for a greenhouse (Flores-Velázquez, 2010).

The presence of turbulence in a fluid is indicated by the fluctuating velocity components and the quantities carried out by the flow, even when the boundary conditions for the problem under study are kept constant. These fluctuations determine the difference between laminar flow and turbulent flow (Figure 3). For most situations, ventilation (effect of temperature, wind or both) measurements and visualization experiments have demonstrated the turbulent air flow inside and outside the greenhouse. Therefore, the phenomenon of turbulence must be taken into account (Norton et al. 2007).

Fig. 3. Velocity vectors on a multi-hood greenhouse of four spans. CFD model considers anti-insect mesh vents (Rico-García, 2008).

If inertial effects are large enough with respect to viscous effects, then the flow can be turbulent. Turbulence means that the instantaneous velocity varies at each point of the flow field. The turbulent nature of the velocity can be explained considering that the rate consists of the sum of two components, a main component (stable) and a fluctuating component. Depending on Reynolds number, laminar or turbulent flow can be modeled. For instance, most turbulence models, such as standard k-ε Model, and Re-Normalized Group Turbulence Model (RNG), to name a few (Rico-García, 2008).
The post-processing stage allows the user to visualize and search for the solution. Figures contours, vectors and graphs can be obtained from analyzing the solution. It is remarkable that figures allow us to observe the full distribution of temperature, speed, pressure and so on the whole flow field (Figure 4).

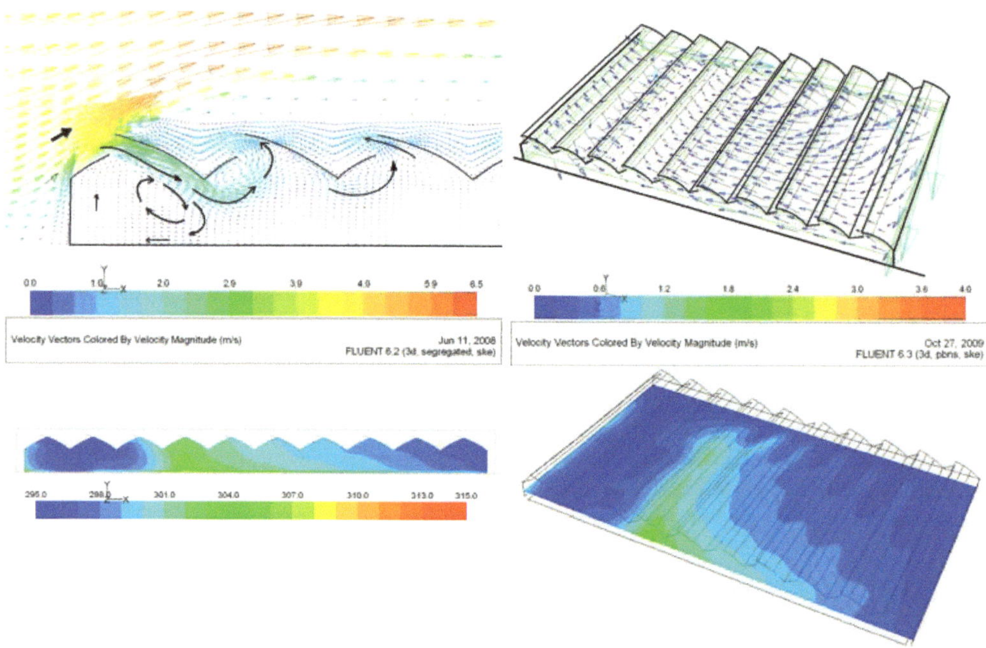

Fig. 4. Wind velocity (m s^{-1}) and temperature (K) representative post processing characteristics at different greenhouses sceneries (Flores-Velázquez, 2010)

As soon as a CDF model has been tested, the computational greenhouse environment can become a powerful climate analysis tool. Nowadays, it is possible to visualize, for instance, the wind distribution along the greenhouse when the income windows are up or down, and also the consequent temperature profiles, among many other possibilities (Figure 5). Even though, in the last decade research on wind behavior inside the greenhouse has been enormous, still, as a fundamental part of the greenhouse environment modeling process, it is necessary to take into consideration the physical verification in order to provide accuracy on the results obtained by numerical simulation (Flores-Velázquez, 2010). Scale models, water and wind tunnels and direct measurements of the climatic variables are some of the main options for verification of the CFD models of the greenhouse climate (Flores-Velázquez, *et al.*, 2011).

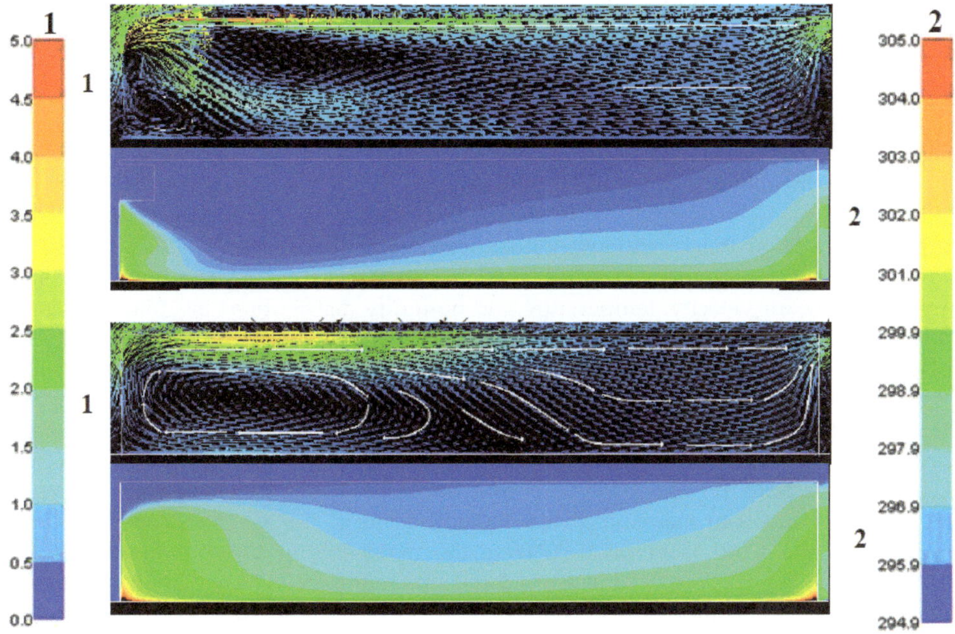

Fig. 5. Comparison of wind speed (m s^{-1}) (1) and profiles of air temperature (K) (2) in the greenhouse with and without the projection of the rectangular window. Wind speed exterior 4 m s^{-1}, soil heat flux 315 W m^{-2} (Flores-Velázquez, 2010).

3. Approaches used in the application of CFD to the greenhouse environment

CFD modeling is used to design facilities that provide suitable climatic conditions for the crops. According to Sase (2006), within a mild climate, appropriate design and control of ventilation are required to ensure effective cooling and uniformity of the environment. It is possible to design an optimal greenhouse by calculating its area, volume and vents area as well as the material properties of the roof (Impron *et al.*, 2007). Rico-García *et al.* (2006), comparing two different greenhouses, showed the importance of its geometry and found that the ventilation rate for a greenhouse with larger vertical roof and windows was better than a multi-span greenhouse. Omer (2009) describes several designs of low energy greenhouses. In agreement with Baeza *et al.* (2008), design changes in the greenhouse, such as size and shape of vents, can improve air movement in the area of crops. Bakker *et al.* (2008) investigated energy balance, and determined that the amount of energy used per unit of output is defined by improvements in energy conversion, environmental control to reduce energy consumption and efficiency of agricultural production. In a study of outdoor areas using the turbulence model Reynolds-averaged Navier-Stokes equations (RANS), van Hoff (2010) found that small geometric modifications can increase the ventilation rate up to 43%. The performance of ventilation in enclosed spaces is affected by the flow of outside air, type of cover, height of the installation and the ventilation opening (Kim *et al.*, 2010). Computational parametric studies on greenhouse structures can help to identify design factors that affect greenhouse ventilation under specific climatic conditions (Romero-Gómez *et al.*, 2008; Romero-Gómez *et al.*, 2010; Flores-Velázquez *et al.*, 2008).

3.1 Windward and leeward wind directions

The wind direction outside the greenhouse is an important factor in defining the flow of air and climate inside the greenhouse system. The boundary conditions of wind speed distribution are deduced from experimental data and wind direction with respect to the longitudinal axis of the greenhouse, which can range from 0 ° to 90 °. Roy and Boulard (2005) simulated the impact of wind at 45 ° and 90 °, showing the influence of wind direction in the air velocity, temperature and humidity distributions inside the greenhouse; a similar result was found by Campen (2003). Rico-García *et al.* (2006) also showed that a greenhouse with larger vertical roof windows works better with a windward condition, whereas the multi-span greenhouse works better with a leeward condition. Therefore, wind direction affects the degree of ventilation. In a experiment carried out by Khaoua *et al.* (2006), four different openings of roof vents obtained ventilation rates from 9 to 26.5 air exchanges per hour for the windward and 3.7 to 12.5 on the leeward wind condition, respectively, which can maintain acceptable and uniform climate conditions for particular cases where the wind is perpendicular to the main axis of the greenhouse. Overhead ventilation to the windward and leeward directions represents a reduction in the ventilation rate by 25% to 45%, compared with only opening to the windward direction (Bournet *et al.*, 2007). Openings to the windward direction generate the highest rate of ventilation; however, the greatest homogeneity of the temperature and wind speed arises from combining windward and leeward roof vents (Bournet and Khaoua, 2007).
Kacira *et al.* (2008) showed that the air temperature inside the greenhouse was higher on the windward side than on the leeward side when roof vents were used. Wind speed had a linear influence on air exchange rates, while the wind direction did not affect them. Majdoubi *et al.* (2009) observed a strong wind air current above a tomato canopy that was fed by a windward side vent and a slow air stream flowing within the tomato canopy space. The first third of the greenhouse, until the end of the leeward side, was characterized by a combination of wind and buoyancy forces, with warmer and more humid inside air that was removed through upper roof vents. There may be a conflict between increasing ventilation and improving uniformity because there is little information on air movement affecting the cooling efficiency and the uniformity of the environment (Sase, 2006). According to Rico-García (2008) the relationship between the thermal gradient and ventilation of gases shows a linear behavior, while the relationship between the combined effect of temperature and wind greenhouse ventilation presents a piecewise linear behavior. The wind pattern in a greenhouse is strongly affected not only by the outside wind velocity but also by the number of greenhouse spans (Flores-Velázquez, 2010). It was found recently, that as the greenhouse has three or four spans roof windows orientation is independent, however, when the greenhouse has five or more spans, side ventilation is dominant over the roof ventilation (Figure 6).

By increasing the surface of the front windows, opening windows and increasing roof capacity exhaust fans can ventilate properly larger greenhouses (Figure 7). It is important to calibrate the fan power, but also to determine a representative inlet area (Flores-Velázquez *et al.*, 2009; Flores-Velázquez, 2010).

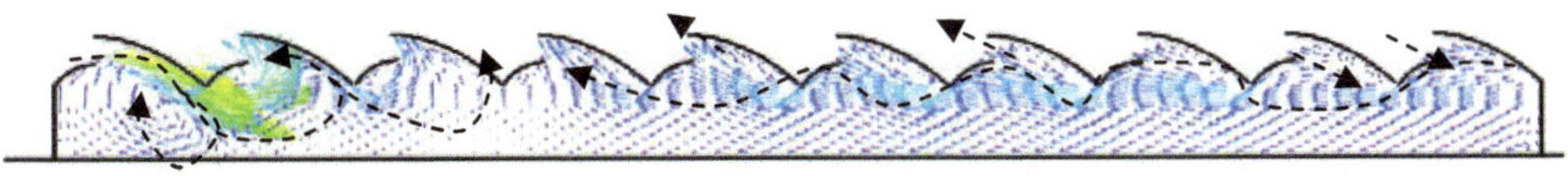

A) Roof windows windward open

B) Roof windows leeward open

C) Roof and side windows windward open

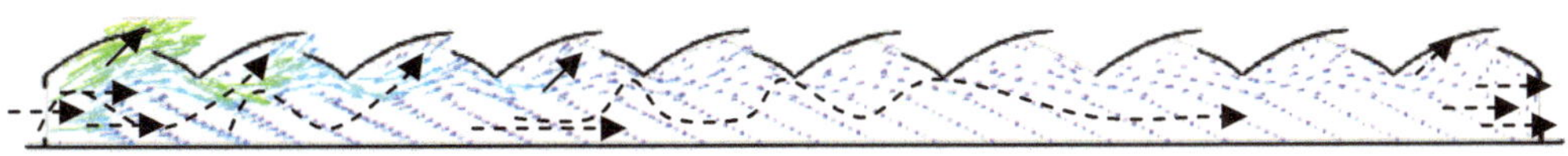

D) Roof and side windows leeward open

Fig. 6. Wind velocity interior vectors with 5 m s^{-1} wind velocity outside the greenhouse, on four open windows sceneries tested.

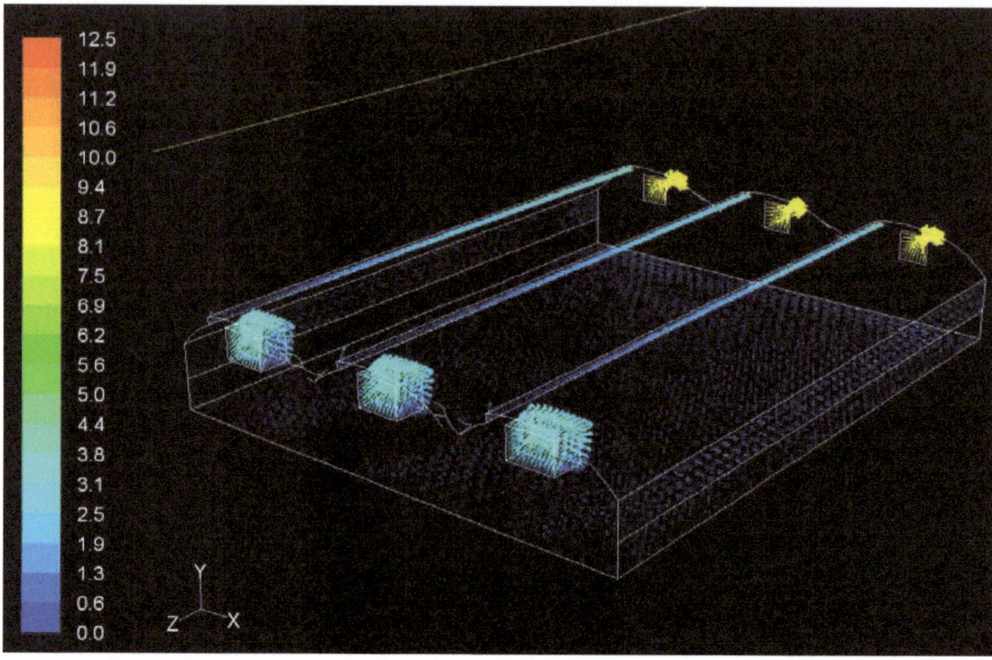

Fig. 7. Wind velocity vectors in a inlet area and outside fans in a three span greenhouse (Flores-Velazquez, 2010)

3.2 Heat exchange tube and natural ventilation

Rouboa and Monteiro (2007) simulated, by using a CFD model, the effects on temperature and wind speed of the introduction of hot water pipes along a greenhouse on nighttime conditions, under three scenarios: natural convection heating (case A), artificial heat pipes (case B) and artificial heat pipes and natural ventilation (case C) by using the turbulence model. Re-Normalization Group (RNG) observed an average increase in air temperature to 2.2 ° C, 6.7 ° C and 3.5 ° C; the turbulence was lower for case A, slightly increasing with the heating system for case B and higher for case C, due to the effect of natural ventilation.

3.3 Forced ventilation

The study of fluid dynamics in ventilation systems application provides elements of natural ventilation. A numerical investigation by Rousseau (2008) on a prototype air-forced unit for crop growth chambers obtained simulations that show a nonlinear relationship between airflow rate and opening vents, showing the mixing zone. Dayan *et* al. (2004) developed another simplified model to demonstrate the calculation of plant temperature when applying forced ventilation for climate control in greenhouses. The use of a device air flow deflector below the roof vents proved to increase air exchange in the area of cultivation effectively. According to the CFD simulations, the combination of the side vent dual configuration has little effect on overall air exchange; nonetheless it increases air movement in the crops and homogenizes temperatures (Baeza *et al.*, 2008). Another investigation by Hughes and Abdul (2010) took into account the effect of the external angle of the ventilation device (wind vent) louvers against the internal pressure and velocity to optimize the device performance. The optimum angle was 35 ° to 40 ° with a wind velocity of 4.5 ms^{-1}. Forced ventilation is an excellent option to abate the high temperature, but mainly in small greenhouses as fans are designed properly (Figure 8). As the length of the greenhouse increases, the overhead natural ventilation becomes a positive and more important complement to mechanical ventilation (Flores-Velazquez, 2010; Flores-Velázquez *et al.*, 2011).

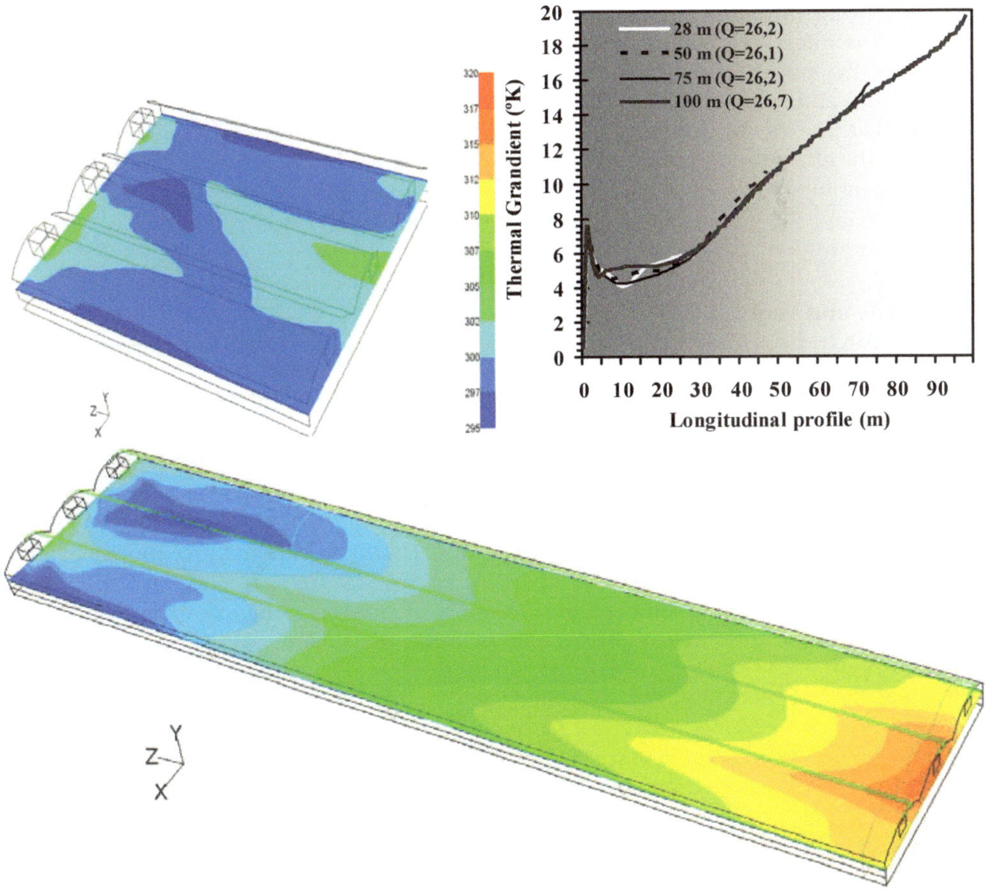

Fig. 8. Three spans mechanically ventilated greenhouse, power fan 25 Pa, Heat flux convection 315 W, top view 2 m high.

3.4 Fog-cooling system

According to Sase (2006), in a fog-cooled greenhouse in combination with natural ventilation, air cooled by fogging above the plants is likely to go down. Kim *et al.* (2007) developed a CFD model to simulate air temperature and relative humidity distribution in a greenhouse with fog-cooling systems, regardless of the presence of plants. Air temperatures and simulated measures ranged from 0.1 ° to 1.4 ° C and relative humidity differences were 0.3-6.0%. The results showed that the best cooling system performance occurs when fog nozzles are within 2.3 meters of the floor and 1.9 m of the side walls with a uniform spacing of 3.7 m and the best location for the injectors is at the entrance of the side openings of the greenhouse.

3.5 Eave cladding type

In order to understand the air mixing properties of the installation, Norton *et al.* (2010b) studied airflow and buoyancy, determining the relationship between eave openings and resistance to airflow and indoor air mixing for ventilation. They quantified the domain based on the effect of opening conditions, eave cladding type, porosity and height and found that the porosity of the eave opening cladding system significantly increases the efficiency of ventilation. They developed another CFD model incorporating shipping space and the cladding of the eave under conditions of opening and modifying its height, to determine the effects and characteristics of ventilation inside. It was found that the cladding of the eave influences the efficiency of ventilation and thermal comfort. They also found that the strength and the height of the eave determine whether it opens to leeward, acting as an air inlet (Norton *et al.*, 2010c).

3.6 Screens and vents

Recent research using CFD models includes further refinement in adaptive meshing areas, in order to maintain a high level of accuracy during modeling making the simulations more reliable (Norton and Sun, 2006). Screens reduce ventilation rate by 33%, according to a study carried out by Kittas *et al.* (2005). In agreement with Harmanto *et al.* (2006), using different screen's sizes over the vent opening has a significant effect, reducing 50% to 35% mesh 40, 78 and 52 and giving rise to a temperature gradient of 1 to 3 ° C with a mesh of 52 as optimal for a tropical greenhouse. Majdoubi *et al.* (2007) found that insect-proof screens significantly reduced airflow, increasing thermal gradients inside the greenhouse by 46%. Using a wind tunnel with screens of different porosity (0.62, 0.52 and 0.4), Teitel *et al.* (2008a) showed that a screen inclined by airflow reduces drag compared to a flow perpendicular to the screen, allowing an increase of 15-30% and 25% in the upper compared with a flat screen. Also, Teitel *et al.* (2009) found that higher speed screens are inclined at 45 ° and decreased to 135 ° tilt.

Ali *et al.* (2009) investigated the effect of roof vents on the temperature and coefficient of heat transfer in naturally ventilated facilities. Better flow patterns and heat transfer from the heated ceiling are observed when the front opening is located closest to the ceiling and the rear opening is located closest to the center. The increase of temperature and humidity as a result of insect-proof screens is particularly evident in the vicinity of the crop canopy (Majdobi *et al.*, 2009). Also it was found that a larger roof vent area can greatly enhance ventilation, while the extension of an insect-proof screen on side walls hardly changes the air exchange rate (Romero-Gómez *et al*, 2008; Romero-Gómez *et al*, 2010).

In hot regions a common practice is the substitution of the plastic cover of the greenhouse by a screen cover. This structure is called a screenhouse or a shadehouse. Although the climate control is rather difficult in such structure, it is an interesting and important system for growing crops. In some respect a screenhouse is better than a greenhouse as it is required to avoid hot air temperatures during specific hours of a day or a season, which is feasible because of the roof ventilation lets a higher air exchange (Figure 9).

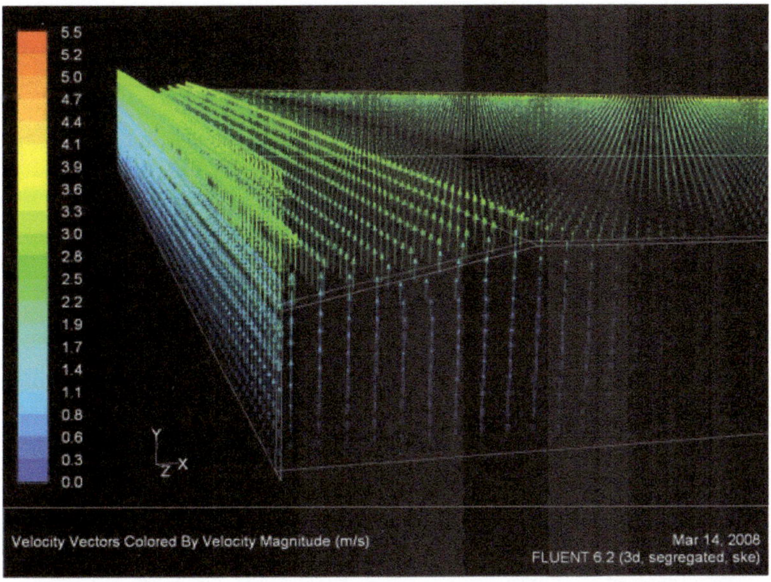

Fig. 9. Y-component wind velocities showing the wind exchange by the roof at entrance of a screenhouse (Flores-Velazquez, *et al.*, 2008).

When several screen porosities are located on side/roof screenhouses a similar behavior is observed, regardless the kind of screen, and a strong reduction of the wind velocity is predicted as the crop effect is taken into account to simulate the CFD model of the screenhouse (Figure 10).

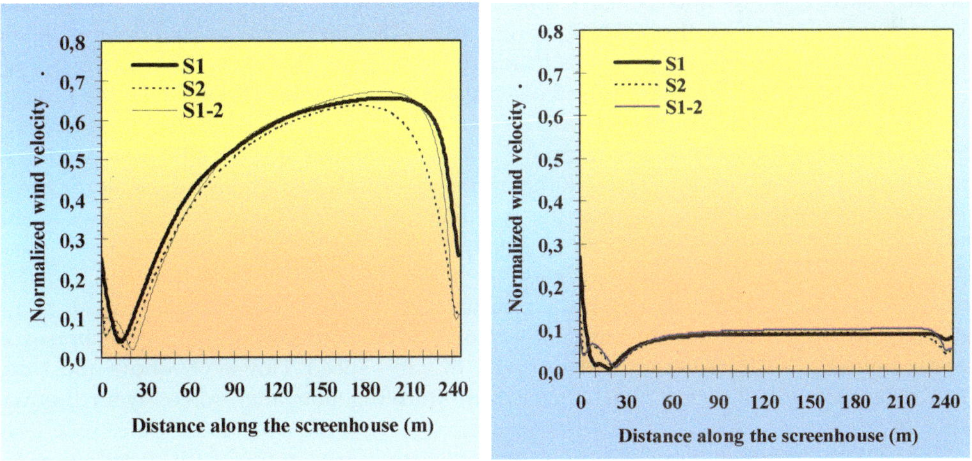

(a) (b)

Fig. 10. Comparison of wind velocities between screenhouses with the same screen (S1 and S2) and screen on the side (S1) and screen on the roof (S2). (a) With crop effect simulated. (b) Without crop effect simulated.

3.7 Solar radiation and temperature

Some studies have used solar radiation and transpiration models based on the heat and water balances of the crop, to investigate the distributions of air temperature and humidity and also the interactions between the crop and the air, in addition to the airflow distribution (Sase, 2006). According to Tablada (2005), the factor of solar protection plays a crucial role in maintaining stable thermal conditions indoors, even if the outside air temperature is higher. The slightly higher air speed on the top floor is insignificant in view of reducing the negative effect of the solar radiation over the roof and facade. The temperature of the greenhouse cover is an essential parameter needed for any analysis of energy transferred in the greenhouse. A sub-model developed by Impron *et al.* (2007) calculated the transmission of radiation through the greenhouse, including the reduction of NIR transmission through the roof. Tong *et al.* (2009) developed a numerical model to determine time-dependent temperature distributions based on hourly measured data for solar radiation, indoor air, soil and outside temperature, taking into account variable solar radiation and natural convection inside the greenhouse during the winter in northern China.

3.8 Temperature and air exchange

The effect of solar and thermal radiation is often taken into account by setting specific wall or heat fluxes at the physical boundaries of the greenhouse. Radiation transfer within the crop itself is still the major concern since it determines the two main physiological crop processes: transpiration and photosynthesis. This challenge is now launched and will probably receive more attention within the next few years (Bournet and Boulard, 2010). Pontikakos *et al.* (2006) analyzed data obtained from a CFD model, and showed that the external boundary temperature is a critical parameter in the pattern of internal greenhouse temperatures and that for specific external temperatures and wind directions, airspeed becomes the crucial parameter. According to Molina *et al.* (2006), opening vents affect the air flow, the ventilation rate and the air temperature distribution in a greenhouse; where the mean air temperature at the middle varied from 28.2 to 32.9°C with an outside air temperature of 26°C, there were regions inside the greenhouse that were 13°C warmer than the outside air. Nebbali *et al.* (2006) used a semi-analytical method to determine the ground temperature profile from weather parameters and other characteristics, to help in evaluating heat flux exchange between the surface and the air. Rico-García *et al.* (2008) showed that ventilation in greenhouses due to the temperature effect produces high air exchange rates; however, those air patterns occur near the openings, causing almost no air exchange in the central zone of the greenhouse due to a stagnant effect that reduces the wind effect throughout the greenhouse. In agreement with the results of Majdoubi *et al.* (2009), convection and radiation are the dominant forms of heat transfer. The measurements show that the difference between the air temperature inside and outside the greenhouse is strongly linked to solar radiation and secondly to wind speed. However, Chow and Hold (2010) obtained the following conclusions from studying buoyancy forces from thermal gradients:

a. Thermal radiation without air involvement changes air temperature distribution by radiating upper zone thermal energy in the wall towards the lower zone wall, which then affects air temperature through conduction and convection;
b. The inclusion of air absorption increases the effect of radioactive thermal redistribution by allowing air to absorb and radiate heat, reducing temperature gradients further;

c. Thermal boundary conditions and heat loads affect the predicted absolute temperature bounds, but do not affect the temperature distribution.

Radiation conditions play an important role in redistributing heat. Atmospheric conditions, especially relative humidity, are important for the calculation of radiation and heat transfer. Flores-Velázquez (2010) recently found that without overhead natural ventilation, it is possible to find linear relationships between temperature increase and the length of the greenhouse respectively, and the fan power that determines the slope of the regression line.

3.9 Turbulence and buoyancy

As computing power has increased the complexity and sophistication of CFD models also have increased. According to Norton and Sun (2006), the standard k-ε turbulence model commonly used in CFD models for greenhouses, in some cases provides inadequate results, and the choice of turbulence models must be based on the phenomena involved in the simulation. Different turbulence models give rise to differences in speed, temperature and humidity patterns, confirming the importance of choosing the model that most closely matches the actual conditions of turbulence (Roy and Boulard, 2005). Teitel and Tanny (2005) showed that the output of the turbulent heat flux is mainly due to cold air entering the greenhouse, which produces hot and cold eddies coming in and out the greenhouse. Roy and Boulard (2005) showed that the effects of wind direction on climate parameters inside the greenhouse are usually simulated by using different turbulence models available, to determine the energy balance between the flow of perspiration and the flow of radiation. Under ventilation parameters based on Bernoulli's theorem, Majdoubi *et al.* (2007), showed that bad ventilation performance is not a result of the low value of the greenhouse wind−related ventilation efficiency coefficient, but rather, that the low rate of discharge due to pressure drop in air flow is generated both by the use of anti-insect screens with small openings as an obstruction due to the orientation of the rows of crops. Moreover, Rouboa and Monteiro (2007) note that the RNG turbulence model is best suited to simulate microclimates in arc-shaped greenhouses.

According to Baxevanou *et al.* (2007), the circulation of air buoyancy effect shows the importance of internal temperature gradients, forced convection resulting from natural ventilation predominates. Rico-García *et al.* (2008) found that applying temperatures as the main driven forces for the buoyancy effect provides a simple way to study ventilation and inner air patterns. Vera *et al.* (2010a) observed that differences in temperature and ventilation rates strongly influence the movement of air, pushing it through openings where space is colder, while creating rising air currents when it is hot. Majdoubi *et al.* (2009) showed that the buoyancy forces induced by air temperature and increased humidity result in loops of air between the crop and the roof windows, which in turn tend to accelerate the pace of removal of heat and water vapor, enhancing indoor climate. Fidaros *et al.* (2010) studied turbulence in Greek greenhouses and found that external temperature variation is very important because internal temperature is determined by convection induced by the input current. The housing area had a higher circulation in the center of the greenhouse near the deck and in the corners of the ground, where the effect of the input current is weak. Defraeye *et al.* (2010) used a RANS turbulence model in CFD simulations to evaluate heat transfer by forced convection at the surface of a cube immersed in a turbulent boundary layer for applications in the atmospheric boundary layer (ABL), where wind speed is not disturbed at a height of 10 m. In a study of airfoil wakes, three turbulence models were simulated by Roberts and Cui (2010); the Reynolds Stress Model (RSM) is superior over the k-ε model, and when a time-dependent solution is necessary, Large Eddy Simulation (LES) is the desired option. However, LES does require the airfoil geometry to be included in the domain because it performs poorly when given only inlet velocities, turbulence kinetic energy and eddy dissipation at the trailing edge of the airfoil. According to Bournet and Boulard (2010), although they have been used for a long time in both the agriculture and environment studies, less empirical approaches to turbulence based on the use of LES have never been applied to greenhouse climate modeling and might perhaps be used to look for a solution to this complex situation.

Inside buildings, it is difficult to maintain a thermally stratified space with low ceilings, such as in offices and houses. Vera *et al.* (2010b) studied buoyancy in enclosed spaces, drawing the following conclusions:

a. Rising air currents and the exchange of humidity are closely related to the temperature difference between the lower and upper space. Low temperature in the upper space promotes the exchange of humidity and air flow through the opening; the hotter you are, the greater the restriction of air and humidity transport.
b. The existence of upward air currents when the space is warmer than the bottom is caused by local conditions such as non-uniform temperature distributions in the upper space and convective warm currents of the base and the humidity source.
c. Compared with conditions without mechanical ventilation, ventilation severely restricts the flow of air through the opening.

The main difficulty in the choice of the model is that greenhouse systems cover a range of length and velocity scales that generally require different modeling approaches (Bournet and Boulard (2010).

3.10 Incorporation crop effects and crop modeling

The effect of plants on greenhouse ventilation has also been studied in the past. For instance; Bournet *et al.* (2007), based on studies by Nederhoff (1985) and Lee and Short (1998), assumed that a crop of 90 cm high and low density decreases between 12 and 15% greenhouse ventilation. Dayan *et al.* (2004) built a representative model of a greenhouse for three vertical segments, horizontally oriented to the directions of energy and vapor transfer between the segments containing plants, considering the external weather. They concluded that Representative Plant Temperatures (RPTs) can be calculated instead of measured. Roy and Boulard (2005) developed a 3D CFD model for the characterization of climatic conditions in a greenhouse, incorporating five rows of ripe tomatoes as a porous medium where the buoyancy, heat and moisture transfer between the crop and air flow inside were considered. The heat and moisture transfer coefficients are deduced from the characteristics of the laminar boundary layer of the leaf, which are calculated with the velocity of flow in the crop. Khaoua *et al.* (2006) found that under external conditions of 1 ms^{-1} air velocity and 30° of temperature, wind speed at crops' height varies according to the modalities of ventilation from the windward 0.1 and 0.5 ms^{-1} for the leeward side, while temperature differences ranged from 2.0 to 6.1 ° C. In a study with tomatoes, Majdoubi *et al.* (2007) found that crop rows oriented perpendicular to air movement reduce the rate of airflow through the cultivation in a greenhouse by 50%. According to Baeza *et al.* (2008), a greenhouse with natural ventilation efficiency must combine an enough number of air changes to remove excess of heat, with good circulation of air through the crop. The effect of the crop was evaluated by Impron *et al.* (2007) using a sub-model to determine its effects on ventilation, the properties of the cover, and crop transpiration. In agreement with Kruger and Pretorius

(2007), the temperature and velocity at the plant level are influenced by the arrangement and number of windows. A study carried out by Sapounas *et al.* (2007) simulated a tomato crop as a porous medium, taking into account the addition of buoyancy to develop a model of the pressure drop of air flow due to the crop. The model depended on the area leaf stage of growth and cultivation, under the RANS turbulence model together with the RNG k-ε turbulence model. The results, validated with experimental measurements obtained at 1.2 m inside the canopy; show that the evaporative cooling system is effective with numerical parameters, providing a useful tool to improve system efficiency. A study performed by Roy *et al.* (2008) on leaf level through an experimental setup based on Münger cells measured the temperature, relative humidity and different heat flows to the leaves of soybeans, obtaining minimum stomatal resistance values ranging from 66 to 200 sm^{-1}.
Teitel *et al.* (2008b) built a small-scale model and found that wind direction significantly affects the ventilation rate and temperature distribution in crops. A study by von Elsner *et al.* (2008) on the effect of near-infrared (NIR) reflecting pigments in microclimate and plant growth found that a temperature drop up to 4 ° C in a young crop is the result of a 18% reduction in the transmission of global radiation in spring. At the same time, during the rainy season, minimizing transpiration differences in temperature and shading reduces water requirements in the plants, and they observed parthenocarpic fruit rot and yield-reducing crop. In a tunnel-type greenhouse, a tomato crop was modeled by Bartzanas *et al.* (2008) by designing a porous medium, where they emphasize the influence of the heating system on greenhouse microclimate. The climatic behavior of the rows of the tomato crop is taken into account using external user defined functions (Baxevanou *et al.*, 2007). According to Majdoubi *et al.* (2009), reorienting crop rows in simple ways improved climatic conditions. Endalew *et al.* (2009) performed CFD modeling of a plant with leaves and branches of the canopy, using turbulent energy equations in porous sub-domains created around the branches. Fidaros *et al.* (2010) simulated a greenhouse tomato crop as a porous medium so as to model radiation transport by discrete ordinates (DO). According to Teitel *et al.* (2010a), when applying the porous medium approach, the Forchheimer equation is often used, which gives rise to erroneous results with respect to the pressure drop through screens. An alternative way to calculate it through several panels of porous media used to simulate screens with realistic geometries. Moreover, the crop exerts a mechanical strain (drag force) on the flow just above but also interacts through the transpiration process with the temperature and humidity distributions (Bournet and Boulard, 2010). A simple model of transpiration of a crop was developed by Sun *et al.* (2010), who related it to the characteristics of ventilation in a greenhouse in eastern China, obtaining a good approximation. In general, there have been enormous efforts devoted to the analysis of ventilation in greenhouses (Norton, 2007); each new study provides new elements not only in the movement of air in the greenhouse but also in the forms it takes due to interactions occurring in the environment, such as position, shape and size of windows, and (one of the most important), the presence of a crop (Flores-Velázquez, 2010).

3.11 Humidity

Roy and Boulard (2005) simulated wind directions of 0 °, 45 ° and 90 ° with respect to the orientation of the greenhouse ridge to determine wind speed, temperature and humidity distributions inside the greenhouse, getting a good approximation for the humidity. In agreement with Demrati *et al.* (2007), models allow estimation, with better accuracy, of water requirements for a banana crop under cover and improved water saving in regions where water is the main limiting factor for agriculture. Roy *et al.* (2008) studied moisture on the surface of leaves at low light levels; crop transpiration and air flow were integrated into a single parameter model of leaf stomatal response to air flow and radiation. Campen (2008) showed that climate through a ventilation system is more homogeneous and the control is more efficient than with the conventional method of steam extraction. Dehumidifiers and cooling reduce the overall difference in humidity between the middle and lower areas of a greenhouse, as demonstrated by Kim (2008) using a 3D model could identify the heterogeneous distribution of relative humidity in a greenhouse. According to Majdobi *et al.* (2009), an increase in air temperature precedes a more moderate increase in specific humidity.

4. Main results of CFD models in greenhouses

Many CFD studies are focused on defining the conditions for a suitable environment. There has been less work on automation and control variables. Investigations that seek for a greater understanding of the interactions among climatic variables are increasing. Studies such as those of Hooff, 2010; Teittel, 2010 and Fidaros, 2010, evaluating geometries, have increased in the last year. Figure 11 shows the frequency of climatic variables studied during the period from the year 2005 to 2009 in the studies of CFD models in greenhouses.

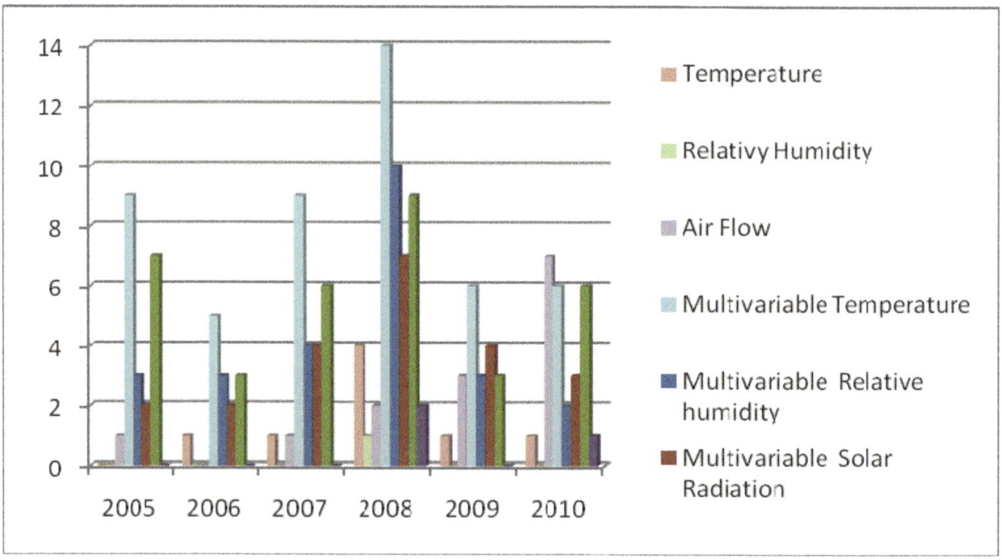

Fig. 11. Frequency of climatic variables analyzed by CFD models applied to greenhoses.

Most studies show multi-variable relationships, of which temperature and air flow are predominant. Humidity has been linked to temperature, while there are still few CO_2 distribution models. Solar radiation is the subject of investigations that evaluate housing, and is also related to the temperature in simulations with a greater degree of realism.

Studies to determine the influence of windward and leeward wind direction indicate that roof vents are of great importance for air renewal, where aperture settings that maximize the number of air renewals are obtained only with windows open to the windward roof (Bournet *et al.*, 2007; Kacira, 2008 and Majdoui, 2009). However, the combination of openings to the windward and leeward sides homogenizes the temperature inside the greenhouse better (Bournet and Khaoua, 2007). There have been many studies to determine a kind of optimal geometric design of greenhouses that encourage improvements in weather conditions and the use of new technologies such as monitoring systems in real time, allowing improvements in automation using Web technology (Pontikakos, 2005). Other studies have focused on the evaluation of misting systems (Kim *et al.*, 2007; Gázquez *et al.* 2008), forced ventilation (Dayan *et al.*, 2004, Baeza *et al.*, 2008 and Hughes and Abdul 2010) looking for energy savings. As simulation technology and computing power have improved, accuracy and realism in research based on CFD models, has increased as well, by defining more detailed models and by the use of textures that define the materials of the facilities.

The use of insect-proof screens in commercial greenhouses is very important as a means of crop protection; even though they reduce natural ventilation, by this, there have been many research efforts to reduce its negative influence (Kittas *et al.*, 2005; Harmanto *et al.*, 2006; Majdoubi *et al.*, 2007; Teitel *et al.*, 2008a). These studies tested different designs in size of the box and tilt and determined the most affected areas within the greenhouse, where the use of porous media allowed its CFD simulation.

Several studies have investigated the influence of solar radiation on temperature and relative humidity (Tablada *et al.*, 2005; Impron *et al.*, 2007, Tong *et al.*, 2009), and the result in crop response (Baxevanou *et al.*, 2007). Other studies evaluated the use of pigments (Elsner *et al.*, 2008) taking into account the convection, and thermal gradients.

Most of the recent studies developed 3D CFD models, some of which reported the use of models of turbulence and buoyancy, which appear more often during the past two years (Fidaros, 2010; Defraeye, 2010; Norton, 2010; Majdoubi, 2009). By taking into account turbulence, CDF models can make simulations more accurate, in turn increasing the processing and memory requirements for computing resources. Norton and Sun (2006) and Roy and Boulard (2005) discuss the importance of choosing the turbulence model that best meets the conditions of the study. Moreover, the concept of buoyancy appears frequently in order to incorporate the effects of growing space on the air flow and temperature gradients into the models (Figure 12).

Many studies consider the growing space, some of which are designed to measure phenomena based on their influence on the development and crop yield. Other studies are focused on the influence of crops on the other elements, such as temperature, relative humidity, CO_2 concentration and air flow, where it is necessary to model the space occupied by the crop by using porous media approach (Fidaros et al ., 2010). Other investigations measure biological phenomena such as evapotranspiration and Photosynthetically Active solar Radiation (PAR) by using indirect measures of climatic variables (Baxevanou, 2007; Sun, 2008). However, some studies do not mention an experimental phase aimed at validating the numerical model. In studies of air flow, the experimental methods mostly used are scaled models and unidirectional anemometry; the tracer gas technique is used less often, as well as three-dimensional anemometry, which is considerably more expensive. Studies that have used new methods to assess ventilation systems, such as those by Lu (2009), Molina (2010), Endalew (2009), van Henten (2008), Mikulka (2010) and Defraeye (2010), have been increasing in the past three years. The main question is the validation of these studies because they mainly concern to real scale greenhouses, whereas the measurements and characterizations have merely been done on scale models.

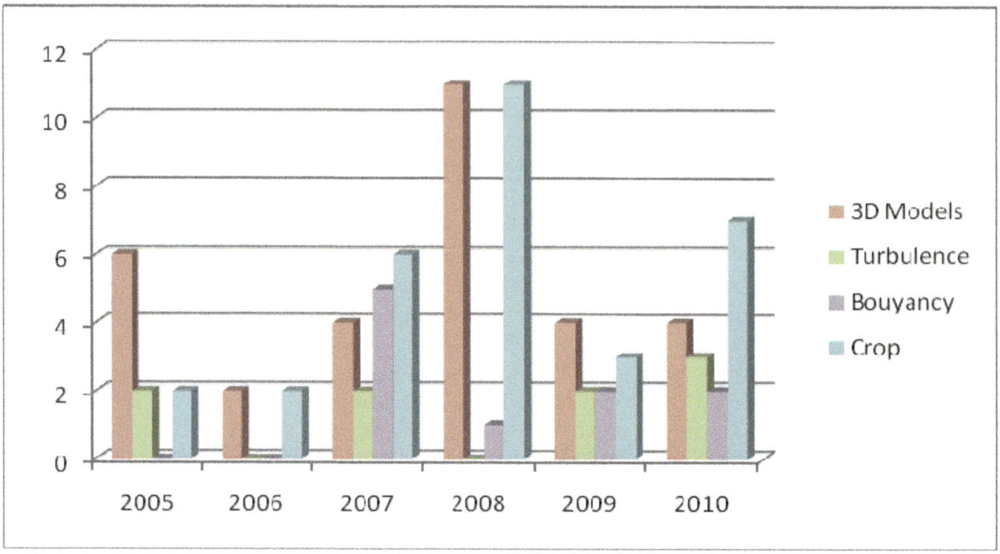

Fig. 12. Degree of realism and accuracy in the CFD models for greenhouse climate.

CFD modeling is an area of knowledge that in recent years has developed enormously through the development of software and hardware, which has contributed to research on natural ventilation a greater understanding of the interactions between the variables that make up the climate inside greenhouses. In the past five years, CFD simulation has become increasingly realistic and detailed, obtaining more accurate solutions. However, their use requires depth and extensive knowledge of climatic variables, fluid dynamics and turbulence. Simulating more accurately requires more processing power, so research tends to use CFD models together with other tools. Further studies are required to incorporate more realistic crops beyond a porous medium, taking into account the role of gas exchange, which is necessary for an understanding of the physiology and phenology of crops. There is still a need to develop high-precision systems in greenhouses, and CFD is a powerful tool for defining parameters with high precision, in order to control better the greenhouse environment.

5. Validation procedures for CFD models of greenhouse environment

5.1 Models and experimental validation

According to Sase (2006), recent progresses in CFD techniques have accelerated a more detailed analysis of air movement in combination with verification tests. However, studies in this area are required in order to address the detailed design of each element involved in the greenhouse climate, highlighting the difficulty involved in the analysis of air movement inside a greenhouse (Flores-Velázquez, 2010). The quality of the CFD models predictions is often evaluated from the agreement with experimental data. Nevertheless, no standard procedure exists yet in order to properly assess the accuracy of the simulations, and the type of comparison often differs from one study to the next (Bournet and Boulard, 2010). Figure 13 summarizes the main approaches used in the recent past to validate the CFD models of the greenhouse environment.

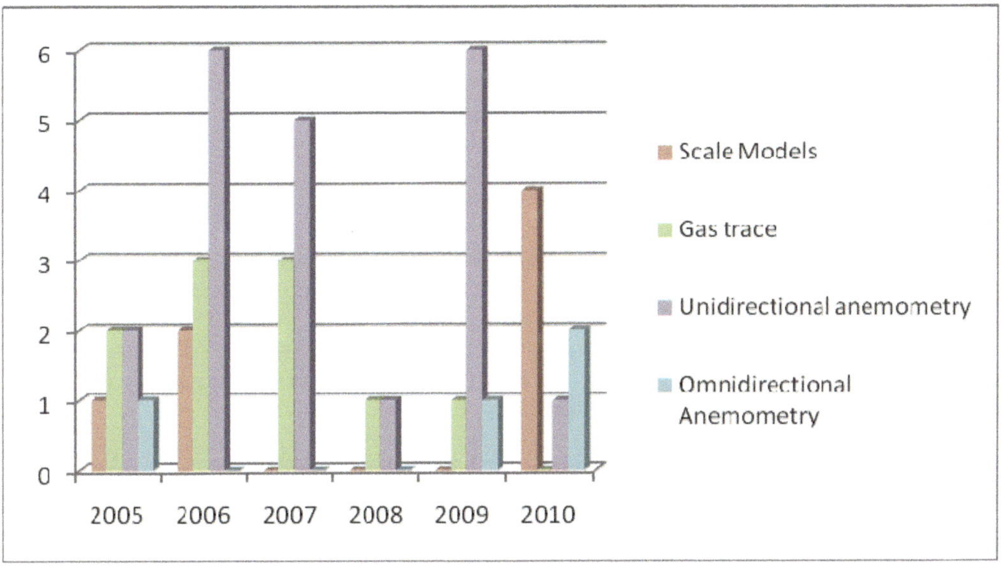

Fig. 13. Frequency of validation procedures used in CFD models applied to greenhouses.

The use of porous media models to simulate the pressure drop across flow boundaries, such as insect screens and fences, is very popular and must be validated experimentally. New technologies such as particle image velocimetry have worked properly to complement predictions; field solutions may include biological responses increasing the realism of the simulations (Norton *et al.*, 2007). According to Rouboa and Monteiro (2007), improvements could be achieved by incorporating nighttime transpiration and optimizing the size of the mesh elements to lower computation time. Recent progress offers the opportunity to build a grid that fits the physical boundaries of the structures studied much more realistically than a Cartesian structured grid, which closely follow the contour of the solid boundaries. However, they require verification of the meshing quality to obtain accurate data and an appropriate computational convergence (Bournet and Boulard, 2010).

5.2 Statistical models

The statistical models developed by Pontikakos *et al.* (2006) are less computationally expensive than the original CFD model, and therefore, they could be used for real-time estimates of temperature and flow rate in a greenhouse.

5.3 Model types

In a small-scale model developed by Teitel *et al.* (2008b), wind direction significantly affected the ventilation rate and temperature distribution in crops. Chen *et al.* (2010) evaluated seven types of models (analytical, empirical, experimental small-scale, pilot-scale, multi-area network and CFD to predict the ventilation rate in crops, obtaining the following conclusions:

a. The analysis of the model can give an overall assessment of a ventilation system if flow can be approximated to obtain a solution.

b. The empirical model has similar abilities to the model of analysis, but develops a database.
c. The small-scale model can be useful to examine the complex problems of ventilation if a similarity of flow can be maintained between the model and reality.
d. The large-scale model is the most reliable in predicting the efficiency of ventilation, but is expensive and time consuming.
e. The multi-zone model is a useful tool for the design of the ventilation of an entire building, but cannot provide detailed flow information in a room.
f. The zone model can be useful when a user has prior knowledge of the flow in a room.
g. The CFD model provides more detailed information on the performance of ventilation and is the most sophisticated.

However, the model must be validated by corresponding experimental data and the user must have depth knowledge of fluid mechanics and numerical technique. Therefore, the choice of an appropriate model depends on the problem to be solved.

5.4 Finite Element vs. Finite Volume

In a study by Molina *et al.* (2010), on the effectiveness of the Finite Element Method (FEM) and Finite Volume Method (FVM) for two-dimensional incompressible turbulent flow in ventilation rates, it was found that the FEM requires twice the computation time and 10 times more memory storage than FVM. FVM software (ANSYS/FLUENT v 6.3.) is the most frequently used CFD package in ventilation research, and only few papers using FEM software (ANSYS/FLOTRAN v. 11.0) have been published. CFD simulations have been compared to experimental data for 12 cases corresponding to three greenhouse types. The experimental greenhouses were chosen to represent a large range of ventilation situations: buoyancy effect in a mono-span greenhouse with adiabatic walls, as well as buoyancy and wind effect in a multi-span greenhouse and ventilation.

6. Looking into the future

Advances in telecommunications such as wireless networking and Internet technology (TCP / IP) facilitate the monitoring of environmental conditions in greenhouses. Pontikakos *et al.* (2005) designed a Web-based application for real-time predictive modeling of temperature and air velocity patterns, which consists of a user interface, interpolation process data generated by CFD and an output interface.

A lighting systems model with different optical properties was developed by Mikulka *et al.* (2010) who shows various settings for the R-FEM method in the CFX environment.

CFD ventilation space still tends to be a slow process today, while the computation time for the ventilation system and control simulation strategy is negligible. Sun and Wang (2010) found that the test method is more effective than the simplified numerical models, which require more powerful computers. Stavrakakis *et al.* (2010) concluded that Artificial Neural Networks coupled with CFD models are a powerful computational tool to evaluate the energy savings of various architectural designs.

Currently, CFD studies that mainly considered natural ventilation, increasing the realism of simulations by adding features such as the modeling of the crop inside greenhouse. However, it is still necessary to model crop physiological processes such as respiration and transpiration that define the terms of CO_2 and relative humidity mainly in crop space (De la Torre-Gea and Rico-García, 2010).

7. References

Ali, A.A.A.; Nabil, M.G. & Mahmoud, A.H (2009). Effect of natural ventilation and wind direction on the thermal performance of a building ceiling, *Proceedings Of ISES World Congress 2007, Vol.* I-V, Springer, No. 2, pp. 436-440.

Anderson, J.D. (1995). *Computational Fluid Dynamics.* ISBN: 978-3-540-85055-7, McGraw-Hill, US.

Baeza, E.J.; Perez-Parra, J.J.; Lopez, J.C. & Montero, J.I. (2008). CFD simulation of natural ventilation of a parral greenhouse with a baffle device below the greenhouse vents. *Acta Horticulturae,* No. 801, pp. 885-892.

Bakker, J.C.; Adams, S.R.; Boulard, T. & Montero, J.I. (2008). Innovative technologies for an efficient use of energy. *Acta Horticulturae,* No. 801, pp. 49-62.

Bartzanas, T.; Tadj, N.; Draoui, B. & Kittas, C. (2008). Numerical and experimental analysis of convective heat transfer in a heated greenhouse. *Acta Horticulturae,* No. 801, pp. 847-854.

Baxevanou, C.; Bartzanas, T.; Fidaros, D. & Kittas, C. (2008). Solar radiation distribution in a tunnel greenhouse. *Acta Horticulturae,* No. 801, pp. 855-862.

Bournet, P.E.; Ould Khaoua, S.A.; Boulard, T.; Migeon, C. & Chassériaux, G. (2007). Effect of roof and side opening combinations on the ventilation of a greenhouse using computer simulation. *American Society of Agricultural and Biological Engineers,* No. 50, pp. 201−212.

Bournet, P.E. & Ould Khaoua, S.A. (2007). Predicted effects of roof vent combinations on the climate distribution in a glasshouse considering radioactive and convective heat transfers. *Acta Horticulturae,* No. 801, pp. 925-932.

Bournet, P.E. & Boulard, T. (2010). Effect of ventilator configuration on the distributed climate of greenhouses: A review of experimental and CFD studies. *Computers and Electronics in Agriculture,* No. 74, pp. 195–217.

Campen, J.B. (2008). Vapor removal from the greenhouse using forced ventilation when applying a thermal screen. *Acta Horticulturae,* No. 801, pp. 863-868.

Chen, Q. (2009). Ventilation performance prediction for buildings: a method overview and recent applications. *Building and Environment,* No. 44, pp. 848-858.

Chen, Q; Lee, K.; Mazumdar, S.; Poussou, S.; Wang, L.; Wang, M. & Zhang, Z. (2010). Ventilation performance prediction for buildings: model assessment. *Building and Environment,* No. 45, pp. 295–303.

Chow, K.; & Hold, A.E. (2010). On the influence of boundary conditions and thermal radiation on predictive accuracy in numerical simulations of indoor ventilation. *Building and Environment,* No 45, pp. 437–444.

Dayan, J.; Dayan, E.; Strassberg, Y. & Presnov, E. (2004). Simulation and control of ventilation rates in greenhouses. *Mathematics and Computers in Simulation,* No. 65, pp. 3-17.

Defraeye, T.; Blocken, B. & Carmeliet, J. (2010). CFD analysis of convective heat transfer at the surfaces of a cube immersed in a turbulent boundary layer. *Int. J. Heat Mass Tran.,* pp.297-308.

De la Torre-Gea, G. & Ríco-García, E. (2010). Crop Under Greenhouse Simulations as Porous Media in Computational Fluid Dynamics Models. *Proceedings of The 5th symposium of engineering, Queretaro State University,* Queretaro, Mexico, November 2010.

Endalew, A.M.; Hertog, M.; Gebrehiwot, M.G.; Baelmans, M.; Ramon, H.; Nicolaï, B.M. & Verboven, P. (2009). Modelling airflow within model plant canopies using an integrated approach. *Computers and Electronics in Agriculture,* No. 66, pp. 9-24.

Fidaros, D.K.; Baxevanou, C.A.; Bartzanas, T. & Kittas, C. (2010). Numerical simulation of thermal behavior of a ventilated arc greenhouse during a solar day. *Renewable Energy,* No. 35, pp. 1380-1386.

Flores-Velázquez, J. (2010). Climate analysis in the main models of greenhouses in Mexico (mesh shade and Baticenital multitunnel) using CFD. PhD Tesis, Almeria University, Spain.

Flores-Velázquez, J. & Montero J.I. (2008). Computational fluid dynamics CFD study of large-scale screenhouse. *Acta Horticulturae,* No. 797, pp. 117-122.

Flores-Velázquez, J.; Montero, J.I; Baeza, E.J.; Lopez, J.C.; Bonachela, S. & Pérez-Parra, J.J. (2009). Analysis of mechanical ventilation in a three span greenhouse using computational fluid dynamics (CFD). *Acta Horticulturae* No. 893, pp. 653-660.

Flores-Velázquez, J.; Mejía, E.; Montero, J.I.; & Rojano, A. (2011). Numerical analysis of the inner climate in a mechanically-ventilated greenhouse with three spans. Acepted in *Agrociencia.*

Gan, G. (2009). CFD modelling of transparent bubble cavity envelopes for energy efficient greenhouses. *Building and Environment,* No. 44, pp. 2486–2500.

Gan, G. (2010). Impact of computational domain on the prediction of buoyancy-driven ventilation cooling. *Building and Environment,* No. 45, pp. 1173–1183.

Hughes, B.R. & Ghani, S.A.A. (2010). A numerical investigation into the effect of windvent louvre external angle on passive stack ventilation performance. *Building and Environment, No.* 45, pp. 1025–1036.

Iglesias, N.; Montero, J.I.; Muñoz, P. & Antón, A. (2009). Study for the night weather and the use of double cover roof as a passive alternative to increase temperature in greenhouses through a model based in computational fluid dynamics (CFD). *Horticultura Argentina,* No. 28, pp. 18-23.

Impron, I.; Hemming, S. & Botb, G.P.A. (2007). Simple greenhouse climate model as a design tool for greenhouses in tropical lowland. *Biosystems Engineering,* No. 98, pp. 79-89.

Kacira, M.; Sase, S.; Ikeguchi, A.; Ishii, M.; Giacomelli, G. & Sabeh, N. (2008). Effect of vent configuration and wind speed on three-dimensional temperature distributions in a naturally ventilated multi-span greenhouse by wind tunnel experiments. *Acta Horticulturae,* No. 801, pp. 393-401.

Khaoua, S.A.O.; Bournet, P.E.; Migeon, C.; Boulard, T. & Chassériaux, G. (2006). Analysis of greenhouse ventilation efficiency based on computational fluid dynamics. *Biosystems Engineering,* No. 95, pp. 83–98.

Katsoulas, N.; Bartzanas, T.; Mermier, M. & Kittas, C. (2006). Effect of vents opening and insect screen on greenhouse ventilation. *Biosystems Engineering.,* No. 94, pp. 427-436.

Kim, K.; Giacomelii, G.A.; Yoon, J.Y.; Sase, S.; Son, J.E.; Nam, S.W. & Lee, I.B. (2007). CFD modeling to improve the design of a fog system for cooling greenhouses. *Japan Agricultural Research Quarterly,* No. 41, pp. 283-290.

Kim, K.; Yoona, J.Y.; Kwonb, H.J.; Hana, J.H.; Sonc, J.E.; Namd, S.W.; Giacomelli, G.A. & Lee, I.B. (2008). 3-d cfd analysis of relative humidity distribution in greenhouse with a fog cooling system and refrigerative dehumidifiers. *Biosystems Engineering,* No. 100, pp. 245-255.

Kim, T.; Kim, K. & Kim, B.S. (2010). A wind tunnel experiment and cfd analysis on airflow performance of enclosed-arcade markets in korea. *Building And Environment,* No. 45, pp. 1329-1338.

Kruger, S. & Pretorius, L. (2007). The effect of ventilator configurations in naturally ventilated greenhouse applications. *Proceedings of 10th Uk Heat Transfer National Conference,* Edinburgh, Scotland, September, pp. 10-11.

Majdoubi, H.; Boulard, T.; Hanafi, A.; Bekkaoui, A.; Fatnassi, H.; Demrati, H.; Nya, M. & Bouirden, L. (2007). Natural ventilation performance of a large greenhouse equipped with insect screens. *American Society of Agricultural and Biological Engineers,* No. 50, pp. 641-650.

Majdoubi, H.; Boulard, T.; Fatnassi, H. & Bouirden, L. (2009). Airflow and microclimate patterns in a one-hectare canary type greenhouse: an experimental and CFD assisted study. *Agricultural and Forest Meteorology,* No. 149, pp. 1050–1062.

Mercado-Luna, A.; Rico-García, E.; Lara-Herrera, A.; Soto-Zarazúa, G.; Ocampo-Velázquez, R.; Guevara-González, R.; Herrera-Ruiz, G. & Torres-Pacheco, I. (2010). Nitrogen determination on tomato (*Lycopersicon esculentum* Mill.) seedlings by color image analysis (RGB). *African Journal of Biotechnology,* No. 9,Vol. 33, pp. 5326-5332.

Mikulka, J; Kriz, T.; Kroutilova, E. & Fiala, P. (2010). Numerical modeling of light sources with R-Fem method in CFX environment. *Piers Proceedings,* March 22-26, pp. 424-427.

Molina-Aiz, F.D.; Valera, D.L.; Pena, A.A.; Alvarez, A.J. & Gil, J.A. (2006). Analysis of the effect of rollup vent arrangement and wind speed on Almería-type greenhouse ventilation performance using Computational Fluid Dynamics. *Acta Horticulturae,* No. 719, pp. 173-180.

Molina, F.D.; Fatnassi, H.; Boulard, T.; Roy, J.C. & Valera, D.L. (2010). Comparison of finite element and finite volume methods for simulation of natural ventilation in greenhouses. *Computers and Electronics in Agriculture,* No. 72, pp. 69-86.

Nebbali, R.; Makhlouf, S.; Boulard, T. & Roy, J.C. (2006). A dynamic model for the determination of thermal boundary conditions in the ground of a greenhouse. *Acta Horticulturae,* No. 719, pp. 295-302.

Norton, T. & Sun, D.W. (2006). Computational Fluid Dynamics (CFD) an effective and efficient design and analysis tool for the food industry: a review. *Trends in Food Science & Technology,* No. 17, pp. 600-620.

Norton, T.; Sun, D.W.; Grant, J.; Fallon, R. & Dodd, V. (2007). Applications of Computational Fluid Dynamics (CFD) in the modeling and design of ventilation systems in the agricultural industry: a review. *Bioresource Technology,* No. 98, pp. 2386-2414.

Norton, T.; Grant, J.; Fallon, R. & Sun, D.W. (2009). Assessing the ventilation effectiveness of naturally ventilated livestock buildings under wind dominated conditions using Computational Fluid Dynamics. *Biosystems Engineering,* No. 103, pp. 78–99.

Norton, T.; Grant, J.; Fallon, R. &, Sun, D.W. (2010a). Optimising the ventilation configuration of naturally ventilated livestock buildings for improved indoor environmental homogeneity. *Building And Environment,* No. 45, pp. 983–995.

Norton, T.; Grant, J.; Fallon, R. & Sun, D.W. (2010b). Assessing the ventilation performance of a naturally ventilated livestock building with different eave opening conditions.

Computers and Electronics in Agriculture, No. 71, pp. 7-21.Norton, T.; Grant, J.; Fallon, R. & Sun, D.W. (2010c). A Computational Fluid Dynamics study of air mixing in a naturally ventilated livestock building with different porous eave opening conditions. *Biosystems Engineering,* No. 106, pp. 125-137.Norton, T.; Grant, J.; Fallon, R. & Sun, D.W. (2010d). Improving the representation of thermal boundary conditions of livestock during CFD modeling of the indoor environment. *Computers and Electronics in Agriculture,* article in press.

Omer, A.M. (2009) Constructions, applications and the environment of greenhouses. *African Journal of Biotechnology,* No. 8, Vol. 25, pp. 7205-7227.

Pontikakos, C.; Ferentinos, K.P. & Tsiligiridis, T.A. (2005). Web-based estimation model of natural ventilation efficiency in greenhouses using 3D Computational Fluid Dynamics. *In Proceedings of the International Congress on Information Technologies in Agriculture, Food and Environment,* October: 12-14.

Pontikakos, C.; Ferentinos, K.P.; Tsiligiridis, T.A. & Sideridis, A.B. (2006). Natural ventilation efficiency in a twin-span greenhouse using 3D Computational Fluid Dynamics. Of The *3rd International Conference on Information and Communication Technologies in Agriculture,* September: 20-23.

Rico-García, E.; Reyes-Araiza, J.L. & Herrera-Ruiz, G. (2006). Simulations of the climate in two different greenhouses. *Acta Horticulturae,* No. 719, pp. 325-332.

Rico-García, E.; López-Cruz, I.L.; Herrera-Ruiz, G.; Soto-Zarazua, G.M. & Castaneda-Miranda, R. (2008). Effect of temperature on greenhouse natural ventilation under hot conditions: Computational Fluid Dynamics simulations. *Journal of Applied Sciences,* No. 8, pp. 4543-4551.

Rico-García, E. (2008). Climate Modeling Greenhouse: Natural Ventilation. PhD Tesis, Queretaro State University, México.

Roberts, R.A. & Cui, J. (2010). Selection of the simulation domain of turbulent flow around and airfoil. *Engineering Aplications of Computational Fluid Mechanics,* No. 4, pp. 441-449.

Rodríguez, N.Y. (2006). Fundamental Equations of Fluid Dynamics. Universidad de la Habana, Facultad de Matemática y Computación, Ciudad de la Habana, Cuba.

Romero-Gómez, P.; Choi, C.Y. & Lopez-Cruz, I.L. (2010). Enhancement of the greenhouse air ventilation rate under climate conditions of central Mexico. *Agrociencia* No. 44, pp. 1-15.

Romero-Gómez, P.; López-Cruz, I.L. & Choi, C.Y. (2008). Analysis of greenhouse natural ventilation under the environmental conditions of central México. *Transactions of the ASABE,* Vol. 51(5), pp. 1753-1761.

Rousseau, T. (2008). Structure design and indirect adaptive general predictive temperature control of a class of passive HVAC. *WSEAS Transactions On Systems and Control,* No. 3, pp. 493-502.

Rouboa, A. & Monteiro, E. (2007). Computational Fluid Dynamics analysis of greenhouse microclimates by heated underground tubes. *Journal of Mechanical Science and Technology,* No. 21, Vol 12, pp. 2196-2204.

Roy, J.C. & Boulard, T. (2005). CFD prediction of the natural ventilation in a tunnel-type greenhouse: influence of wind direction and sensibility to turbulence models. *Acta Horticulturae,* No. 691, pp. 457-464.

Roy, J.C.; Vidal, C.; Fargues, J. & Boulard, T. (2008). CFD based determination of temperature and humidity at leaf surface. *Computers and Electronics in Agriculture*, No. 61, pp. 201-212.

Sapounas, A.A.; Nikita-Martzopoulou, C.; Bartzanas, T. & Kittas, C. (2007). Aspects of CFD modeling of fan and pad evaporative cooling system in a greenhouses. *International Journal of Ventilation*. No. 6, pp. 379-388.

Sase, S. (2006). Air movement and climate uniformity in ventilated greenhouses. *Acta Horticulturae*, No. 719, pp. 313-324.

Seo, I.H.; Lee, I.B.; Moon, O.K.; Kim, H.T.; Hwang, H.S.; Hong, S.W.; Bitog, J.P.; Yoo, J.I.; Kwon, K.S.; Kim, Y.H. & Han, J.W. (2009). Improvement of the ventilation system of a naturally ventilated broiler house in the cold season using computational simulations. *Biosystems Engineering*, No. 104, pp. 106-117.

Stavrakakis, G.M.; Zervas, P.L.; Sarimveis, H. & Markatos, N.C. (2010). Development of a computational tool to quantify architectural-design effects on thermal comfort in naturally ventilated rural houses. *Building and Environment*, No. 45, pp. 65–80.

Sun, H.; Yang, J. & Wang, X. (2010). CFD based determination of crop transpiration in local greenhouses in eastern china. *On 4th International Conference, Bioinformatics and Biomedical Engineering*, pp. 1-4.

Sun, Z. & Wang, S. (2010). A CFD-based test method for control of indoor environment and space ventilation. *Building and Environment*, No. 45, pp. 1441–1447.

Tablada, A.; Blocken, B.; Carmeliet, J.; De Troyer, F. & Verschure, H. (2005).The influence of courtyard geometry on air flow and thermal comfort: CFD and thermal comfort simulations. *Proceedings of 22nd Conference on Passive and Low Energy Architecture*, No.1, pp. 75-80.

Teittel, M.; & Tanny, J. (2005). Heat fluxes and airflow patterns through roof windows in a naturally ventilated enclosure. *Flow, Turbulence and Combustion*, No. 74, Vol. 1, pp.: 21-47.

Teitel, M.; Liron, O; Haim, Y. & Seginer, I. (2008a). Flow through inclined and concertina-shape screens. *Acta Horticulturae*, No. 801, pp. 99-106.

Teittel, M.; Ziskind, G.; Liran, O.; Dubovsky, V. & Letan, R. (2008b). Effect of wind direction on greenhouse ventilation rate, airflow patterns and temperature distributions. *Biosystems Engineering*, No. 101, pp. 351-369.

Teitel, M.; Dvorkin, D.; Haim, Y.; Tanny, J. & Seginer, I. (2009). Comparison of measured and simulated flow through screens: effects of screen inclination and porosity. *Biosystems Engineering*, No. 104, pp. 404-416.

Teittel, M. (2010). Using Computational Fluid Dynamics Simulations to determine pressure drops on woven screens. *Biosystems Engineering*, No. 105, pp. 172-179.

Tong, G.; Christopher, D.M. & Li, B. (2009). Numerical modelling of temperature variations in a chinese solar greenhouse. *Computers and Electronics in Agriculture*, No. 68, pp. 129-139.

Van Hoff, T. & Blocken, B. (2010). Coupled urban wind flow and indoor natural ventilation modelling on a high-resolution grid: a case study for the amsterdam arena stadium. *Environmental Modelling & Software*, No. 25, pp. 51 – 65.

Vera, S.; Fazio, P. & Rao, J. (2010a). Interzonal air and moisture transport through large horizontal openings in a full-scale two-story test-hut: part 2 - CFD study. *Building and Environment* No. 45, pp. 622–631.

Vera, S.; Fazio, P. & Rao, J. (2010b). Interzonal air and moisture transport through large horizontal openings in a full-scale two-story test-hut: part 1 - experimental study. *Building and Environment*, No. 45, pp. 1192–1201.

5

Computational Fluid Dynamics (CFD) and Discrete Element Method (DEM) Applied to Centrifuges

Xiana Romani Fernandez, Lars Egmont Spelter and Hermann Nirschl
Karlsruhe Institute of Technology, Campus South, Institute for Mechanical Process Engineering and Mechanics, Karlsruhe
Germany

1. Introduction

The simulation of fluid flow in e.g. air channels, pipes, porous media and turbines has been the emphasis of many research projects and optimization processes in the academic and industrial community during the last years. A key advantage of the calculation of fluid flow through a given geometry when compared to the experimental determination is the efficient identification of disadvantages of the current state and the fast cycle of change in geometry and analysis of its effects. This saves time and costs and furthermore opens up new fields of research and development. There are, however, few drawbacks of the new optimisation and research methods. The numerically obtained results need validation before it can be stated whether they are correct or misrepresent the true state. In some cases, such as the optimisation of existing systems, the experience of the user may be sufficient to evaluate the calculated flow pattern. In the case of the prediction of flow in new geometries or process conditions, the numerical result can be validated only by measuring the flow pattern at representative process conditions (flow velocity, pressure, and temperature). For non-moving geometries, many measurements have been conducted in order to determine the flow behaviour in various geometries and for different process conditions. These measurements are readily available for the evaluation of new simulations. However, only little research has been conducted to evaluate the flow conditions in centrifuges, especially at high rotational speeds. The reason for the lack of experimental data lies in the difficult conditions for the measurement in the centrifugal field with high circumferential velocities and pressure gradients. For low rotational speeds, the flow patterns have been determined by an electrolytic technique (Glinka, 1983) and visualised by adding ink to the flowing liquid (Bass, 1959; 1959; 1962).

The emphasis of the work presented is the evaluation of the efficiency of the acceleration geometry and the prediction of the axial flow profile in the centrifuge. In centrifugation technology, the prediction of the separation efficiency is often restricted to the cut size, the smallest particle size that can settle on the bowl wall at certain operating conditions. In a centrifuge, the highest velocities occur in tangential direction but, due to the throughput, a secondary flow in the axial direction appears. The tangential velocity creates the centrifugal force acting on the particles and the flow in the axial direction determines the residence time of particles in the bowl. Thus it is necessary to know the flow patterns in a centrifuge in order to calculate its separation capability. Often the predicted cut size of a centrifuge diverges from the real separation efficiency due to the fact that the assumptions of the analytical models are not strictly valid for industrial process conditions. The deviation is caused by the complex shape and internal assemblies of industrial solid bowl centrifuges, which create complex flow patterns. Hence estimating these flow patterns, as well as measuring them, is complicated.

In rotating machinery, the fluid flow is mostly turbulent, but in centrifuges the degree of turbulence is difficult to determine. Furthermore it may vary throughout the different zones of the geometry. For example, at the inlet of a continuous centrifuge, high shear rates and intensive mixing occur due to the high tangential velocity gradient. Hence a solver that incorporates equations to solve turbulent flow has to be used to obtain accurate predictions. Behind the inlet zone, the liquid flows undisturbed until the continuous phase leaves the centrifuge via an overflow weir or another discharge system. In this zone, the degree of turbulence depends on the geometry of the outlet system and the velocity gradients at the outlet. This example shows how complicated it is to gain the basis for the selection of an appropriate simulation method. Thus it is advisable to choose a model that is as precise as possible, but with an affordable computational demand. Various models have been developed to predict turbulent flow, but the codes were initially not designed to solve the Navier-Stokes equation in a rotating reference frame. Modifications were made over the years to include rotating systems and phenomena, e.g. the modified k-ε model "k-ε-RNG". Since then, only little work has been conducted to evaluate the accuracy of the various turbulence models when applied to centrifuges. The work presented compares several models such as the k-ε RNG, the k-ω, the Reynolds Stress Model (RSM) and the large-eddy simulation (LES) for predicting flow pattern in centrifuges. Furthermore the computational demand is evaluated and compared for different mesh sizes and types. The nomographs presented help to estimate the computational time for a given problem and computer system depending on the model and mesh.

Since with increasing fill level the sediment interacts with the flow pattern and reduces the capacity of the centrifuge, the sediment build-up must be included in the calculation of the separation efficiency and the determination of the flow pattern for tubular centrifuges. For calculation of the settling particles, diverse approaches are available. In this study, a Lagrange formulation has been chosen for calculation of the particle trajectories. In order to investigate the sediment build-up, particle–particle interactions must be considered. For this purpose, it is appropriate to apply the Discrete Element Method (DEM) (Cundall & Strack, 1979), originally developed for calculating the flow of bulk granular materials and then extended to the entire particle processing technology. This method accounts for particle-particle and particle-wall interactions by means of various contact models. However this method is limited because it does not account for hydrodynamic forces (drag, lift, torque), which play a vital role in e.g. centrifuges. Hence the simulation of the multiphase flow in centrifuges requires a coupling of the DEM with computational fluid dynamics (CFD). The flow pattern and particle trajectories must be calculated in an alternating scheme. First, the flow pattern is determined using CFD and afterwards, the motion of the particles is calculated using DEM adding the hydrodynamic forces to the contact forces.

The advantage of the numerical simulation of the complex multiphase flow in centrifuges is that the flow, the particle trajectories and their deposition are represented accurately. It allows a detailed description of the flow and particle behaviour for various operating

conditions, which is of high importance for the design and optimization of centrifuges and other apparatus in solid-liquid separation.
This work focuses on the prediction and validation of the results of flow patterns in solid bowl centrifuges of different kind. The centrifuges operate within a range of 1000 up to 40000 rpm and with throughputs between 0.1 and 6 l/min. The article highlights the issues that will occur if flow behaviour has to be calculated in a rotating confinement using commercial software. Detailed results show the variation of the flow patterns achieved using different models and the possibilities to run a sanity check for the calculated flow pattern.

2. Solid bowl centrifuges

The centrifugal separation of particles in a suspension is one of the most common problems appearing in industrial processes such as waste water, mining and mineral processing, biotechnology, solid-fuel, pharmaceutical, chemical and food industry. Solid bowl centrifuges are widely used in the pharmaceutical and fine chemicals industry. The tubular bowl centrifuge has been providing the basis for all centrifuge designs. A scheme of a typical state of the art tubular bowl centrifuge is shown in Figure 1.

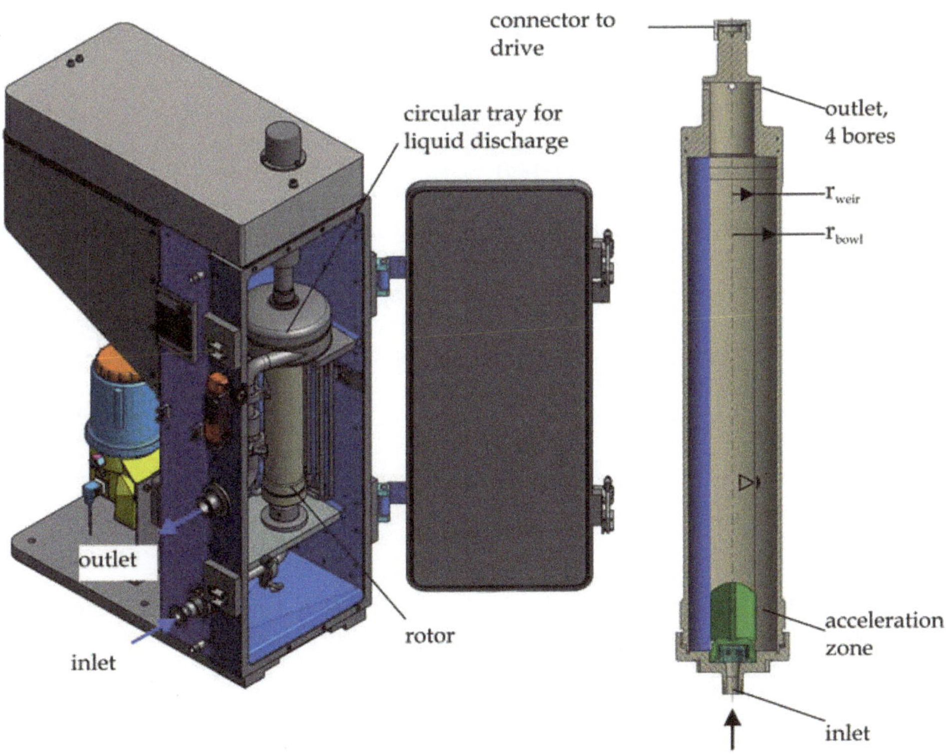

Fig. 1. Tubular bowl centrifuge (CEPA Z41G)

Starting with a rotating cylinder, modifications were made, leading to the disc stack centrifuge, screen scroll centrifuge and various other types. The basic design still has its applications. Due to the lean cylinder, high rotational speeds are possible. One of the first high speed centrifuges was the Sharples superfuge (Saunders, 1948; Taylor, 1946) that has been modified and adapted by various companies. Applications range from laboratory purposes to the production of pharmaceuticals in an industrial scale. New centrifuge designs were developed allowing the extension of the techniques to the industrial manufacture of new products such as vaccines (Anderson et al., 1969; Brantley et al., 1970; Perardi & Anderson, 1970; Perardi et al., 1969). The application of most high-speed centrifuges is limited by the solids capacity of the rotor. Due to the lean design, the solids capacity usually does not exceed a few kilograms. For most products that are processed in such centrifuges, the small capacity does not increase the production costs significantly because the value of the final product exceeds the cost of labour and machinery by an order of magnitude. This is the reason why most research has been conducted in the field of process optimization, e.g. in sterilisation, product recovery, and operator safety. Optimisation of the flow pattern in the tubular centrifuge has not yet been the focus of research.

The advantages of this type of centrifuge are the high centrifugal numbers and the simple geometry as compared to e.g. disc stack centrifuges. Tubular bowl centrifuges reach centrifugal numbers up to 40000 g. Their stability is due to their relatively high length to diameter aspect ratio in the range of approximately 4 to 7. There are other types of solid bowl basket centrifuges with conical or tubular bowls but smaller length to diameter aspect ratios, usually below 0.75 (Leung, 1998) such as overflow centrifuges. Sometimes radial or vertical compartments as well as blades are included to improve the separation efficiency.

Up to now, the flow pattern in existing centrifuges has been analysed and compared to some theories describing the fluid flow in rotating geometries (Glinka, 1983; Golovko, 1969; Gösele, 1968; 1974; Schaeflinger & Stibi, 1991; Schaeflinger, 1991; Trawinski, 1959). Using the determined axial flow pattern, the authors developed various analytical models for calculation of the cut size. Since it was not possible to measure or predict the tangential velocity, corrections were made to take into account insufficient tangential acceleration. These empirical corrections were often only valid for a specific centrifuge.

Moreover, the cut size depends not only on the tangential and axial flow but also on the density difference $\Delta\rho$ between the solids and the liquid, the viscosity of the suspension μ and the geometry of the rotor as stated in Equation (1). The geometric parameters are the length of the rotor L, the radius of the overflow weir r_{weir} and the radius of the bowl r_{bowl}:

$$x = \sqrt{\frac{18 \cdot \mu \cdot v_{ax} \cdot \ln\left(\frac{r_{bowl}}{r_{weir}}\right)}{\Delta\rho \cdot L}} \cdot \frac{2 \cdot \pi \cdot n}{60} \tag{1}$$

Modifications were made because the observed separation efficiency of the test rig often was not in good agreement with simple theoretical approaches. These modifications included certain levels of turbulence, boundary layer flow and inhomogeneous flow in the acceleration and discharge zones (Bass, 1959; 1959; 1962; Sokolow, 1971; Zubkov & Golovko, 1968; 1969). The different flow patterns in centrifuges are explained in detail in Chapter 2.1. The assumption that the entire length L is available for the deposition of particles is only valid for an ideal tangential acceleration of the feed so that no turbulent inlet area is present and a rotor without any settled particles that may interact with the flow pattern. In fact, the length L is reduced by up to 30 % due to the effects of the inlet. With increasing fill level, the sediment interacts with the flow pattern and reduces the free length L so that the cut size of the centrifuge increases (Spelter et al., 2010). Therefore the sediment build-up has to be included in the calculation of the separation efficiency and the determination of the flow pattern for tubular centrifuges.

2.1 Flow pattern

There are two limiting cases for the flow pattern in solid bowl centrifuges. Either the entire volume is passed by the feed or the main axial fluid flow takes place in a thin layer on top of a stagnant but rotating pool. The two different flow patterns are schematically shown in Figure 2. When a boundary layer flow is present, the geometric constant r_{bowl} in Equation 1 has to be replaced by the radius where the stagnant pool begins ($r_{boundary\ layer}$).

When starting building centrifuges and evaluating the separation performance, it was assumed that the proportion of the volume in which significant axial fluid velocities occur, further referred to as active volume V_{ac} or area A_{ac}, is nearly as high as the entire volume of the centrifuge, V_{ac}=100 % (Horanyi & Nemeth, 1971). Other researchers realised that the active volume of the centrifuge is significantly smaller than the capacity of the centrifuge (V_{ac}<<100 %) and developed the boundary layer theory (Bass, 1959). Today, most researchers who work with solid bowl centrifuges support the boundary layer theory.

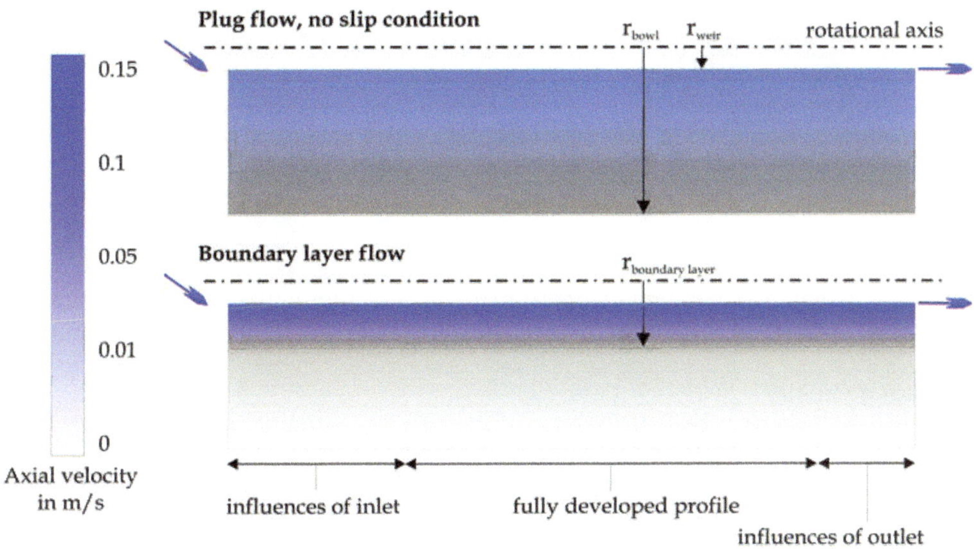

Fig. 2. Flow pattern in solid bowl centrifuges

Recent studies by the authors have confirmed the validity of this theory for industrial solid bowl and tubular bowl centrifuges (Romani Fernandez & Nirschl, 2009; Spelter & Nirschl, 2010; Spelter et al., 2010). Nevertheless it is important to pay attention to the acceleration geometry of the centrifuge and the influence of the settled particles on the flow pattern. In many centrifuges, the feed enters the rotor via a cone or a jet hitting a distributor plate. The liquid is not accelerated to the tangential velocity of the pool surface and thus hits the surface with a certain velocity gradient. The rotating pool accelerates the incoming liquid to the angular velocity. The pressure gradient in the rotating liquid stabilises the flow pattern and reduces the mixing of the feed and the rotating pool in radial direction. Thus a stratification of the liquid in radial layers occurs from the inlet and the liquid flows axially at the liquid-gas interface in a boundary layer. The flow is eased by the reduced friction at the interface between liquid and gas. The flow pattern will be different if the liquid is not fed to the pool surface or if intensive mixing occurs at the inlet. This is the case for some tubular centrifuges, as long as no sediment is present. Figure 3 a) shows the inlet geometry of a tubular bowl centrifuge as it was used in previous studies. The metal structure accelerates the incoming liquid to a certain rotational speed and significant mixing of incoming and rotating liquid occurs. The rigid body rotation is reached after a few centimetres behind the inlet. Figure 3 b) shows the acceleration geometry after 5 minutes of centrifugation. The settled yeast cells block a significant section of the inlet geometry and thus impede intensive mixing and acceleration. This has been observed for various minerals and biological products (Stahl et al., 2008) and is detectable by measuring the residence time and separation efficiency in the centrifuge at different fill levels. The growth of the sediment will change the flow pattern significantly if it reaches the active area. This may not be the case for industrial scale solid bowl centrifuges but has to be included in the calculations of the separation efficiency in tubular bowl centrifuges.

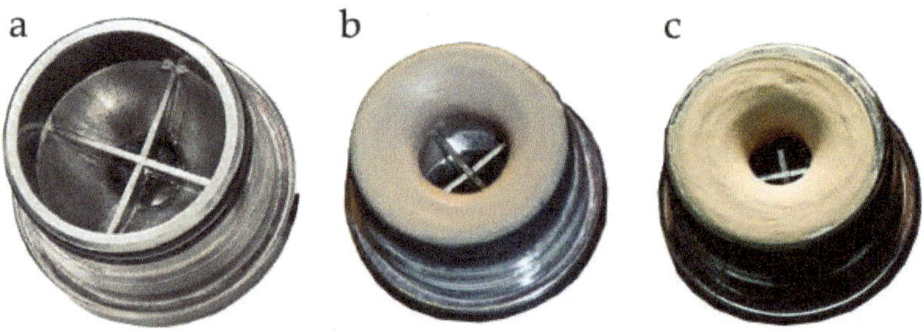

Fig. 3. Inlet geometry of tubular bowl centrifuge - a) empty rotor; b) with sediment after 5 min of centrifugation at 15000 rpm; c) as "b" but at 40000 rpm

2.2 Rotor geometry

Other inlet systems have been analysed in a recent research project (Spelter, et al. 2011). The liquid is accelerated by rotating inlet pipes and the outlet of these pipes is below the pool surface. This minimises the lag of the tangential velocity of the incoming liquid significantly and the trapping of air in the fluid. This feeding system may be used in existing centrifuges in order to enhance the separation efficiency by reducing the axial section where the tangential velocity gradient reduces the effective centrifugal force on the settling particles. Figure 4 shows a simplified scheme of the rotor of the test rig. The test rig is operated with two different rotor setups. Either the centrifuge runs using both cylinders, further referred to as tandem setup, or without the inner glass cylinder, further referred to as overflow setup. The tandem setup is similar to the one applied in high-performance centrifuges used for the production of vaccines. In these centrifuges, the outer rotor is often made of titanium and the inner one of plastic, e.g. Noryl®. The liquid enters the cylinder halfway between inner and outer cylinder at a radius of 42 mm at the top. The inlet is marked with an arrow and the height of the inlet is further referred to as H=0 mm. Both cylinders and the feed pipes rotate with the same revolutions per minute. There is no air in the system as the void is entirely filled with water. The liquid leaves the centrifuge at two outlet bores at a radius of 36 mm. In the overflow setup, only the outer cylinder is used and thus an air core is present. The liquid leaves the centrifuge via the two outlet bores at 36 mm and via two additional outlet bores at a radius of 30 mm, which adjust the minimum pool level. The two bores at 36 mm can be sealed. The height of the outlet is at H=170 mm, the overall length of the cylinder is 200 mm.

Depending on the pressure drop at the outlet bores, the water level rises above the radius of the weir. The superelevation is the equilibrium of pressure drop, rotational speed and throughput. With higher rotational speed and lower throughput the elevation is close to zero. With increasing throughput and lower rotational speeds, the centrifuge can be entirely filled. In the test rig, the maximum elevation is limited by a bore (radius 12.5 mm) that is necessary for the measurement of the flow pattern with Laser Doppler Anemometry (Spelter, et al. 2011). Once this radius is reached by the liquid, it overflows the circular weir without hindrance.

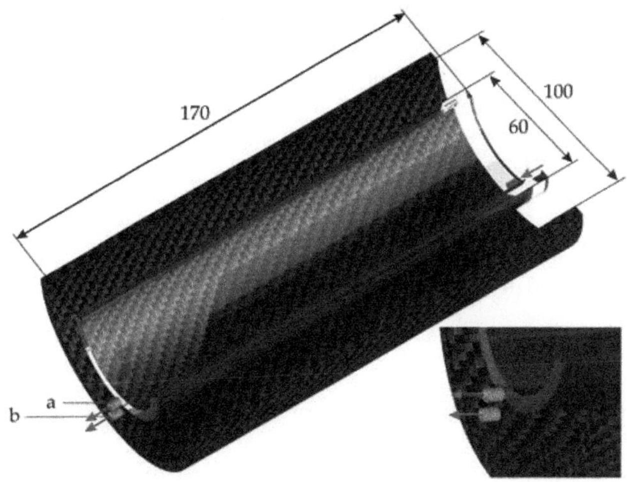

Fig. 4. Simplified scheme of the setups of the centrifuge test rig - outer rotor made of carbon fibre reinforced plastic, inner one made of tempered glass - a) additional outlet for overflow setup; b) outlet for tandem setup

3. Mathematical formulation of the fluid

To describe the flow pattern, it is necessary to determine the velocity components in all of the spatial directions considered as well as the pressure and, in the case of compressible flow, the density at any time and position in the domain. This chapter summarizes the mathematical methodology used to calculate these parameters.

The conservation equations of the fluid mechanics, conservation of mass, momentum and energy, can be obtained by means of a balance around an infinitesimally small fluid element. The continuity equation, which is derived from the mass conservation, describes the temporal change of mass in the volume element as the net rate of flow mass into the element across its boundaries. In the case of the incompressible flow considered here, the continuity equation simplifies to

$$\nabla \cdot \vec{v} = 0 \,. \tag{2}$$

In an analogous way, the momentum equation can be determined based on the conservation of momentum in all of the spatial directions considered. Newton's second law states that the rate of change of momentum of a fluid element equals the sum of the forces acting on the volume element. These forces can be surface forces, such as viscous forces and shear and normal stresses, or mass forces, such as gravity and centrifugal forces. For incompressible and Newtonian fluids in turbulent regime, the resulting momentum conservation equation follows to be

$$\rho\left(\frac{\partial \vec{v}}{\partial t} + (\vec{v} \cdot \nabla)\vec{v}\right) = -\nabla p + \mu \nabla^2 \vec{v} + \nabla \cdot \tau_t + \vec{F} \,, \tag{3}$$

where τ_t represents the tensor of turbulences and F other additional forces that can be included expressed as volumetric forces. All terms from Equation (3), the Reynolds-averaged Navier-Stokes equation (RANS), are discretized and calculated for each volume cell with the exception of τ_t, which is modelled. The RANS equation together with the continuity equation form an equation system of four differential equations with four unknown variables, the flow velocities u, v, w in x, y and z direction, respectively and the pressure p, which can be numerically calculated. This equation will be used for simulation of the centrifugal flow when only one phase is considered. For the cases where the liquid pool rotates around an air core, a multiphase approach must be used in order to simulate both, the air and the liquid. The presence of solid particles is ignored for flow simulation purposes, which is an acceptable assumption for low solid concentrations of the suspensions, as is the case in the present study.

3.1 Two-phase approach

The Volume of Fluid method (VOF) (Hirt & Nichols, 1981) is an interface tracking method that simulates the gas-liquid multiphase flow. This method is a simple and efficient formulation designed to track the interface of two phases that do not interpenetrate with a relatively small number of interfaces. Hence it is often used to simulate the gas-liquid multiphase flow in industrial devices (Brennan, 2003; Brennan et al., 2007; Li et al., 1999; Mousavian & Najafi, 2009). VOF introduces a variable α which takes values from zero to one and represents the volume fraction of one of the phases in each cell. A continuity conservation equation is solved for each phase

$$\frac{\partial}{\partial t}\left(\alpha_q \rho_q\right) + \nabla \cdot \left(\alpha_q \rho_q \vec{v}_q\right) = 0 \,. \tag{4}$$

Furthermore the volume fraction of each phase α_q in any cell has to obey

$$\sum_{q=1}^{n} \alpha_q = 1 \,. \tag{5}$$

Using the VOF method, a single momentum equation, the Reynolds-averaged Navier-Stokes equation, is solved throughout the domain, and the resulting velocity field is shared among the phases. The dependency on the volume fraction is implemented by using volume-averaged values for the density $\overline{\rho}$ and viscosity $\overline{\mu}$, as explained in Equations 6 - 8.

$$\overline{\rho}\left(\frac{\partial \vec{v}}{\partial t}+(\vec{v}\cdot\nabla)\vec{v}\right)=-\nabla p+\overline{\mu}\,\nabla^2\vec{v}+\nabla\cdot\tau_t+\vec{F} \tag{6}$$

$$\overline{\rho}=\sum_{q=1}^{n}\alpha_q\,\rho_q \tag{7}$$

$$\overline{\mu}=\sum_{q=1}^{n}\alpha_q\,\mu_q \tag{8}$$

3.2 Turbulence models

Due to high velocities and complex flow areas, most of the flow patterns of interest in process engineering, aviation, and automotive engineering become unstable above certain Reynolds numbers and, thus turbulences emerge. Turbulent eddies appear at very different time and length scales and are always three-dimensional. The kinetic energy stored in the rotational movement of the large eddies is passed to smaller eddies, until they disappear by energy dissipation. A turbulent flow is characterized by a chaotic and random fluctuation of the flow properties velocity and pressure. The conservation equations are able to capture the physics of the fluctuation motion, but in order to solve all the vortex structures of a turbulent flow, an extremely fine grid and very small time steps would be necessary. The memory and the computing power required for the so-called Direct Numerical Simulation (DNS) can only be provided by supercomputers. For that reason, turbulence is usually modelled with different mathematical approaches. A large-eddy simulation (LES) performs a filtering of the fluctuating flow quantities. The small eddies have a common behaviour and are nearly isotropic, while large eddies are more anisotropic and their behaviour is strongly influenced by the geometry of the domain, the boundary conditions and the body forces. Thus larger eddies are directly calculated, while fine structures are modelled. As a result, the computational demand decreases regarding to the DNS but it is nonetheless high in comparison with the approaches modelling all the turbulence scales.

The flow variables can be represented as the sum of a time-averaged size and an additional fluctuation:

$$u=\overline{u}+u';\qquad v=\overline{v}+v';\qquad w=\overline{w}+w';\qquad p=\overline{p}+p'. \tag{9}$$

The substitution of these expressions in the momentum conservation equation for incompressible flows and the subsequent temporal averaging yield to the Reynolds-averaged conservation equations. In the modified continuity equation, the velocity component is just replaced by the time-averaged variable. In the Reynolds-averaged Navier-Stokes equation (RANS), the fluctuation velocities appear in an additional tensor, the Reynolds stresses tensor τ_t:

$$\tau_t = -\rho \begin{pmatrix} \overline{u'u'} & \overline{u'v'} & \overline{u'w'} \\ \overline{v'u'} & \overline{v'v'} & \overline{v'w'} \\ \overline{w'u'} & \overline{w'v'} & \overline{w'w'} \end{pmatrix}. \tag{10}$$

The description of a turbulent flow with the RANS equation requires modelling of these tensor terms. The most complex classical turbulence model is the Reynolds Stress Model, RSM, (Launder et al., 1975), which uses transport equations to model each of the elements of the stress tensor and in addition an equation for the dissipation rate of the turbulences. This means that five additional transport equations are required in two-dimensional flows and seven in three-dimensional flow. This way, the directional effects of the Reynolds stress field can be taken into account. This model is recommended for flows with complex strain fields, such as highly swirling flows, or significant body forces. However, the computational cost of RSM is often unaffordable.

An effective method to reduce the modelling effort is to apply the Boussinesq approach, Equation (11). This approach introduces the eddy viscosity or turbulent viscosity μ_t, which represents the momentum and energy transport by diffusion by the eddies or turbulent fluctuations. Thus the Reynolds stresses can be calculated as the viscous stresses and only this new variable μ_t must be modelled.

$$-\rho\,\overline{u'v'} = -\rho\,\overline{v'u'} = \mu_t\left(\frac{\partial \overline{u}}{\partial y} + \frac{\partial \overline{v}}{\partial x}\right) \tag{11}$$

One of the most common turbulence models for the determination of μ_t is the k-ε model (Launder & Spalding, 1974). This model introduces two variables to calculate μ_t, the turbulent kinetic energy k, and its dissipation rate ε:

$$\mu_t = \rho C_\mu \frac{k^2}{\varepsilon}, \tag{12}$$

where C_μ represents a semi-empirical dimensionless constant. The temporal and spatial changes of both variables are described with a transport equation which includes not just the convection and diffusion but also the creation of turbulence and its dissipation by using several source and sink terms. This advantage justifies the additional expense of the transport equation turbulence models compared to algebraic turbulence models, which presume that the turbulence only depends on local conditions. The disadvantage of the k-ε model is that it assumes the turbulences to be isotropic, which is not strictly true in most of the flows. This model, well established and most widely validated, has already been used to simulate the flow in cyclones and centrifuges; although in its development this model assumes a fully turbulent flow, which is only partially applicable to centrifuges.

The k-ε renormalization group model, k-ε RNG, (Yakhot & Orszag, 1986) is an extension of the standard k-ε model that takes into account the effect of swirl on the turbulence by means of an extra source term in the transport equations. Thus the k-ε RNG model exhibits a higher accuracy for swirling flows. The RNG procedure systematically removes the small scales of motion from the governing equations by expressing their effects in terms of larger scale motions and a modified viscosity (Versteeg & Malalasekera, 2007). While the standard k- ε model is appropriate for high-Reynolds-number flows, the RNG theory accounts for low-Reynolds-number effects. Thus this model is more reliable for a wider class of flows than the

standard k-ε model. The k-ε RNG model has been succeeding for previous simulations of the flow in centrifuges by other groups and in our own research group (Romani Fernandez & Nirschl, 2009; Spelter & Nirschl, 2010).
The k-ω model (Wilcox, 1988), like the k-ε model, uses two equations to describe turbulence. Instead of the dissipation rate ε, the specific dissipation or turbulent frequency ω=k/ε is introduced in this model. The turbulent viscosity is then given by Equation (13). The accuracy of the model depends on the undisturbed velocity of the fluid outside the boundary layer which is subject to large fluctuations. This dependence can lead to significant errors in the calculation (Bardina et al., 1997). Due to the strong dissipation of the turbulence on the wall, no special treatment of the boundary layer at the wall is necessary. This feature is desired for the exact calculation of the flow near the wall.

$$\mu_t = \rho \frac{k}{\varpi} \tag{13}$$

Comparing the turbulence models, the k-ω model, as a two-equation model, has a similar range of strengths and weaknesses as the k- ε model. RSM is complex, but it is generally accepted as the simplest kind of model with the potential to describe all mean flow properties and Reynolds stresses without case-by-case adjustment. RSM is by no means as comprehensively validated as the k-ε model and due to the high computational cost it is not as widely used in industrial flow simulations as the k-ε model. LES, due to the inherent unsteady nature, is much more computationally expensive than the k-ε and k-ω models. However, compared to RSM it was proved that it requires only twice the computational demand of RSM (Ferziger, 1977) as cited in (Versteeg & Malalasekera, 2007). This is not a big difference taking into account the solution accuracy and the ability of LES to resolve certain time-dependent features. In order to investigate the applicability of the different turbulence models to the simulation of the flow pattern in centrifugal field, different approaches, the k-ε RNG model, the k-ω model, the RSM, and the LES, were used in this work.

3.3 Boundary conditions, discretization schemata and solver

In order to simplify the simulation, the air is considered as an incompressible gas. This is a reasonable assumption for the operation conditions of atmospheric pressure and for a non-temperature dependent problem. The fluid is defined as an incompressible Newtonian one with the density and viscosity of water.
The inlet velocity is set to match the desired volume flux. Using a turbulence model, also the turbulence quantities of the inlet flow must be specified. The input of fixed values for k and ε or k and ω are difficult to estimate. Thus these parameters are usually obtained as a function of the turbulence intensity and a characteristic length which must be set at the boundaries (Paschedag, 2004). For the multiphase simulations, a value of one was given to the volume fraction of water at the inlet. A static pressure of 101325 Pa (1 atmosphere) is set as ambient condition for the outlet, defined as a pressure outlet. Sometimes, there is a backflow through the pressure outlet in order to obey the mass conservation. Therefore it is recommended to set realistic conditions for a possible backflow to avoid convergence problems. For the cases presented, only backflow of air is allowed.
It is possible to reduce the computational demand by dividing the geometry into periodic sections. The flow pattern is calculated in one of these segments with periodic conditions at

the intersecting planes. The periodic surfaces were defined with a periodic boundary condition which uses the flow conditions at one of the periodic surfaces as its cells were the adjacent cells of the other periodic plane. The rotational periodicity must be taken into account and the rotation axis must be defined. Due to the rotational nature of the system, it is helpful to use a rotating reference frame to perform the calculations. The side and top walls of the bowl, as well as the inlet accelerator, all defined with no-slip condition, rotate with the same angular velocity as the reference frame.

Another problem emerging from the use of turbulence models is the treatment of turbulent quantities at the walls. Despite the turbulent flow, there always exists a laminar sublayer at the wall followed by a transient region until the flow can be considered as turbulent. A common approach is to create a first layer of cells next to the wall with a width including the entire viscous sublayer and the transition layer. In this layer, standard wall functions (Paschedag, 2004), most widely used for industrial flows, can be applied. The wall functions define values for the velocity field and the turbulent quantities suitable as boundary conditions for the solution in the second cells layer, where the turbulence is fully developed. This approach is particularly suitable for flows with larger spatial extent in which the thickness of the boundary layer at the wall is small compared to the extension of the whole domain. To choose the thickness of the first layer, the criterion of the y+ value (Equation 14) is often used. The value of y+ is a function of the first layer thickness y, the shear stress at the wall τ_w, and the physical properties of the fluid density ρ and viscosity μ. Values of y^+ between 30 and 100 allow the use of standard wall functions. For details concerning the exact formulation of the y^+ and standard wall functions, the reader is referred to (ANSYS, 2009) which is based on (Launder & Spalding, 1974).

$$y^+ = \frac{\rho\, y}{\mu}\sqrt{\frac{|\tau_w|}{\rho}} \tag{14}$$

The k-ε model and the RSM need these wall functions in order to calculate the turbulent properties near the wall. The k-ω model does not require wall-damping functions, because the value of turbulence kinetic energy at the wall is set to zero and for the turbulent frequency ω a hyperbolic variation at the near-wall grid point is applied (Wilcox, 1988).

The discretization schema used for the velocity and the turbulent variables is set to first-order upwind, due to the convective nature of the flow in centrifugal field and to avoid convergence problems. For the volume fraction, the Geo-Reconstruct method (ANSYS, 2006) is applied in order to obtain a sharp interface between both phases. For the interface between two fluids, this method assumes a linear slope, which is calculated with a piecewise linear approach. The variable pressure was discretized with the PRESTO! (PREssure STaggering Option) scheme recommended for high speed rotating flows (ANSYS, 2009). If velocities and pressure values are both defined at the centres of the cells, a highly non-uniform pressure field can act like a uniform field in the discretized momentum equations (Versteeg & Malalasekera, 2007). A solution to this problem is to use a staggered grid to calculate the pressure via a discrete continuity balance. In the staggered grid, the values of pressure in the centre are known and these are the values at the interfaces in the normal grid.

The temporal discretization of the transport properties was performed using a first-order implicit method. However, using the VOF model, the time step for the variable volume fraction is refined with respect to the general time step based on the maximal Courant number allowed for this variable. The Courant number, also called Courant-Froudel-Lewy number (CFL), Equation 15, is a dimensionless number that compares the time step of the calculation to the characteristic time needed by a fluid element to cross a cell. In some explicit time discretization methods, the criterion CFL<1 (Oertel & Laurien, 2003) is imposed to achieve stability in the numerical calculation. An explicit schema limited by a Courant number of 0.25 was applied for the volume fraction.

$$CFL = \frac{\Delta t}{\Delta x_{cell} / v_i} \tag{15}$$

The simulations were performed with the commercial software ANSYS FLUENT with the versions 6.3.26, 12.0.16, 12.1.2 and 13.0, which are comparable. An overview of the type of solver used by ANSYS can be found in (Kelecy, 2008). The solver used is an unsteady, segregated pressure-based solver. The segregated pressure-based algorithm solves the conservation equations sequentially. Because these equations are coupled, the solution loop must be carried out iteratively in order to obtain a converged numerical solution. Pressure and velocity were coupled either with the SIMPLE algorithm, Semi-Implicit Method for Pressure Linked Equations, or with the PISO schema, which stands for Pressure Implicit with Splitting of Operators and is available for unsteady calculations. The PISO pressure-velocity coupling algorithm has one predictor step and two corrector steps, extending the pressure correction procedure of the commonly used SIMPLE algorithm with a further corrector step (Versteeg & Malalasekera, 2007). The SIMPLE algorithm is a guess-and-correct procedure which uses a Poisson's equation, developed from the continuity condition, for the pressure correction.

4. Simulated flow in solid bowl centrifuges

The previously published simulations of a tubular bowl centrifuge were carried out two-dimensionally (Spelter & Nirschl, 2010); hence it was not possible to include the real acceleration geometry in the model. For the work presented, all simulations were run three-dimensionally, including the acceleration and discharge geometry, and were compared with experimental data. The flow pattern was predicted for a tubular bowl centrifuge with a feeding system similar to the one used in a recent published work (Spelter, et al. 2011). The experimental data has been obtained by Laser Doppler Anemometry, a non-invasive technique for the determination of flow velocities.

4.1 Prediction of the flow pattern (k-ε RNG model, transient solution) - tandem setup

Figure 5 shows the tangential velocity profiles for four different rotational speeds at a constant throughput of 1 l/min. There is no significant deviation between the predicted tangential velocity and the rigid body rotation (RBR). The measurement of the tangential velocity for various rotational speeds and throughputs at different heights was in good agreement with the rigid body rotation as well (Spelter, et al. 2011). Thus it can be stated that the simulation is in good agreement with the experimentally determined tangential velocity profiles. Therefore it is possible to optimise existing acceleration geometries by using a feeding system similar to the one proposed. By calculating the present state and modifying the geometry, the changes in the acceleration efficiency can be evaluated by means of CFD. Furthermore the calculation will help in the case of a scale-up of the centrifuge. The acceleration efficiency and the shear in the feed zone can be calculated before a pilot machine or a full scale centrifuge is built. With the knowledge of the flow behaviour, the geometry can be optimised, leading to an improved apparatus with less cost in development and testing.

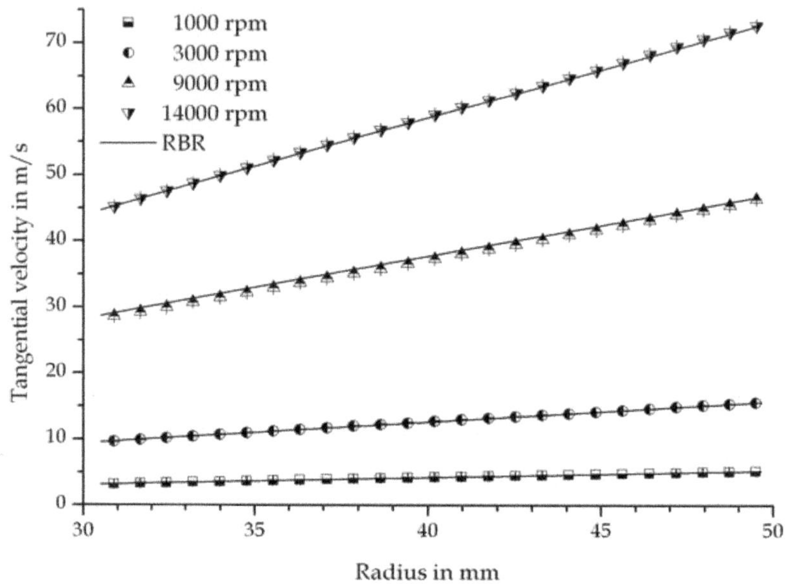

Fig. 5. Comparison of calculated tangential velocity and rigid body rotation, 1 l/min

The tangential velocity of the fluid creates a pressure in the cylinder in addition to the stress in the rotor created by the rotating cylinder itself. For the case that the rotor is designed as an ideal cylinder, the pressure distribution and thus the tensile stress can be calculated analytically (Sokolow, 1971; Stahl, 2004). However in many industrial centrifuges, the rotor has a non-cylindrical shape. For these designs, the stress in the rotor is determined via finite element methods (FEM), but the arbitrary load by the rotating liquid has to be calculated separately. The geometry of the rotor can be imported from various CAD programs such as ProEngineer, Autodesk Inventor, Catia, etc.. The pressure on the different planes in the bowl is then accessible by calculating the pressure distribution for the highest rotational speed that occurs in the centrifuge. The combination of FEM and CFD leads to a complete virtual development and saves time and costs. The simulations will never substitute completely experimental evaluation of the separation efficiency and mechanical integrity of the centrifuge, but the scale-up factor may be significantly increased. The predicted pressure distributions in the cylinder are shown in Figure 6 for 9000 rpm (a) and for a range of rotational speeds in (b). The outlet is located at the top end of the cylinder. The predicted pressures in Figure 6 b) are compared with the analytically calculated pressure distribution for the different rotational speeds. All values are given in bar.

The gauge pressure must be taken into account for dimensioning the rotor. Furthermore it is possible to increase the density of the liquid to model the arbitrary load of a sediment on the bowl wall.

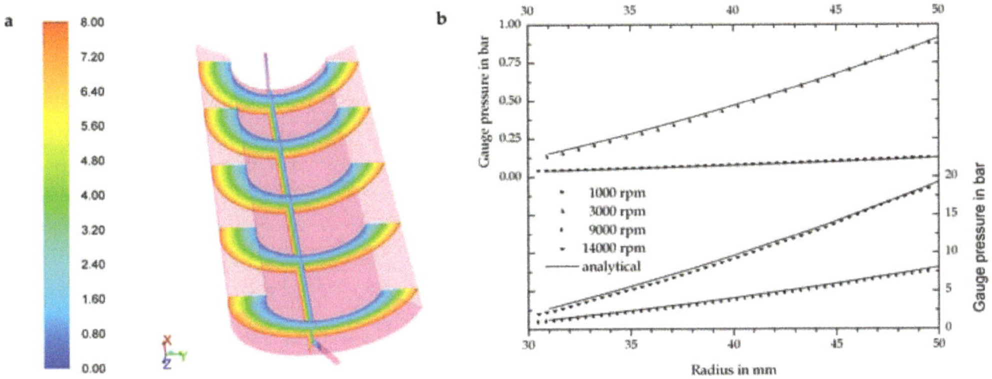

Fig. 6. Pressure distribution, all values in bar: a) 9000 rpm, pressure b) 1000-14000 rpm and comparison to analytical values

The axial velocity profiles are displayed as contour plots in Figure 7 for 1000 rpm (a) and 3000 rpm (b) at a constant throughput of 1 l/min. The inlet is located at the bottom while the liquid leaves the centrifuge at the top. The incoming liquid enters the centrifuge as a jet that does not widen up significantly towards the outlet. At 1000 rpm, the main axial flow takes place at a small circular cross-section that is not displaced in angular direction on the path towards the outlet bores. With increasing rotational speed, a small angular displacement of the axial flow path is predicted, as shown in Figure 7 b). This effect has been observed by Laser Doppler Anemometry as well.

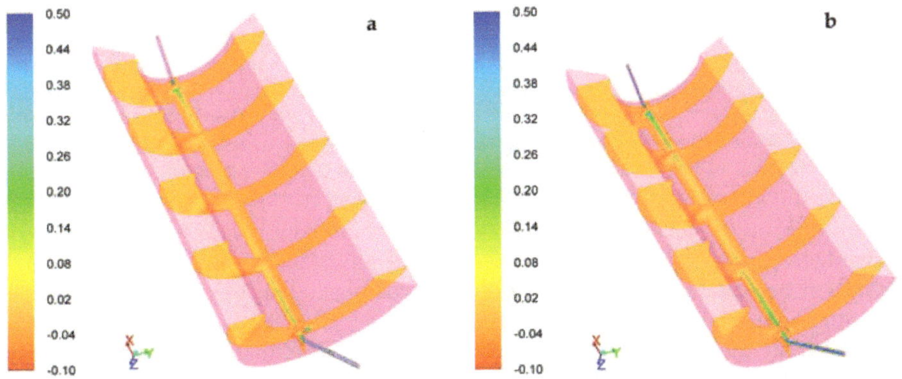

Fig. 7. Axial velocity in m/s, throughput 1 l/min: a) 1000 rpm b) 3000 rpm

However, the predicted axial flow pattern does not depict the experimentally determined flow behaviour accurately. It has been observed that the inlet jet widens within the first 30 % of the rotor (Spelter, et al. 2011) until a parabolic profile is developed that is distributed over the whole circumference. The predicted axial flow profile close to the inlet is in good agreement with the experimental data but the change of flow pattern in direction of the angle of rotation is not correct. The axial flow profiles averaged over the length of the centrifuge between 1000 and 14 000 rpm for two angles of rotation and a constant throughput of 1 l/min are shown in Figure 8. The standard deviation is plotted in both diagrams. The scatter of the data is very high for the angle where the inlet is located, which is 0°. The diagram on the right side of Figure 8 shows the axial flow profiles for an angle of rotation of ±45° displacement to the inlet. There is no significant flow towards the exit of the rotor, only recirculating currents are predicted. The standard deviation of the data is small when compared to the angle of 0°. The origin of the high standard deviation of the data displayed in the diagram on the left in Figure 8 is caused by the change in flow pattern from the feed towards the liquid discharge. The radius of the inlet is higher than the radius of the outlet. Hence the main flow is shifted towards the rotational axis. These effects are displayed in Figure 9 for 1000 rpm (left) and 3000 rpm (right).

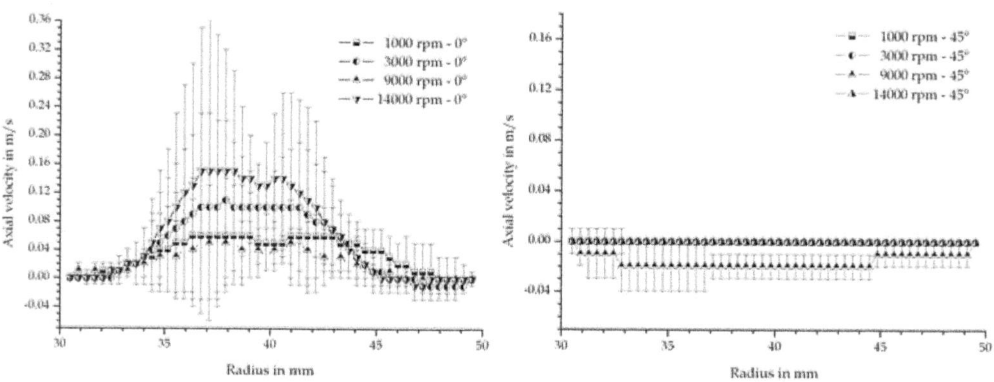

Fig. 8. Axial velocity, mean values from throughout the rotor: left: radius of inlet; right: angular position ± 45°

The maximum of the axial velocity is close behind the inlet (H=0mm) at a radius of 42 mm, which corresponds to the inlet radius. Towards the liquid discharge, the maximum shifts to a radius of 37 mm, where the outlet pipes are located. The change in flow pattern is indicated by the arrows in Figure 9. This behaviour was detected for all rotational speeds and different throughputs. The changing flow profile explains the high standard deviation of the averaged profiles shown in the diagram on the left in Figure 8.

The deviation between simulation and experimental data with respect to the flow profile over the full angle of rotation may be explained by the turbulent model used. The modelling does not calculate the turbulences in detail, but estimates the degree of turbulence and thus extenuates the fluctuations at the inlet and outlet. Furthermore the flow between inlet and outlet may not be entirely turbulent, as mentioned in the introduction.

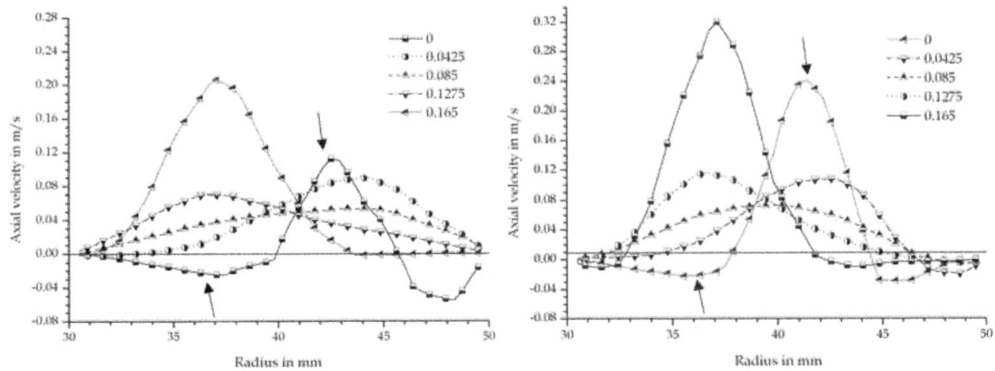

Fig. 9. Axial velocity, radius of inlet for different distances from inlet: left: 1000 rpm; right: 3000 rpm

4.2 Comparison of different turbulence models

It is presumed that the deviations between observed flow pattern and predicted flow behaviour are caused by the turbulence model used. Some authors already compared different models that simplify turbulent effects by correlation factors and additional equations like the k-ε RNG model. They stated that the k-ε RNG model is a reasonable compromise between computational demand and validity of the results (Wardle et al., 2006). Other authors relied on the more extensive Reynolds Stress Model (RSM) or the large-eddy simulation (LES) (Mainza et al., 2006; Wang & Yu, 2006). Nowakowski et al. reviewed the milestones in the application of computational fluid dynamics in hydrocyclones (Nowakowski et al., 2004). The most common models used are the k-ε and recently the RSM and LES model. For the evaluation of the differences in the prediction of the flow behaviour in centrifuges between several turbulence models, the flow pattern was calculated for various rotational speeds and throughputs, using the k-ε RNG, k-ω, Reynolds Stress Model and large-eddy simulation. A representative comparison of the obtained results by using the different models is shown in Figure 10. The plots illustrate the differences in the calculated tangential and axial velocity halfway between inlet and outlet at the angular position of the inlet.

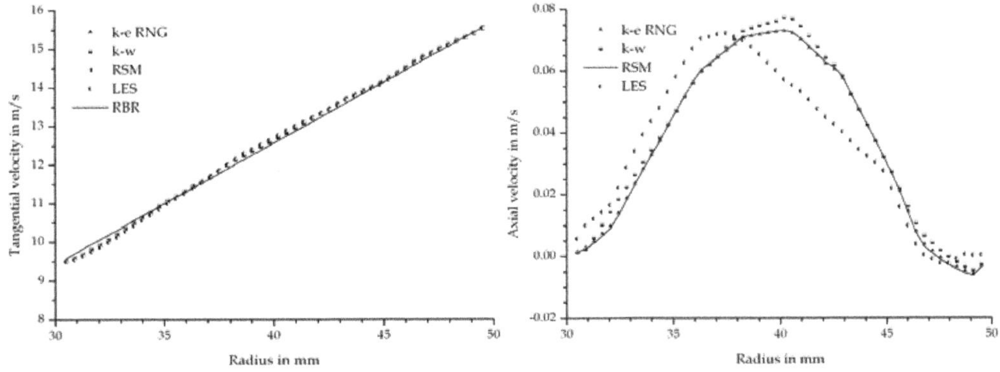

Fig. 10. Tangential (left) and axial velocity profiles (right), throughput 1 l/min, k-ε RNG, k-ω, RSM and LES turbulence model

The comparison reveals no significant deviation of the predicted flow pattern. The differences in the tangential velocity are barely detectable. All calculations are in good agreement with the rigid body rotation and thus with the experimentally determined profiles (Spelter, et al. 2011). Using the LES model, the obtained axial flow profile differs from the other models by predicting an earlier shift of the main axial flow towards the radius of the outlet at 36 mm. Because of the high computational demand and instability of the RSM, the simulation with this turbulence model was started with a solution obtained with the k-ε RNG model. Nevertheless the differences are small and all calculated axial flow pattern are in disagreement with the measured flow profile. The mixing and expanding of the inlet jet is significantly understated. By not questioning and validating the simulations, the user of the CFD software is led to false conclusions which may result in low performance of the centrifuge due to disadvantageous design. The classical evaluation of the simulation does not indicate a computational error. The mass balance between the inlet and the outlet is fulfilled as stated in Table 1, the residuals are sufficiently small (10^{-3} to 10^{-5}) and the flow patterns do not change with increasing flow time. Furthermore the predictions are reproducible with different turbulence models. All these facts suggest an accurate calculation. Only the comparison of the main velocities with experimentally determined data shows that the prediction misrepresents the true case in respect to the axial flow profile.

Model	k-ε RNG	k-ω	RSM	LES
% dV	0.0395	0.0015	0.0022	0.0172

Table 1. Deviation of mass balance dV for different turbulence models in % at 3000 rpm and 1 l/min

4.3 Comparison of two and three dimensional modelling

The analysis of the acceleration geometry is only possible in three dimensional modelling, because, as mentioned in the introduction, the inlet geometry cannot be correctly reproduced in the two dimensional case. The results presented in chapter 4.1 and 4.2 show a short-circuit flow between inlet and outlet. The comparison between the results obtained when calculating three and two dimensional are shown in Figure 11. The predicted tangential velocity profiles are in good agreement with the rigid body rotation for both cases

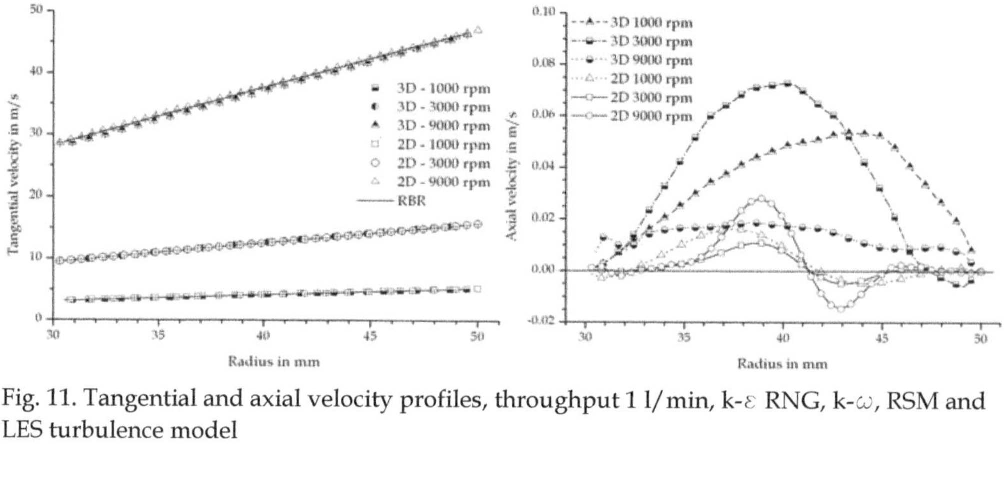

Fig. 11. Tangential and axial velocity profiles, throughput 1 l/min, k-ε RNG, k-ω, RSM and LES turbulence model

at various rotational speeds. The water is fed through an annulus in the two dimensional model and the tangential acceleration takes place via the friction at the wall at both sides of the annulus. According to the prediction, this feeding geometry is sufficient for an acceleration of the incoming liquid up to 9000 rpm. The differences between the cases are distinguishable in the axial flow profiles which are shown in the diagram on the right side of Figure 11. The profiles presented are the values halfway between inlet and outlet. Due to the symmetry condition, the flow is homogeneously distributed along the whole circumference for a specific radius. Therefore the maximum axial velocity is lower than in the three dimensional case due to the conservation of mass. There are recirculating currents next to the inlet in the two dimensional cases which increase in magnitude with rotational speed. In the three dimensional case, recirculating flows are located next to the inlet as well, but displaced few degrees in the angle of rotation.
Nevertheless the predicted axial velocity profiles in the two dimensional cases are comparable to the ones obtained in the three dimensional ones in respect to the radial distribution of the flow. The inlet jet remains until the water leaves the centrifuge via the circular weir, so that no plug-shaped profile - resulting in a high active volume -, as observed in the experiments, is formed.

4.4 Calculation of the flow pattern (k-ε RNG model, VOF method, transient solution) - overflow setup

The overflow setup was simulated by using the previously described Volume of Fluid method. The centrifuge is filled with water at the beginning of the simulation. As an initial condition, the water is already spinning with the rigid body rotation. This assumption is justifiable because in the experiments, the centrifuge was slowly filled with water and ran for at least 30 minutes in steady state prior to the measurements. The initial condition reduces the necessary time until the steady state is reached significantly. Figure 12 shows

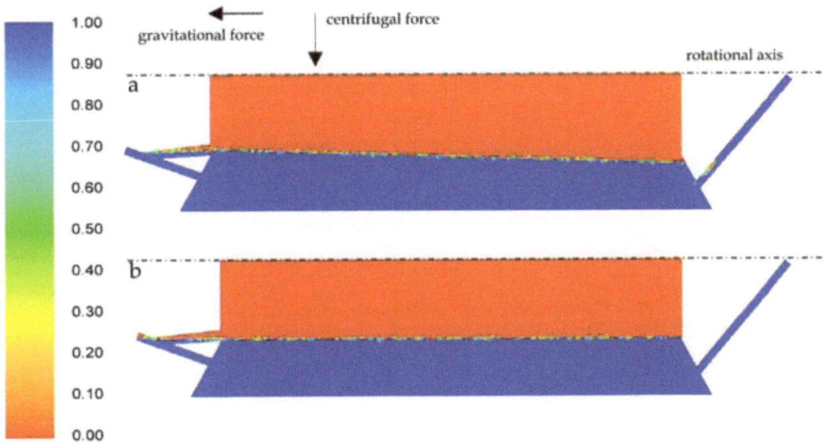

Fig. 12. Profiles of water-air interface for a throughput of 1 l/min: a) 1000 rpm; b) 9000 rpm

the distribution of the two phases for 1000 (a) und 9000 rpm (b) for a throughput of 1 l/min. Due to the gravitational and centrifugal force, a conical shape of the water surface is developed at 1000 rpm. The centrifugal number outbalances the gravitational force with increasing rotational speed, thus the water phase approaches the shape of a cylinder, as shown in Figure 12 b).

The tangential velocity profiles for 1000 and 9000 rpm are shown in Figure 13. The diagram on the right side shows the tangential velocity values averaged over the length of the centrifuge for the water phase only. The computed values are compared to the rigid body rotation. The contours of the tangential velocity at 9000 rpm are shown on the left side of Figure 13. The calculated velocities are in good agreement with the analytical values and with the measurements. At 9000 rpm, the tangential velocity of the water exhibits a small lag when compared to the rigid body rotation. The magnitude of the deviation is within the accuracy of the measurements, so that no statement whether this difference is the true case or not, is possible.

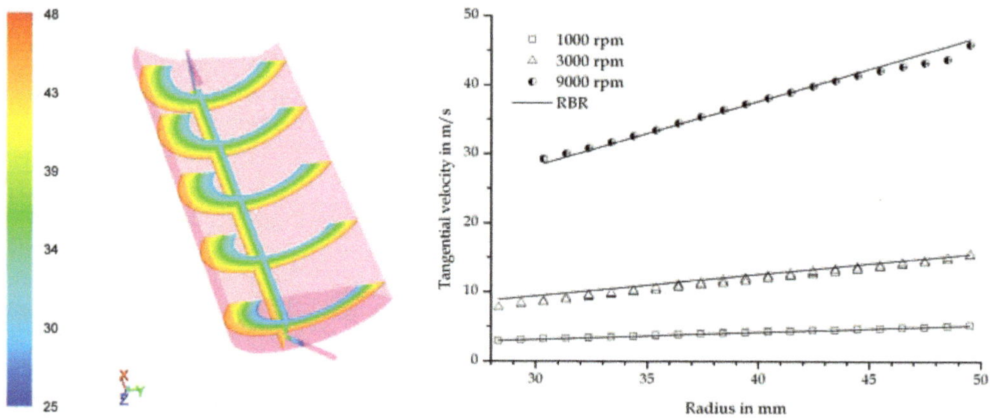

Fig. 13. Tangential velocity profiles, water phase, throughput of 1 l/min; left: contours at 9000 rpm; right: for different rotational speeds and different axial positions

Figure 14 shows the axial velocity profiles for 9000 rpm at a constant throughput of 1 l/min. The axial flow pattern is similar to the one predicted in the single-phase model. The inlet jet

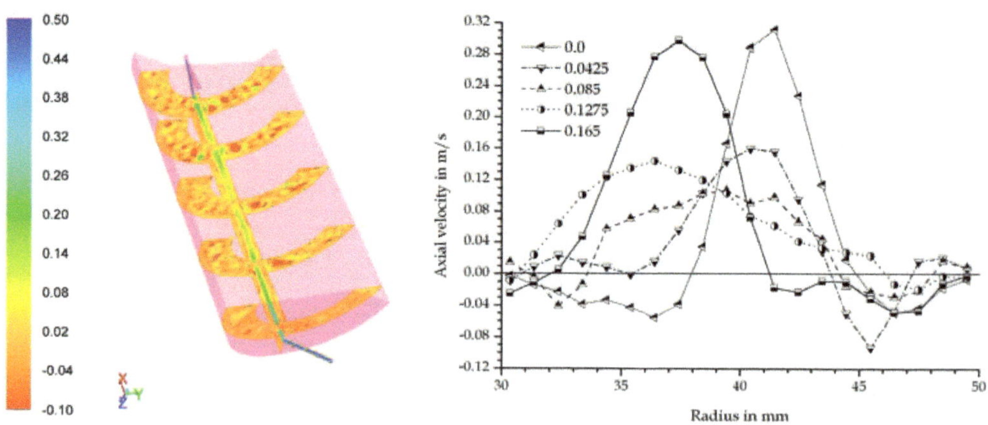

Fig. 14. Axial velocity profiles at 9000 rpm and 1 l/min throughput; left: contours; right: different distances from the inlet

does not widen so that a short-circuit flow towards the outlet is formed, as shown in the contour plot on the left side of Figure 14. There is no significant expansion of the flow profile neither in angular nor in radial direction. The peak of the axial velocity shifts from a radius of 42 mm close to the inlet towards a radius of 36 mm. This behaviour was also predicted in the single-phase model. Nevertheless the simulation misrepresents the true case, in which a significant widening of the inlet jet was observed.

4.5 Influence of the inlet geometry on the flow pattern

In order to investigate how the acceleration of the feed affects the flow pattern, a centrifuge with a simpler geometry and a low performance feed accelerator was simulated. The main differences when compared to the tubular bowl centrifuge are the lower length-to-diameter ratio of 0.8 and the feed accelerator. The feed enters a rotating disc and leaves it radially with a certain angular velocity. The water exits the centrifuge through eight boreholes distributed along the circumference at the top of the bowl. Figure 15 shows a scheme of the centrifuge geometry on the left and the results of the volume fraction of air on the right. Water is fed axially through the inlet to the accelerator; there, the water changes its direction and gains in tangential velocity. The transfer of rotational movement occurs just at the plate surfaces, defined as no-slip boundaries, via momentum diffusion. Then, the water exits the accelerator as streams that travel through the air core and enter the rotating liquid pool. The radius of the interface changes 35 % along the height of the centrifuge from the inlet to the upper part of the centrifuge because of the low rotational speed (454 g).

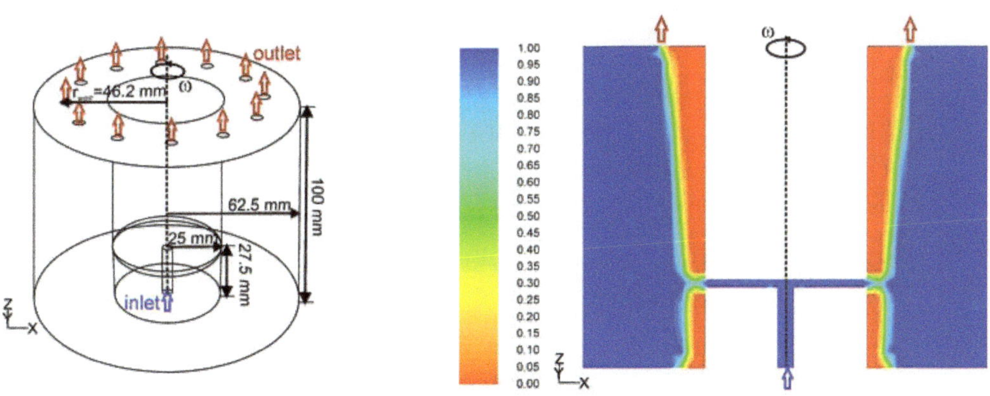

Fig. 15. Left: geometry of the centrifuge; right: contours of volume fraction of air at 2550 rpm and 5.8 l/min

The main flow occurs in the direction of the rotation of the bowl. Both, the liquid pool and the air core rotate. The tangential velocity of the water jet from the inlet accelerator (height 0.0275 m) is lower than the tangential velocity of the bowl. This causes a drag of the whole rotating liquid pool regarding the angular velocity of the bowl. The values of the tangential velocity of the liquid averaged over the circumference are plotted for different heights in Figure 16. The averaged velocity of the liquid in this geometry is approximately 50 % below the rigid body rotation. The values at the wall have to match the rigid body rotation of the bowl to fulfil the no-slip boundary condition. The steep decrease of the tangential velocity near the wall was found to be smoother for a 6 times finer grid. However, the computational effort for the multiphase simulation including the particles with the finer grid was unaffordable. The deviation with respect to the rigid body rotation decreases for lower flow rates because a smaller amount of liquid has to be accelerated. The simulation of this geometry shows how important is to accelerate the feed properly to the angular velocity of the liquid pool before entering it in order to reach the angular velocity of the bowl in the rotating liquid and thus the highest centrifugal acceleration possible.

In contrast to the other centrifuge, a layer with high axial velocity at the gas-liquid interface, as expected by the boundary layer model, has been observed in the simulations as depicted in the diagram on the right in Figure 16. Near the impact height of the inlet water jet two whirls with opposed directions are formed. This can be seen at the negative axial velocity values at the interface for the height 0.02 m, just below the inlet, and the positive axial velocities at he interface for the height 0.0275 m, just above the inlet. Upsides the height of the inlet, a homogeneous axial boundary layer is formed at the interface with a recirculation layer at the bowl wall. This layer exhibits a similar averaged thickness δ as the one proposed by the theoretical model of Gosele (Gösele, 1968) but much thicker as the one calculated following the theory of Reuter (Reuter, 1967) as seen in Table 2. Indeed, it occupies the whole liquid pool until a recirculation layer is formed close to the bowl wall. This is due to the lower tangential velocities which cause a small pressure gradient in the radial direction reducing the stratification of the flow and allowing the fluid to move in the axial direction. The velocity values oscillate with the angle, especially near the inlet at 0.02 m and 0.0275 m and the outlet at 0.0975 m, respectively. Thus the standard deviation is higher at these positions.

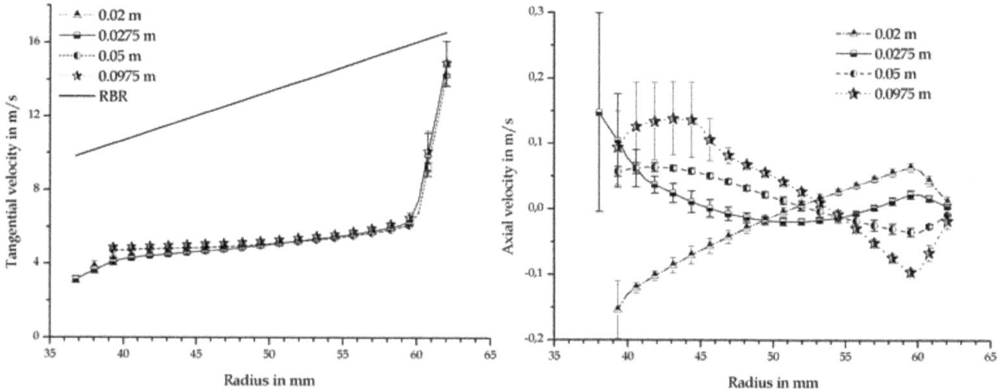

Fig. 16. Left: tangential velocity versus radial position; right: axial velocity versus radial position at 2550 rpm and 5.8 l/min

$\delta_{simulated}$ (mm)	δ_{Reuter} (mm)	$\delta_{Gösele}$ (mm)
14.1	1.71	15.8

Table 2. Thickness of the axial boundary layer δ obtained in the simulations and calculated analytically at 2550 rpm and a throughput of 5.8 l/min

5. Mathematical formulation of the particles

There are several multiphase models available for simulation of a particulate phase. A summary of the different simulation methods for multiphase flows in CFD can be found in (van Wachem & Almstedt, 2003). The multiphase models can be divided into two groups; the Euler-Euler and the Euler-Lagrange methods.

The Euler-Euler method considers all phases as continuous ones. For each phase, a momentum conservation equation is solved. In the momentum equation of the dispersed phases, the kinetic theory for granular flows (Lun & Savage, 1987) is used. The intensity of the particle velocity fluctuations determines the stresses, viscosity, and pressure of the solid phase. The interaction between phases is considered by means of momentum exchange coefficients. This numerical method has the advantage that it is suitable for high volume concentrations of the disperse phase. The disadvantage is that the properties of each dispersed particle cannot be simulated and a particle size distribution cannot be considered. Moreover, the stability of the calculation depends on the choice of model parameters and often convergence problems appear. For the simulation of sedimentation and thickening processes there are also some Euler-Euler models based on the Kynch's kinematics sedimentation theory (Kynch, 1952), which were reviewed by Bürger in (Bürger & Wendland, 2001). These models are all based on the measurement of fundamental material properties of the suspension to be separated as explained by (Buscall, 1990; Buscall et al., 1987; Landman & White, 1994). These multiphase models are based on mass balances for the fluid and the disperse phase. The fluid flow is not calculated but they consider the relative velocity between fluid and particles. On this basis, these models are unsuitable for studying the particle behaviour in solid bowl centrifuges where the liquid flow plays a crucial role in particle settling. These models usually describe the sedimentation and consolidation in batch processes in simple-geometry centrifuges. An extension of these models is necessary to describe the sediment build-up in two and three dimensions correctly.

The Euler-Lagrange method solves the continuum conservation equations just for the characterization of the continuous phase. It describes the dispersed phase as mass points and determines the particle trajectories by integrating a force balance for each individual particle. This way, instead of a partial differential equation, only an ordinary differential equation must be solved for the dispersed phase. The weakness of the Euler-Lagrange method is that it is only valid for diluted suspensions if a one-way or a two-way coupling method between continuous and disperse phase is applied. With the one-way coupling, the motion of the discrete phase is calculated based on the velocity field of the continuous phase but the particles have no effect on the pressure and velocity field of the continuous phase. By using a two-way coupling, the equations of continuous and discrete phase are calculated alternatively. The momentum conservation equation of the continuous phase has a source term to consider the momentum exchange with the particulate phase. The maximum value for the volume concentration of the disperse phase is between 5% (Schütz et al., 2007) and 10-12% (ANSYS, 2009), above this value the interaction between particles should be considered. Thus there is a third method, four-way coupling, which involves the interaction within the disperse phase: impacts between particles, particle-wall interactions, van der Waals forces, electrostatic forces, etc. Thus the restriction for the volume fraction is not valid anymore. However, if the number of particles increases, the computational effort rises significantly. Usually the characteristics of the problem and the information that should be obtained from the simulation decide which multiphase model and coupling methodology should be used.

The Discrete Element Method (DEM) (Cundall & Strack, 1979) is a Lagrangian method originally developed for the simulation of bulk solids and then extended to the entire field of particle technology. This method solves the Newton's equations of motion for translation and rotation of the particles. In these equations, impact forces of particle-particle and particle-wall interactions are considered. Other forces such as electrostatic or Brownian forces can also be implemented if necessary. In order to describe the collisions of particles, the soft-sphere approach was chosen. When two particles collide, they actually deform. This deformation is described by an overlap displacement of the particles. A schematic representation of the forces on the particles in the soft-sphere model is depicted in Figure 17. This approach uses a spring-dashpot-slider system to calculate the behaviour of the particles during the contact time. The soft-sphere model allows contacts of a particle with more neighbouring particles during a time step. The net contact force acting on a particle is added in the case of multiple overlapping. This model for contact forces was first developed by Cundall (Cundall & Strack, 1979), who proposed a parallel linear spring-dashpot model for the normal direction and a parallel linear spring-dashpot in a series with a slider for the tangential direction. A comparison of the performance of several soft-sphere models can be found in (Di Renzo & Di Maio, 2005; Stevens & Hrenya, 2005). There, also the Hertz-Mindlin (Mindlin, 1949) non-linear soft-sphere model used in the present work is analysed. This model proposes a normal force as a function of the normal overlap δ_n, the Young's modulus of the particle material and the particle radii. The damping force in normal direction depends on the restitution coefficient, the normal stiffness, and the normal component of the relative velocity between the particles. The tangential force depends on the tangential overlap δ_t, on the shear modulus of the particle material, and on the particle radii; while in the damping force in tangential direction the tangential stiffness and the tangential component of the relative velocity between the particles are considered.

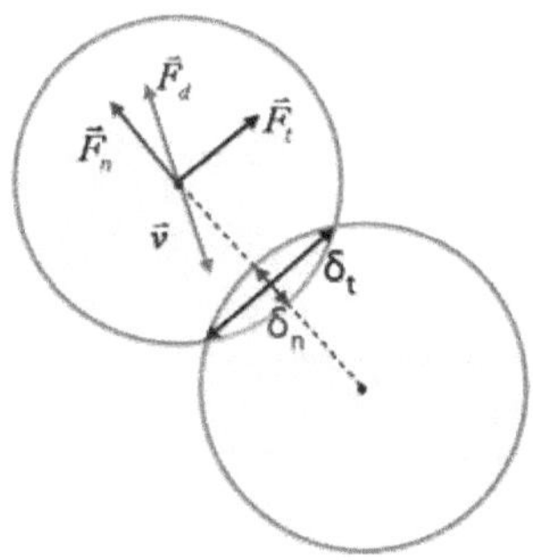

Fig. 17. Contact forces between two particles following the soft-sphere model

While the DEM model was originally used in environments without fluid, increasing efforts have been made to extend this model to the mixture of fluid and particles. For that, a coupling of the DEM calculations with the fluid flow simulation is necessary to be able to include the hydrodynamic forces in the particle simulation. An important aspect in the calculation of particles in a fluid is the energy dissipation into the surrounding fluid. In viscous fluids, there are, in contrast to air, two additional effects that lead to energy losses. The drag force leads to lower particle velocities and when a particle is about to hit a surface or another particle, a deceleration occurs. This is due to the drainage of the fluid, which is located in the gap between the two particles or the particle and the surface. In his work

(Gondret et al., 2002), Gondret presents the coefficient of restitution as a function of the Stokes number St (Equation 16), which describes the ratio of the dynamic response time of a particle to flow changes and the characteristic flow time. He determined empirically a critical Stokes number of about 10. Below this value, a particle does not bounce when it hits a wall and instead remains in contact with it.

$$St = \frac{1}{9}\frac{\rho_s v\, x}{\mu} \tag{16}$$

In a centrifugal field, the hydrodynamic forces (drag, lift, torque) play a crucial role and must be included in the calculation of the particle trajectories. Therefore the fluid flow and particles movement have to be computed alternatively. A scheme of the coupling between the CFD software FLUENT and the DEM software EDEM can be seen in Figure 18. At first, the flow will be determined with CFD, then for each position in the computational domain, the velocity, pressure, density and viscosity of the flow are transferred to DEM in order to calculate the hydrodynamic forces and include them in the balance. Afterwards, the fluid flow will be computed again considering the influence of the particles as an additional sink term in the momentum equations. This sink term include the force transmitted to the particles by the fluid.

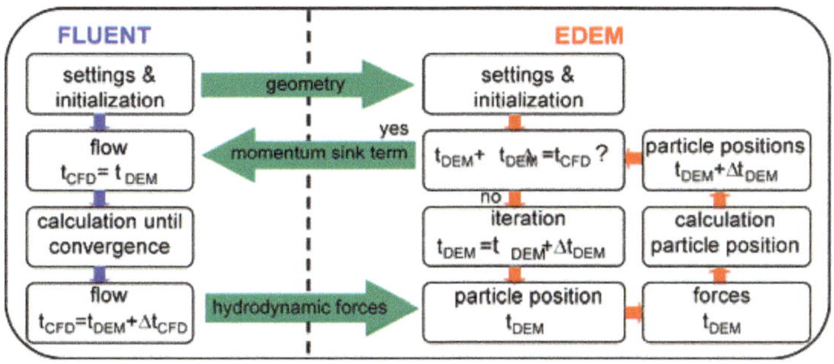

Fig. 18. Coupling between CFD and DEM

Some authors have already simulated such complex multiphase flows in other separation equipment such as filters and hydrocyclones (Chu et al., 2009; Chu & Yu, 2008; Dong et al., 2003; Ni et al., 2006). Nevertheless due to the high velocity gradients and complex interactions between the phases, the simulation of multiphase flow and in particular the sediment behaviour in centrifugal field is still an unsolved challenge. The advantage of this approach is that an accurate description of the flow, the particle paths and the sediment behaviour can be obtained. However, the amount of simulated particles and its size is limited due to the computational demand.

5.1 Particles simulation parameters

The calculation in CFD must be run as an unsteady case when coupled with DEM. The DEM time steps are typically smaller than the CFD time steps in order to correctly capture the contact behaviour. The choice of time steps in DEM simulation is of great importance in order to achieve stability in the calculation but does not perform very long calculations, although still representing the real particle behaviour. Therefore it is necessary to analyse the forces acting on the particles to determine the different time scales. In the case presented, the Hertzian contact time - the total time of contact in Hertz theory of elastic collision -, the particle sedimentation time - the time required for a particle to settle one diameter -, and the particle relaxation time - the time for a particle to adapt to the surrounding fluid flow - are considered. On the basis of these, the smallest time step must be chosen so that all physical phenomena are represented. In the simulations presented, the Hertzian contact time is the smallest time scale and, thus a time step $\Delta t_{DEM} = 10^{-6}$ s was chosen in the DEM software, while in the CFD a time step $\Delta t_{CFD} = 10^{-4}$ s was sufficient.

The size of the particles was defined as a logarithmic normal distribution with a mean diameter of 200 µm and a standard deviation of $\sigma = 10$ µm. Simulations with smaller particles lead to divergence of the calculation despite of the small time steps set. Thus simulations with lower densities of the particles and rotational speeds were performed to obtain similar settling velocities as with small quartz particles (mean diameter of 2.07 µm, σ=2.15 µm) which were used in the experiments. In order to reproduce the experimental concentration of approximately 0.5 vol. %, 100,000 particles per second were injected at the inlet of the centrifuge for the simulation. In future work, the comparison of predicted and experimentally determined sediment build-up will show whether the simulation methodology leads to correct predictions.

For the contact models, a parameter setting is needed to represent the real particle behaviour. The information needed is mainly about the physical properties of the particle system because the spring and damper constants are calculated as a function of the shear modulus and the Poisson's ratio of the particle and wall materials. The coefficient of restitution depends on the Stokes particle number. The friction coefficient is an empirically determinable property, which can be found in the literature for various material combinations. Some authors have already determined this friction coefficient for sand and glass particles for DEM (Asaf et al., 2007; Li et al., 2005). Here, shear experiments of a sediment made of quartz particles were performed in a Jenike shear cell under normal pressure in order to calculate the friction coefficient as was done by (Hartl & Ooi, 2008).

Parameter	Value
Particle radius (µm)	200
Particle density (kg/m³)	1016
Shear modulus particle (Pa)	$2.2 \cdot 10^{8}$
Shear modulus wall (Pa)	$7.0 \cdot 10^{10}$
Poisson's ratio particle (-)	0.25
Poisson's ratio wall (-)	0.30
Restitution coefficient (-)	0.010
Static friction coefficient (-)	0.787
Rolling friction coefficient (-)	0.100

Table 3. Parameters used in the DEM simulation

6. Prediction of the sediment build-up by means of DEM

The geometry of the centrifuge in which the particle trajectories and the sediment build-up were simulated is shown in Figure 15. For the simulation purpose of predicting the

sediment build-up it is not possible to use periodic boundaries as was done for the simulation of the fluid. Moreover, we chose this geometry with the VOF model for air and liquid in order to investigate the interaction between the multiphase model and the use of DEM to calculate the particles. The calculations were made with the software ANSYS FLUENT 12.0.16 and EDEM 2.3, which include a coupling feature for both programs.
The coupled simulation was performed with a rotational speed of 2550 rpm. For this rotational speed, the Stokes settling velocity of the particles is by two orders of magnitude higher than the Stokes settling velocity of the quartz particles used in the experiments. The results of this simulation are depicted in Figure 19.

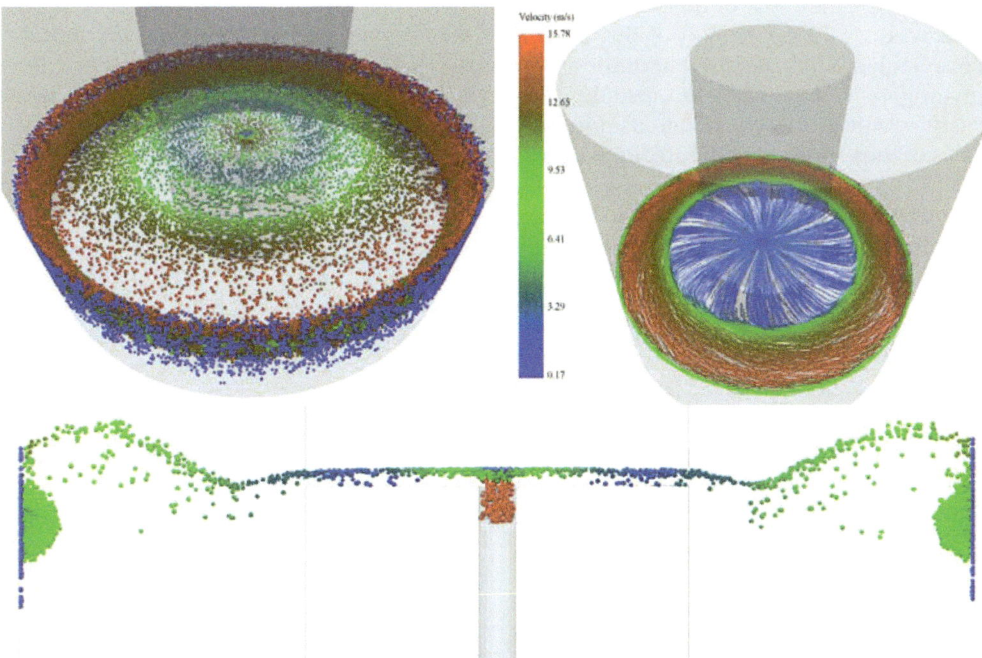

Fig. 19. Particles and path lines simulated with EDEM coupled with a two-phase simulation of a centrifuge in FLUENT and coloured by velocity magnitude

The particles are injected at the inlet pipe with the same axial velocity as the fluid. They hit the top of the feed accelerator and gain in radial velocity. Afterwards, they move through the air core until they hit the liquid pool. There, under the effect of the centrifugal force, the particles perform spiral path lines, as shown in Figure 19 on the right, until reaching the wall or already settled particles. The lower plot in Figure 19 shows a vertical plane of the centrifuge with the particles in the region near the feed accelerator where the sediment is built. Due to the drag force of the axial flow in the rotating pool, the particles move first to higher axial positions. Then, near the wall, they move backwards because of the recirculation layer. The evolution of the sediment can be observed in Figure 20 as a function of the calculated particles for two different rotational speeds of the liquid pool. For the case with lower rotational speed, 686 rpm, the first settled particles deposit at the wall and are pushed to lower axial positions because of the recirculation layer near the wall and due to the lower normal pressure, 3 kPa, acting on the sediment, which allows the influence of the gravity. Then, particles deposit on the sediment and it growths from the wall to the inner part of the centrifuge in radial direction. The sliding of the sediment in this case is due to the relationship of low normal pressure and high shear stress acting on it. In the case of higher rotational speed in the liquid pool, 1937 rpm, the particles travel radially direct to the bowl wall and the sediment formed is flat. The particles settling in the next time steps slide to the side of the sediment which grows along the wall rather than in the radial direction.

In this study the value of the friction coefficient between particles and between particles and wall was constant. Changes of this parameter would also lead to changes in the sediment behaviour under certain conditions of normal pressure and shear stress. A further investigation in this field, including the experimental determination of the friction coefficient for different materials and particle size distributions, is needed. Here only the basis of the coupling methodology was studied.

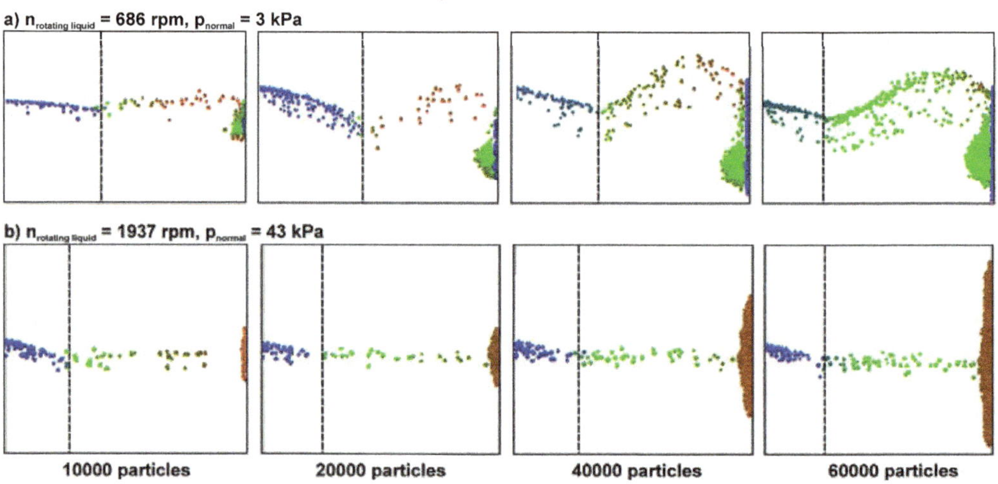

Fig. 20. Sediment built at the bowl wall as a function of the particles simulated with EDEM coupled with a two-phase simulation of a centrifuge in FLUENT and coloured by velocity magnitude for two different rotational speeds

In order to ensure that the coupling between both programs works and that the hydrodynamic forces are correctly passed from FLUENT to EDEM, the particle trajectories calculated with the coupling and with FLUENT were evaluated. Although the time steps and the mathematical method used in FLUENT to calculate the particles is different from the time steps and algorithm used in EDEM, the results are comparable as shown in Figure 21. Another different is that in FLUENT, the particles were calculated using a static flow field at a certain time step whereas in EDEM, the flow field was calculated alternating to the position calculation of the particles. Thus in Figure 21 on the left, the particles calculated in FLUENT acquire higher axial positions when compared to the particles calculated with EDEM. The reason for this disagreement is the difference in the axial flow pattern, where the velocities oscillate until a quasi-steady state is achieved. However, qualitatively both particle trajectories are in good agreement. The tangential velocity of the particles is represented in Figure 21 on the right. The tangential velocity of the particles is almost zero in the regions where there is no liquid. Then, they are accelerated in tangential direction until reaching the bowl wall. There, a deviation between FLUENT and EDEM appears. The particles in EDEM hit the sediment and the tangential velocity diminishes because the wall rotation is not taken into account here, but in FLUENT, no particle accumulation is considered and, thus all particles are calculated until they reach the wall. Once they have reached the wall, they are taken from the simulation domain.

The location and quantity of the particles are stored as a variable in FLUENT representing the volume fraction of the particles, as can be seen in Figure 22 on the left. However, for the flow calculations they are considered as points where a momentum exchange occurs. The drag force of the flow acting on the particles is subtracted from the momentum equation diminishing the flow in regions where the particles accumulate. Figure 22 on the right represents the volume fraction of air in the centrifuge. The effect of the particles on the flow can be seen as the water jet coming from the inlet accelerator falls down in the axial position until it reaches the liquid pool due to the particles weight and the momentum sink term.

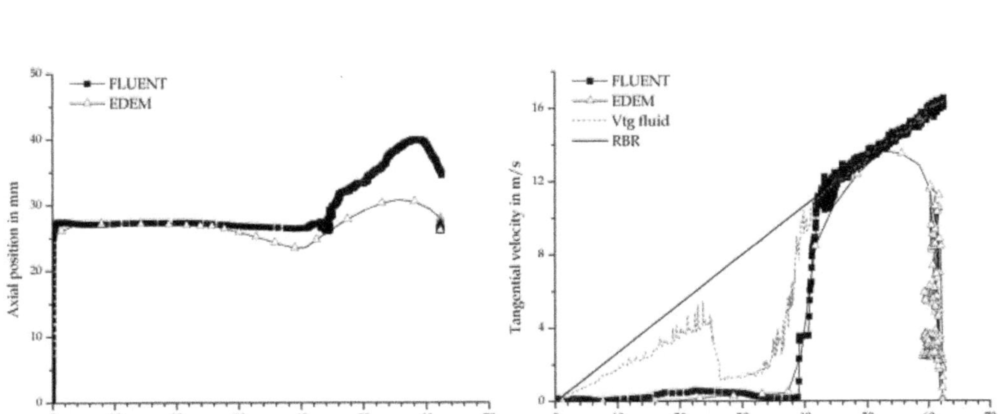

Fig. 21. Comparison of the particle trajectories obtained in FLUENT and EDEM: left: axial position versus radius; right: tangential velocity of the particles and the fluid versus radius

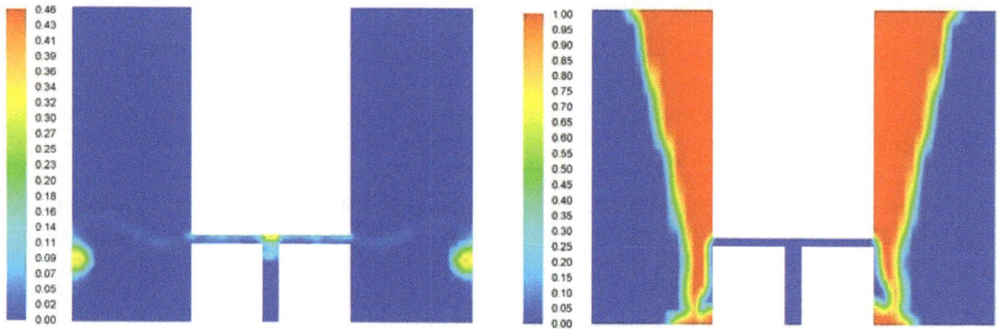

Fig. 22. Contours of the volume fraction in the FLUENT simulation domain; left: particles; right: air

The effect of the sediment on the flow pattern is shown in Figure 23, where the tangential and axial velocity values are represented. The tangential velocity values are lower than in the simulation without particles because of the momentum exchange between the fluid and the particles. Furthermore the tangential velocity is lower in the lower part of the centrifuge, where most of the particles can be found. The interface radius is not constant over the height of the centrifuge, as can be observed in Figure 22 on the right. Therefore the values of the tangential and axial velocity of the water in Figure 23 begin for lower radii in the lower part of the centrifuge. The form of the axial flow pattern does not change qualitatively except for the height where the particles accumulate, around 0.02 m. There, the axial velocity takes approximately the value zero due to the exchange of momentum between fluid and particles. At the wall, where the sediment is located, there is no axial flow in contrast to the simulation without particles. This fact shows how important is to consider the sediment build-up in the multiphase flow simulation of a centrifuge. However this also carries an increase of the computational effort.

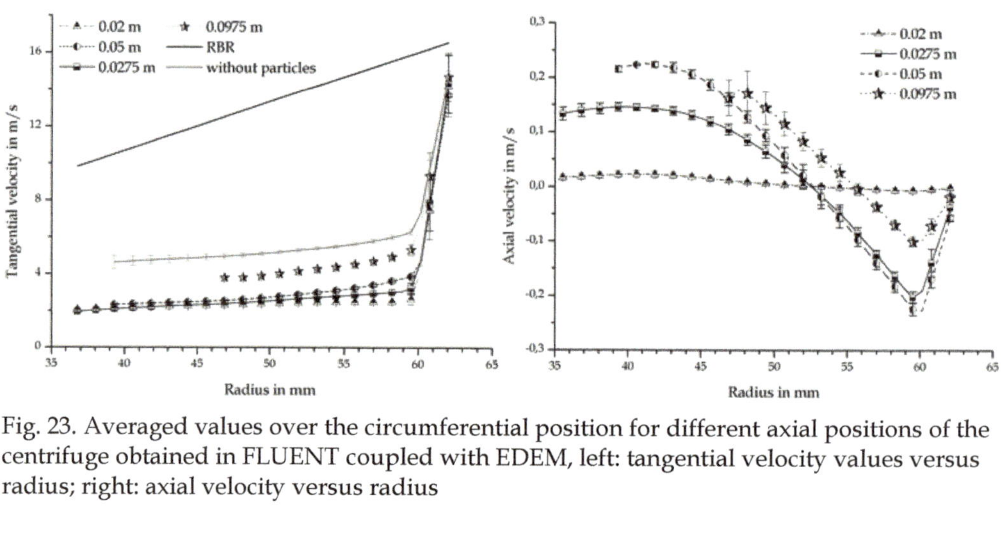

Fig. 23. Averaged values over the circumferential position for different axial positions of the centrifuge obtained in FLUENT coupled with EDEM, left: tangential velocity values versus radius; right: axial velocity versus radius

7. Evaluation of the computational demand for predicting flow in centrifuges

The prediction of flow pattern in centrifuges requires a high computational demand that increases with the maximum velocity in the system. Accordingly to Equation 17, the highest velocity is reached at the bowl wall and increases linearly with the rotational speed.

$$v = \frac{2 \cdot \pi \cdot n}{60} \cdot r \tag{17}$$

Figure 24 shows the computational demand in $h_{sim}/(h_{flow\ time}\ 3.6\ 10^{-3})$ (dimensionless) against circumferential velocity in m/s for different simulations which include two and three dimensional problems, one and two-phase simulations, and different models to predict turbulent flow. The simulations were performed with Win xp 32 bit desktop computers using an Intel Core 2 Duo E4500 2.2 GHz processor, 2 GB DDR 2 memory running at 266 MHz (2 x 1GB Kingston) and a ASrock motherboard (dual channel memory). All data are averaged values. For reasons of readability, the standard deviation is not stated, because it is too small to be included in the plot. Evaluating the diagram on the left side of Figure 24, it is remarkable how costly the VOF method for simulating two-phase flow for the two and three dimensional cases is. Moving from one to two phases causes the same increase in computational demand as moving from a two to a three dimensional grid and simultaneously increasing the cell number by one order of magnitude. However, the size of the grid does not necessarily increase the computational time. The convergence in a fine grid is faster and the influence of the rotating geometry overcomes the one of the grid when

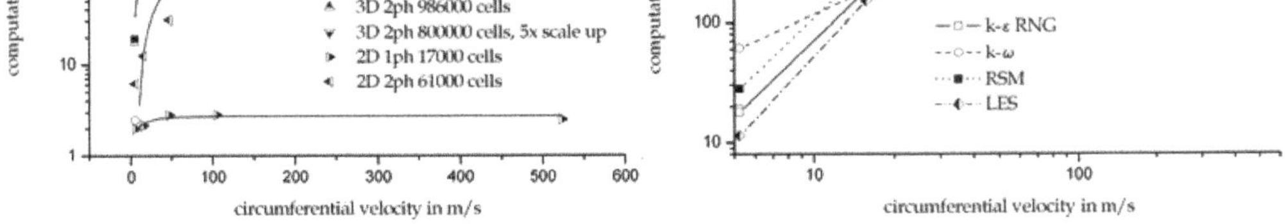

Fig. 24. Correlation of computational demand in $h_{sim}/(h_{flow\ time}\ 3.6\ 10^{-3})$ and circumferential velocity: for one and two-phase modelling, two and three-dimensional geometries, and different sizes of the geometry (left); dependence on turbulent model used (right), 715000 cells, three-dimensional and single-phase model

simulating centrifuges. This is concluded by comparing the three dimensional, single-phase calculations with a grid size of 215000 cells with the simulations with a grid of 715000 cells. The same effect is observed when comparing the three dimensional, two-phase simulations with different cell sizes and different size of geometry. All data is well fitted with a single function for each of the two compared sets of calculations. Looking at the differences between the turbulent models, all simulations exhibit a comparable computational demand. The RSM is the model with the highest computational demand whereas the large-eddy simulation needs less computational resources than all other models evaluated. This is unexpected since the LES is usually rated as more demanding than the RSM. This is explainable due to the single equation modelling of turbulence in the subgrid, the low degree of turbulence in the case presented and the fine mesh that was constant in all cases presented in the diagram on the right side of Figure 24. The k-ε RNG and the k-ω simulations require comparable computational resources. This is in agreement with the theory, because both models are similar to each other except of the modelling of the near wall region.

8. Conclusions

The results presented show that the prediction of the flow pattern in centrifuges by means of computational fluid dynamics give good results for evaluation of the tangential acceleration efficiency. The predictions are in good agreement with the measurements conducted with Laser Doppler Anemometry. Nevertheless the axial fluid profiles do not depict the true case accurately. The calculated axial velocity profiles give reasonable results close behind the inlet, but the widening of the inlet jet is not distinguishable in the simulations. The axial velocity profile is essential for the calculation of the cut size and the sediment distribution in the centrifuge, thus any significant deviation from the true case is not acceptable. It was demonstrated that the Reynolds Stress Model (RSM) and the large-eddy simulation (LES) do not increase the accuracy of the predictions when compared to the results obtained using the k-ε RNG and k-ω model, but increase the computational demand in case of the RSM. All models predict a short-circuit flow between inlet and outlet. The flow pattern does not change with increasing rotational speed or throughputs, only the maximum fluid velocity increases with higher throughputs as expected. This behaviour was observed in single and two-phase simulations.

The geometry of the feed accelerator is decisive for the axial flow pattern developed in the centrifuge. A radial feed impacting on the surface of the liquid pool leads to an axial boundary layer flow similar to the one predicted and experimentally verified by other authors such as Glinka (Glinka, 1983). The design of the feed accelerator and its efficiency is essential for a good centrifugal performance. If the liquid is not accelerated to the angular velocity of the bowl, the rotating pool exhibits a significant lag in tangential velocity when compared to the rigid body rotation. The proposed feeding system via rotating inlet bores leads to an efficient tangential acceleration of the fed suspension.

Particles with a diameter of 200 μm were calculated for the first time in a centrifuge using a coupling between the simulation of the flow in FLUENT and the simulation of the particles in EDEM. The results show a good agreement between particles simulated in FLUENT only considering the hydrodynamic forces and the results from EDEM, which implies that the hydrodynamic forces are correctly transferred from one program to the other. Furthermore it proves that the coupling method can be applied for rotating geometries such as centrifuges. The particle trajectories and the sediment build-up agree qualitatively with the expected results. Particles travel through the air core and enter the liquid pool, where they describe spiral paths until they reach the wall or already settled particles to form a sediment. The volume occupied by the particles has not been considered in the calculation of the fluid but the effect of the particles is taken into account by a sink term in the momentum equation. Thus in the flow pattern obtained in FLUENT, the fluid velocity is approximately zero where the sediment is located. The consideration of the particle volume in the calculation of the fluid is a gap in the methodology. A simulation with three Euler phases - air, liquid and particles - calculated in EDEM could be a possible but computationally very expensive solution. The method is complex and there is still some effort needed to reduce the particle size without significantly increasing computing time. A possibility to reduce the computational time is to divide the computational domain into regions with different time steps according to the controlling forces and phenomena.

There is a substantial increase in computational demand with the increase in rotational speed. This limits the predictions to low rotational speeds when standard workstation computers are used for the calculations. If supercomputers are accessible as it will be the case in research at universities, predictions for high rotational speed are possible. But before advancing to higher rotational speeds it is important to improve the accuracy of the simulations of flow in centrifugal force fields and validate the models with experimental data, especially in terms of the axial velocity profile.

In order to run a sanity check for the simulations in FLUENT or EDEM, it is not sufficient to evaluate the residuals and mass balance. For new methodologies or boundary conditions, for example when simulating centrifuges, it is necessary to compare different turbulent models, transient and stationary calculation methods and most important to validate the simulation methodology proposed with experimental data. Once validated, geometry and parameters can be changed and the advantage of CFD, which is a fast change to evaluation cycle, can be used.

9. Notation

C_2	resistance coefficient, m^{-1}
C_μ	constant of the k-ε turbulence model, dimensionless
dV	difference in mass balance, %
$\vec{F}$	volumetric force, N m^{-3}
$\vec{F}_d$	drag force, N
$\vec{F}_t$	tangential force, N
$\vec{F}_n$	normal force, N
k	turbulent kinetic energy, m^2 s^{-2}
L	length of the rotor of the centrifuge, m
n	revolutions per minute, rpm
p	pressure, Pa (N m^{-2})
p_c	permeability, m^2
r_{bowl}	bowl radius, m
r_{weir}	weir radius, m
$r_{boundary\ layer}$	radius of boundary layer, m

S_i	sink term, N m^{-3}
t	time, s
u	velocity in x direction, m s^{-1}
v	velocity in y direction, m s^{-1}
v_{ax}	axial velocity, m s^{-1}
$\vec{v}$	velocity, m s^{-1}
w	velocity in z direction, m s^{-1}
x	particle diameter, m
x_{cell}	cell length, m
y^+	characteristic width of viscous sublayer, dimensionless

Greek letters

α	volume fraction, dimensionless
δ	thickness of the axial boundary layer, mm
δ_n	normal overlap, m
δ_t	tangential overlap, m
ε	turbulent kinetic energy dissipation rate, $m^2 s^{-3}$
ε_s	sediment porosity, dimensionless
μ	dynamic viscosity, kg m^{-1} s^{-1}
μ_t	turbulent viscosity, kg m^{-1} s^{-1}
ρ	density, kg m^{-3}
ρ_s	solids density, kg m^{-3}
$\Delta\rho$	density difference between liquid and solid, kg m^{-3}
τ_t	turbulent stress tensor, Pa (N m^{-2})
τ_w	wall shear stress, Pa (N m^{-2})
ν_i	Poissons ratio, dimensionless
ϖ	specific turbulent kinetic energy dissipation rate, s^{-1}

Indices

′	fluctuating component of a variable
q	phase

Abbreviations

LES	large-eddy simulation
RBR	rigid body rotation
RSM	Reynolds Stress Model
VOF	Volume of Fluid
St	Stokes number

10. Acknowledgments

We acknowledge support by Deutsche Forschungsgemeinschaft and Open Access Publishing Fund of Karlsruhe Institute of Technology.

11. References

Anderson, N. G., Waters, D. A., Nunley, C. E., Gibson, R. F., Schilling, R. M., Denny, E. C., Cline, G. B., Babelay, E. F. & Perardi, T. E. (1969). K-Series Centrifuges I.

Development of the K-II continuous-sample-flow-with-banding centrifuge system for vaccine purification. *Analytical Biochemistry*, 32, 3, 460-494

ANSYS. (2006). Chapter 23.3.2 Volume Fraction Equation, In: *User's Guide Ansys Fluent 6.3*

ANSYS. (2009). Chapter 4.12.2 Standard Wall Functions, In: *Theory Guide Ansys Fluent 12.0*

ANSYS. (2009). Chapter 18.4.1 Discretization of the Momentum Equation, In: *Theory Guide Ansys Fluent 12.,*

ANSYS. (2009). Chapter 23. Modeling Discrete Phase, In: *User's Guide Ansys Fluent 12.0*

Asaf, Z., Rubinstein, D. & Shmulevich, I. (2007). Determination of discrete element model parameters required for soil tillage. *Soil & Tillage Research*, 92, 1-2, 227-242

Bass, E. (1959). Strömungen im Fliehkraftfeld I. *Periodica Polytechics chem. Ingenieurwes*, 3, 321-340

Bass, E. (1959). Strömungen im Fliehkraftfeld II - Absetzsicherheit von Röhrenzentrifugen. *Periodica Polytechics chem. Ingenieurwes*, 4, 41-61

Bass, E. (1962). Strömungs- und Absetzvorgänge in Röhrenzentrifugen. 249-264

Brantley, J. N., Willis, D. D., Breillatt, J. P., Gibson, R. F., Patrick, L. C. & Anderson, N. G. (1970). K-series centrifuges : IV. Measurement and control of temperature. *Analytical Biochemistry*, 36, 2, 434-442

Brennan, M. S., M., N. & Holtham, P. N. (2007). Multiphase modelling of hydrocyclones - prediction of cut-size. *Minerals Engineering*, 20, 395-406

Bürger, R. & Wendland, W. L. (2001). Sedimentation and suspension flows: Historical perspective and some recent developments. *Journal of Engineering Mathematics*, 41, 2-3, 101-116

Buscall, R. (1990). The sedimentation of concentrated colloidal suspensions. *Colloids and Surfaces Papers Presented at the Society of Chemical Industry Meeting*, 43, 1, 33-53

Buscall, R., Mills, P. D. A., Stewart, R. F., Sutton, D., White, L. R. & Yates, G. E. (1987). The rheology of strongly-flocculated suspensions. *Journal of Non-Newtonian Fluid Mechanics*, 24, 2, 183-202

Chu, K. W., Wang, B., Yu, A. B. & Vince, A. (2009). CFD-DEM modelling of multiphase flow in dense medium cyclones. *Powder Technology*, 193, 3, 235-247

Chu, K. W. & Yu, A. B. (2008). Numerical simulation of complex particle-fluid flows. *Powder Technology*, 179, 3, 104-114

Cundall, P. A. & Strack, O. D. L. (1979). Discrete Numerical-Model for Granular Assemblies. *Geotechnique*, 29, 1, 47-65

Di Renzo, A. & Di Maio, F. P. (2005). An improved integral non-linear model for the contact of particles in distinct element simulations. *Chemical Engineering Science*, 60, 5, 1303-1312

Glinka, U. (1983). Die Strömung in Überlaufzentrifugen - Neue Ergebnisse mit einem elektrolytischen Markierungsverfahren. *Verfahrenstechnik*, 17, 5, 315-318

Golovko, Y. D. (1969). Liquid Lag in Tubular Centrifuge Rotors. *Khimicheskoe i Neftyanoe Mashinostroenie*, 6, 13-14

Gondret, P., Lance, M. & Petit, L. (2002). Bouncing motion of spherical particles in fluids. *Physics of Fluids*, 14, 2, 643-652

Gösele, W. (1968). Schichtströmung in Röhrenzentrifugen. *Chemie IngenieurTechnik*, 40, 13, 657-659

Gösele, W. (1974). Strömungen in Überlauf-Zentrifugen und ihr Einfluß auf die Sedimentation. *Chemie Ingenieur Technik*, 46, 2, 67

Hartl, J. & Ooi, J. Y. (2008). Experiments and simulations of direct shear tests: porosity, contact friction and bulk friction. *Granular Matter*, 10, 4, 263-271

Hirt, C. W. & Nichols, B. D. (1981). Volume of Fluid (VOF) method for the dynamics of free boundaries. *Journal of Computational Physics*, 39, 1, 201-225

Horanyi, R. & Nemeth, J. (1971). Theoretical Investigation of Clarification Process in a Tube Centrifuge. *Acta Chimica Academiae Scientarium Hungaricae*, 69, 1, 59-75

Kelecy, F. J. (2008). Coupling momentum and continuity increases robustness. *ANSYS Advantage*, 2, 2, 49-51

Kynch, G. J. (1952). A theory of sedimentation. *Transactions of the Faraday Society*, 48, 2, 166-176

Landman, K. A. & White, L. R. (1994). Solid/liquid separation of flocculated suspensions. *Advances in Colloid and Interface Science*, 51, 175-246

Launder, B. E., Reece, G. J. & Rodi, W. (1975). Progress in development of a Reynolds-Stress turbulence closure. *Journal of Fluid Mechanics*, 68, APR15, 537-566

Launder, B. E. & Spalding, D. B. (1974). The numerical computation of turbulent flows. *Computer Methods in Applied Mechanics and Engineering*, 3, 2, 269-289

Leung, W. (1998). *Industrial Centrifugation Technology*, McGraw.Hill, New York

Li, Y., Zhang, J. P. & Fan, L. S. (1999). Numerical simulation of gas-liquid-solid fluidization systems using a combined CFD-VOF-DPM method: bubble wake behavior. *Chemical Engineering Science*, 54, 21, 5101-5107

Li, Y. J., Xu, Y. & Thornton, C. (2005). A comparison of discrete element simulations and experiments for 'sandpiles' composed of spherical particles. *Powder Technology*, 160, 3, 219-228

Lun, C. K. K. & Savage, S. B. (1987). A simple kinetic theory for granular flow of rough, inelastic, spherical particles. *Journal of Applied Mechanics-Transactions of the Asme*, 54, 1, 47-53

Mainza, A., Narasimha, M., Powell, M. S., Holtham, P. N. & Brennan, M. (2006). Study of flow behaviour in a three-product cyclone using computational fluid dynamics. *Minerals Engineering*, 19, 10, 1048-1058

Mindlin, R. D. (1949). Compliance of elastic bodies in contact. *Journal of Applied Mechanics-Transactions of the Asme*, 16, 3, 259-268

Mousavian, S. M. & Najafi, A. F. (2009). Numerical simulations of gas-liquid-solid flows in a hydrocyclone separator. *Archive of Applied Mechanics*, 79, 5, 395-409

Ni, L. A., Yu, A. B., Lu, G. Q. & Howes, T. (2006). Simulation of the cake formation and growth in cake filtration. *Minerals Engineering*, 19, 10, 1084-1097

Nowakowski, A. F., Cullivan, J. C., Williams, R. A. & Dyakowski, T. (2004). Application of CFD to modelling of the flow in hydrocyclones. Is this a realizable option or still a research challange? *Minerals Engineering*, 17, 661-669

Perardi, T. E. & Anderson, N. G. (1970). K-series centrifuges : III. Effect of core taper on particle capture efficiency. *Analytical Biochemistry*, 34, 1, 112-122

Perardi, T. E., Leffler, R. A. A. & Anderson, N. G. (1969). K-series centrifuges II. Performance of the K-II rotor. *Analytical Biochemistry*, 32, 3, 495-511

Reuter, H. (1967). Sedimentation in der Überlaufzentrifuge. *Chemie-Ing. Tech.*, 39, 9, 548-553

Romani Fernandez, X. & Nirschl, H. (2009). Multiphase CFD Simulation of a Solid Bowl Centrifuge. *Chemical Engineering & Technology*, 32, 5, 719-725

Saunders, E. (1948). Nomograph for Particle Size Determination with the Sharples Supercentrifuge. *Analytical Chemistry*, 20, 4, 379-381

Schaeflinger, U. & Stibi, H. (1991). Centrifugal separation of a mixture in a rotating bucket II. *Chemical Engineering Science*, 46, 8, 2143-2152

Schaeflinger, U., Stibi, H. (1991). Centrifugal separation of a mixture in a rotating bucket. *The American Society of Mechanical Engineers*, 118, 173-182

Schütz, S., Piesche, M., Gorbach, G., Schilling, M., Seyfert, C., Kopf, P., Deuschle, T., Sautter, N., Popp, E. & Warth, T. (2007). CFD in der mechanischen Trenntechnik. *Chemie Ingenieur Technik*, 79, II, 1777-1796

Sokolow, W. J. (1971). *Moderne Industriezentrifugen*, VEB Verlag Berlin, Leipzig

Spelter, L. E., Schirner, J. and Nirschl, H., (2011). A novel approach for determining the flow patterns in centrifuges by means of Laser-Doppler-Anemometry. Chemical Engineering Science, In Press, Corrected Proof

Spelter, L. E. & Nirschl, H. (2010). Classification of Fine Particles in High-Speed Centrifuges. *Chemical Engineering & Technology*, 33, 8, 1276-1282

Spelter, L. E., Steiwand, A. & Nirschl, H. (2010). Processing of dispersions containing fine particles or biological products in tubular bowl centrifuges. *Chemical Engineering Science*, 65, 14, 4173-4181

Stahl, S., Spelter, L. E. & Nirschl, H. (2008). Investigations on the Separation Efficiency of Tubular Bowl Centrifuges. *Chemical Engineering & Technology*, 31, 11, 1577-1583

Stahl, W. H. (2004). *Industrie-Zentrifugen*, DrM Press, Männedorf

Stevens, A. B. & Hrenya, C. M. (2005). Comparison of soft-sphere models to measurements of collision properties during normal impacts. *Powder Technology*, 154, 2-3, 99-109

Taylor, A. R. (1946). Concentration of the Rabbit Papilloma Virus with the Sharples Supercentrifuge. *Journal of Biological Chemistry*, 163, 1, 283-287

Trawinski, H. (1959). Die äquivalente Klärfläche von Zentrifugen. *Chemiker Zeitung Chemische Apperatur*, 83, 18, 606-612

van Wachem, B. G. M. & Almstedt, A. E. (2003). Methods for multiphase computational fluid dynamics. *Chemical Engineering Journal*, 96, 1-3, 81-98

Wang, B. & Yu, A. B. (2006). Numerical study of particle–fluid flow in hydrocyclones with different body dimensions. *Minerals Engineering*, 19, 1022-1033

Wardle, K. E., Allen, T. R. & Swaney, R. (2006). Computational Fluid Dynamics (CFD) Study of the Flow in an Annular Centrifugal Contactor. *Separation Science and Technology*, 41, 2225-2244

Wilcox, D. C. (1988). Reassessment of the scale-determining equation for advanced turbulence models. *Aiaa Journal*, 26, 11, 1299-1310

Yakhot, V. & Orszag, S. A. (1986). Renormalization-group analysis of turbulence. *Physical Review Letters*, 57, 14, 1722-1724

Zubkov, V. A. & Golovko, Y. D. (1968). Liquid Flow in the Extraction Head of a Tubular-Centrifuge Rotor. *Khimicheskoe i Neftyanoe Mashinostroenie*, 12, 18-19

Zubkov, V. A. & Golovko, Y. D. (1969). Flow of Liquid in Rotor Outlet of a Tubular Centrifuge. *International Chemical Engineering*, 9, 3, 403-406

6

Air Movement Within Enclosed Road-Objects with Contra-Traffica CFD-Investigation

M. Muhasilovic[1,2], A. Mededovic[1],
E. Gacanin[1], K. Ciahotny[3] and V. Koza[3]
[1]IPSA-Institute Sarajevo, Sarajevo
[2]The CIM-Collaboration Program of the GiZ German International Technology Agency
[3]The VSCHT – Institute for Chemical Technology, Prague
[1]Bosnia and Herzegovina
[2]Germany
[3]Czech Republic

1. Introduction

Accidents in closed-space traffic object do force for further investigation of a flow phenomena - that might be a consequence of such large-scale events. Benefit that almost directly would be coming out of this investigative approach is an optimal method for artificial ventilation, that must find mirroring in a annual statistics on covered roads[1, 2].
For such an investigative task, Computational Fluid Dynamics, the CFD offers ever stronger growing engineering tool. Once[3] passing through it´s first sophisticated developments[4], the zone-model approach was not capable for solving the problems like those were treated by the field models for CFD-research. Generally, field-models are based on the full solution of the fundamental physical laws of energy-conservation, where the computational domain of explored large-scale phenomenon is divided in thousands of smaller control volumes; where mathematical mechanism after it´s discretization are "translated" into programme-steps for computer handling. Today[5, 6], the CFD-society enjoys this software-development that started in 1960-ies and was certainly followed by the hardware development between mid ´80-s and mid ´90-s. Together, this computational mechanism [7-11] both hardware and software, can cope with computational domains with few thousands cells offering very satisfying results [1, 12] were accomplished attempts are done in both validating and exploring area of CFD. Modern field-model codes engaged in CFD-research, supported by today´s powerful hardware, can cope with domains made out of several hundred thousands cells. Some computational codes[13] applied for tunnel-fires, are taking in account simple one-step chemistry for combustion modelling, and the reaction rate is won from a the eddy break-up mixing model[14]. This approach is suitable for turbulent diffusion flames as well - one of a characteristic of large-scale fires, where the rate of reaction is controlled by a mixing of fuel with oxidant (air). In such a step, solved would be (the time-averaged[15]), turbulence-modelled conservation-equations for mass, momentum, energy and species of combustion. The k-ε turbulence model with extra source terms accounting for the effect of

buoyancy in turbulent mixing would be applied in such numerical approach. Similar principle[2] in those sophisticated software, one can find[16] in the closed set of coupled equations that are again discretised and solved onto a three-dimensional finite-volume Cartesian mesh. In the latest phases of on-going software development several general purpose computer-programs for analysis of fluid flow - were developed[17] - that can handle both steady and transient flows. Anyhow, all of these research attempts (after their validation[18]), can document a good capability of the numerical approaches used in investigative treating both the reactive flows as well as large-scale fluid flow - especially in enclosed traffic objects.

2. Mathematical model

(In several reports, proceedings and papers, there was "no time" to mention the mathematic and numerical mechanism - due to the characteristic of those publishing media. Now, the opportunity here n will be used for the presentation of involved mathematic tool).

Applied explanation of the flow phenomena in the study[19] is explained by the Reynolds Averaged Navier-Stokes (RANS) equations, having turbulence treated by the k-ε model[20], representing the major characteristic of the applied CFD-investigation-tool of the FLUENT; and handling the buoyancy by applying the Boussinesq approximation. This approach, is not affected by fluctuation of initial conditions, offering more accurate presentation of the time dependent flow - particularly the distribution of the gaseous combustion products[21]. Since the investigations in the observed underground-space facilities have shown that the Mach Number was varying in the tolerance of the order between 0.003 and 0.022; such a flow can be presumed as incompressible[22]. So assumed as incompressible, the fluid while crossing the reaction front, doesn´t undergo thermal-caused expansion and the reaction makes no impact onto flow-velocity. Further assumption, to have a planar propagation front of combustion in a motionless fluid, leads to the application of the Boussinesq approximation[23] without external forces[24]. Boussinesq buoyancy model is corresponding to the infinitely small difference of density between combusted fluid and fuel. Here, the flow velocity obeys the incompressible Navier-Stokes equation with a temperature-dependent force term[24]. The change of temperature $0 \leq T \leq 1$ is described by an advection-reaction-diffusion equation. For this to-be-investigated incompressible gaseous reactive flow at low velocity, the governing equations of the combustion-induced flow read:

$$\frac{\partial \overline{v}_j}{\partial x_j} = 0 \tag{1}$$

$$\frac{\partial \overline{v}_i}{\partial t} + \frac{\partial \left(\overline{v}_i \overline{v}_j\right)}{\partial x_j} = -\frac{1}{\rho}\frac{\partial \overline{p}}{\partial x_i} + \frac{1}{\rho}\frac{\partial \overline{\tau}_{ij}}{\partial x_j} - g_i \alpha \Delta \overline{T} \tag{2}$$

$$\frac{\partial \overline{T}}{\partial t} + \frac{\partial \left(\overline{T}\overline{v}_j\right)}{\partial x_j} = \frac{\partial}{\partial x_j}\left(\frac{\lambda}{\rho c_p}\frac{\partial \overline{T}}{\partial x_j}\right) + \frac{1}{z}R(T) \tag{3}$$

Here, we recognise $\overline{v}_i$, $\overline{v}_j$ as average velocity components, $\overline{T}$ as averaged local temperature, ρ as density, t as time and x_i, x_j as space coordinates. The R(T) = ¼ T (1 - T) stands for reaction rate[24] where the reciprocal value of reaction time-scale is represented by 1/T and ρ is here again a constant average density. Temperature T will be used as expression for reaction-progress-variable as well, whose purpose is to distinguished burned, unburned and partially burned state, providing facile interpretation of flame propagation. The term $-g_i \alpha \Delta \overline{T}$ denotes buoyancy, treated according to the Boussinesq approximation, where $\Delta \overline{T}$ is showing the difference between local and reference temperature. The symbol g denotes the gravity and α is coefficient of the thermal linear deformation. The model for the unknown Reynolds stress[25], $\partial \overline{\tau}_{ij,s}$ is related to the local strain rate:

$$\tau_{ij} = (\tau_{ij})_N + (\tau_{ij})_T \tag{4}$$

where we distinguish between Newtonean stress $(\tau_{ij})_N = 2\mu \overline{S}_{ij}$ featuring molecular viscosity; and turbulent Reynolds stress $(\tau_{ij})_T = 2\mu_T \overline{S}_{ij}$ featuring turbulent viscosity recognising again $\overline{S}_{ij}$ as local strain stress rate reads:

$$\overline{S}ij \equiv \frac{1}{2}\left(\frac{\partial v_i}{\partial x_j} + \frac{\partial v_j}{\partial x_i}\right) \tag{5}$$

Here we recognise the turbulent viscosity: $\mu_T = C_\mu \frac{k^2}{\varepsilon}$ (6)

The applied k-ε model[26] is a two-equation eddy viscosity model [27, 28] uses transport equations for these two variables[29]. One of these equations governs the distribution through the field of k, the local kinetic energy of the fluctuating motion. The other one is explaining a turbulence characteristic of different dimensions, the energy dissipation rate ε [30].

$$\frac{\partial k}{\partial t} - \nabla \cdot \left(C_\mu \frac{k^2}{\varepsilon} \nabla k \right) = C_\mu \frac{k^2}{\varepsilon} P_d - \varepsilon - \lambda_v N^2 \tag{7}$$

$$\frac{\partial \varepsilon}{\partial t} - \nabla \cdot \left(C_\varepsilon \frac{k^2}{\varepsilon} \nabla \varepsilon \right) = C_1 k P_d - \frac{\varepsilon}{k}\left(C_3 \lambda_v N^2 + C_2 \varepsilon \right) \tag{8}$$

The P_d, the production term for the squared shear frequency is:

$$P_d = \frac{1}{2}\left(\left\| \nabla V + \nabla V^T \right\| \right)^2 \tag{9}$$

where $\left\| \nabla V + \nabla V^T \right\|$ is the 2-norm matrix. The constants are given: $C_1 = 0.126$, $C_2 = 1.92$, $C_\mu = 0.09$, $C_\varepsilon = 0.07$. The variable N denotes the Brunt-Väisälä frequency, with the N^2, related following squared buoyancy:

$$N^2 = -\frac{v_v}{\sigma_t} \frac{g}{\rho_0} \frac{\partial \rho}{\partial z} \tag{10}$$

Here is constant $\sigma_t = 1$ and following expressions of eddy coefficients v_v, λ_v should include the stability parameters to account for the turbulence damping in the stratified fluid flows:

$$v_v = S_u \frac{k^2}{\varepsilon} + v \ , \ \lambda_v = S_b \frac{k^2}{\varepsilon} + \lambda_b \tag{11}$$

$$S_u = \frac{0.108 + 0.0229\alpha_N}{1 + 0.47\alpha_N + 0.0275\alpha_N} \ , \ S_b = \frac{0.177}{1 + +0.043\alpha_N} \tag{12}$$

where λ_b =10^{-6} and the following stability coefficients α_N and C_3 can be expressed as:

$$\begin{aligned} \alpha_N &= \frac{k^2}{\varepsilon^2} N^2 \\ C_3 &= -0.4; N^2 > 0 \\ C_3 &= 1; N^2 < 0 \end{aligned} \tag{13}$$

During these investigations, the wall shear stress was obtained from the logarithmic law of wall ("wall function") for the distance from the wall, y, the von Karman constant κ=0,41 and the constant C=9,793:

$$\frac{\overline{v}}{v_\tau} = \frac{1}{\kappa} \ln C \frac{v_\tau y}{v} \tag{14}$$

Energy transfer within the computational domain of such non-premixed turbulent flow in tunnel cavity is governed by next equation:

$$\frac{\partial H}{\partial t} + \nabla \cdot (\overline{v} H) = \nabla \cdot \left(\frac{\gamma_t}{c_p} \nabla H \right) + Q_h \tag{15}$$

where contribution to the energy budget, characterised through turbulent thermal conductivity γ_t is performed through the heat conduction and diffusion combined in term $\nabla \cdot \left(\frac{\gamma_t}{c_p} \nabla H \right)$, accompanied by the heat-increase component Q_h due to the chemical reaction rate R of a species n and it's molar mass M:

$$Q_h = -\sum_n \frac{h_n^0}{M_n} R_n \tag{16}$$

In the governing energy equation is total enthalpy H based onto particular enthalpy of the nth species H_n and it's mass fraction Y_n

$$H = \sum_n Y_n H_n \tag{17}$$

with

$$H_n = h_n + h_n^0 \tag{18}$$

having so species n characterised through it's thermal capacity $c_{p,n}$ and n^{th} enthalpy $h_n = \int_{T_{ref}}^{T} c_{p,n} dT$ with referent temperature $T_{ref} = 293,15K$ that makes an impact on the formation enthalpy $h_n^0 \left(T_{ref,n} \right)$.

The radiative transfer equation (RTE) for an absorbing, emitting and scattering medium at the position $\vec{r}$ in the direction $\vec{s}$ reads:

$$\frac{dI(\vec{r},\vec{s})}{ds}+(a+\sigma_s)I(\vec{r},\vec{s})=ax^2\frac{\sigma T^4}{\pi}+\frac{\sigma_s}{4\pi}\int_0^{4\pi}I(\vec{r},\vec{s}\,')\Phi(\vec{s},\vec{s}\,')d\Omega' \tag{19}$$

The applied Discrete Ordinate approach for treating the radiation considers the RTE in the direction $\vec{s}$ as a field equation, allowing the modelling of non-gray radiation, using the gray-band model. So, for the spectral intensity $I_\lambda(\vec{r},\vec{s})$ the RTE can be written as:

$$\nabla\cdot(I_\lambda(\vec{r},\vec{s})\vec{s})+(a_\lambda+\sigma_s)I_\lambda(\vec{r},\vec{s})=a_\lambda x^2 I_{b\lambda}+\frac{\sigma_s}{4\pi}\int_0^{4\pi}I_\lambda(\vec{r},\vec{s}\,')\Phi(\vec{s},\vec{s}\,')d\Omega' \tag{20}$$

Here is $\vec{r}$ a position vector and $\vec{s}$ a direction vector. I denotes the radiation intesity, i.e. radiation intensity I_λ for a certain wawe-lenght λ. The radiation intensity for the black body reads $I_{b\lambda}$. Scattering coefficient is represented by σ_s and Stefan-Bolzman-constant reads σ =5.672x10^{-8}W/m^{2}K^4. Φ stands for phase function, T is local temperature and x is refraction index. Path length is s and $\vec{s}\,'$ denotes direction vector. Ω' stands for solid angle and a is absorption coefficient, i.e. absorption coefficient for a particular wave-length a_λ.
So will the RTE be integrated over each wave-length-interval λ resulting in transport equation for entity $I_\lambda\Delta\lambda$, containing the radiant energy of the wave-length Δλ. The total intensity $I(\vec{r},\vec{s})$ in each direction $\vec{s}$ at some position $\vec{r}$ is computed as:

$$I(\vec{r},\vec{s})=\sum_n I_{\lambda,n}(\vec{r},\vec{s})\Delta\lambda_n \tag{21}$$

The weighted-sum-of grey-gases model (WSGGM), that was applied in this study for handling the irradiance in the complicated phenomenon of the confined combustion[31] presents a reasonable compromise between the oversimplified grey gas model and a complete model which takes into account particular absorption bands. Within the Discrete Ordinate approach, where the soot generation[32] was described by the Magnussen model[28], the basic assumption of the WSGGM means that the emissivity E of the any nth gray gas characterised by p_{SUM}, the sum of it´s partial pressures p and the absorption κ within the computational domain over a distance d is presented by:

$$E=\sum_{n=0}^{i}a_{E,n}(T)\left(1-e^{-\kappa_n p_{SUM} d}\right) \tag{22}$$

Here $a_{E,n}$ denotes the emissivity weighting factor for the nth gray gas and $\left(1-e^{-\kappa_n pd}\right)$ is emissivity of the nth gray gas. The temperature dependence of $a_{E,n}$ is commonly approximated by the relation:

$$a_{E,n}=\sum_{n=1}^{J}b_{E,n}T^{n-1} \tag{23}$$

where polynomial coefficient $b_{E,n}$, depending on the gas temperature emissivity is obtained experimentally. The coefficients $b_{E,n}$ and κ_n are slowly changing for broad range of the pressure ($0.001bar \leq p \leq 10bar$) and temperature ($600K \leq T \leq 2400K$) which simplifies relation (22) to:

$$E = \sum_{n=1}^{J} a_{E,n} \kappa_n p_{SUM} \tag{24}$$

for

$$a = \sum_{n=1}^{J} a_{E,n} \kappa_n p \;\; ; \;\; s \leq 10^{-4}\text{m} \quad \text{and} \quad a = -\frac{\ln(1-E)}{s} \;\; ; \;\; s \geq 10^{-4}\text{m} \tag{25}$$

The combustion - the chemistry development is explained by fast chemistry assumption including the prePDF [33] and in the ideal stoichiometric conditions the reaction runs as foolows:

$$C_7H_{16} + 11(O_2 + 3{,}76N_2) \xrightarrow{ignition-energy} 7CO_2 + 8H_2O + 41{,}36N_2 + \Delta H \tag{26}$$

However, in a case of unfulfilled combustion, soot is influencing the radiation absorption. Employed generalised model estimates the effect of the soot onto the radiative heat transfer by determining an effective absorption coefficient for soot. In this case, absorption coefficient is a sum of the one for the pure gas a_g and that one for the pure soot a_s :

$$a_{s+g} = a_s + a_g \tag{27}$$

The a_g is obtained from presented WSGGM approach and

$$a_s = b_1 \rho_m \left[1 + b_T (T - 2000) \right] \tag{28}$$

with $b_1 = 1232{,}4 \, \text{m}^2/\text{kg}$ and $b_T \cong 4{,}8 \cdot 10^{-4} \, 1/\text{K}$, where ρ_m denotes soot density.

3. Numerical approach

Discretizing the governing equations in both space and time[28, 34] while choosing the numerical method of the standard finite volumes [22, 34, 35], the CFD-tool is "adapted" for the time-dependent modelling of the explored event. The spatial discretization of time-dependent equations within here-employed segregated solution method are linearized in an implicit way. Since variables described the flow by their value at the centre of a single mesh-cell in explored computational domain, the system of linear equations occurs, containing mentioned unknown variable at the cell centre as well as their unknown values in neighbour cells. A scalar transport-equation within such method is also used to discretize the momentum equations; in the same mode for the pressure-field (if cell-face mass-fluxes were known). This applies for velocity-field as well. In case that the pressure-field and face mass fluxes are not known, applied commercial software of *FLUENT* [28] uses a co-located scheme, whereby the local values for pressure and velocity are stored at centres of a single mesh-cell. A need for interfacial values includes an application of an interpolation scheme to compute pressure and velocity out of mesh-cell values. The integration over the observed control-volume, where is a mesh-cell in a computational domain, can be performed yielding the discretised equation for the mass-flux through a control-surface - a mesh-face on the mentioned mesh-volume. Procedure of the segregated solver that follows as a mechanism in this research, the continuity equation, is used as an equation for pressure as well[28].

However, the pressure does not appear for an incompressible flow in explicit way, since the density is not directly related to pressure. However, through the application of the *SIMPLE*-algorithm-family[34] during this investigation (within the commercial software package of *FLUENT*) the won values for pressure were introduced into the continuity equation and in this way supported the pressure-velocity coupling. This was done through the *NITA*-algorithm for the unsteady.. Further, the *SIMPLE* uses a relationship between velocity and pressure corrections to enforce mass conservation and to obtain the pressure field. So, executing these numerical steps, the equations can express the state for each other cell in the computational grid, the mesh. This will result in a set of algebraic equations with a coefficient-matrix. In this way the segregated solver is "updating" variable-field by considering all the cells of the domain in the same time, solving the governing equations sequentially (segregated one from another). What follows are the iterations of the solution-loop before one wins the converged solution. Subsequently, the next field of another variable will be solved by again considering entire cells at the same time, etc. Finally, determining of the boundary conditions relied on some previous done studies [19, 36] that were basis for the estimation.

4. Objects of investigation - Two different tunnel-shapes

Today´s major two shapes of the underground traffic (road) facilities that are having their cross-section as a "horse-shoe" and rectangular shape - we explored here. And according to the final use, these traffic covered object, were taken as very frequently used ones. Additionally, those tunnels (one, having bifurcation - is going under Slovenian Capital of Ljubljana, and another is situated in a curvy valley of the river Bosna, between Bosnian cities of Travnik and Zenica) are "offering" interesting geometry for a CFD-based investigation; such is air-movement and the movements of air-pollutants and other gaseous products (in accidental situations).

4.1 The "Sentvid" tunnel

As mentioned, the northern tube of tunnel "Sentvid" (undergoing the city of Ljubljana) was investigated with the CFD-approach, where the "chimney-effect" was expected (coming from bifurcated part of the elevating trek, which comes out, up to the city-streets) Although well performed acquisition of data during the 3,5MW-fire-experiment, Slovenian Road Administration (DARS) was not able and/or allowed to organize the test for longitudinal ventilation in case of 40MW-fire event. In this scenario, during it´s functioning, the artificial ventilation should response by the 2nd minute, securing so the area of accident, by suppressing the temperatures that can provoke concrete-destruction. Therefore, a CFD-based approach towards this important question was here an optimal choice. Particularly because of the fact that the "Sentvid" tunnel is having a change from it´s rectangular cross-section´s shape (in it´s old part) into it´s new part that has "horse-shoe" shape of the cross-section (when one is "rising" from it´s western exit to Ljubljana). Further characteristic of this investigated north-tube is that the western exit (towards city of Kranj) is lying geodetically lower then it`s Entrance from Ljubljana. So is the natural air-flow always confronted with the traffic-flow which is always against it.

Going about 50m in the north tube away from the interchange between the old-part of the "Sentvid" towards it´s new ("horse-shoe"-shaped) part - the presumed accidental fire-place is situated. In order to produce thermal load of 40MW heat release, from the surface that corresponds to the dimensions of the used pan in the experiment (1m x 2m) the mass flux of the inflammable agent (the heptane) was set to 0.889kg/s (as same as during the CFD-based investigation of the artificial accidental fire-case.

The northern tube of "Sentvid", with the slope of 2,2 %, has two major main-curves with a Radius of 4000m and 1500m, where the "horse-shoe" cross-section determines the first 1080m of the tunnel. Starting as a three-lane tunnel, after the bifurcation (on the 720th meter of its length, going from Ljubljana-entrance) the "main stream" of the tunnel "continues" as a two-lane traffic communication towards Kranj-exit; and the bifurcated "horse-shoe-shaped" tunnel-line elevates and exites to the point of about 12m above the road-level of the main tunnel-stream, with the length of a further 400m and a change from an one-lane to the two-lane "horse-shoe" cross-section. Therefore, the Aspect-Ratio changed in this tunnel: for three-lanes part Ap =1,707 to the rectangular and other horse-shoe-shaped part with two lanes with Ap = 1,32. Tunnel-entrance as open (pressure) boundary was used for initializing computational values for the velocity and pressure in the computational-domain since the global temperature was set to the 300K. The tunnel housing, tunnel road and tunnel walls, were presumed to be heat transparent.

4.2 The "Vranduk" tunnel

Another tunnel which was object of interest and it is in a well active use in the road-network through the Highlands of Bosnia and Herzegovina - is a motorway-tunnel of "Vranduk-2" that is binding two Bosnian cities of Travnik and Zenica. "Vranduk-2" was chosen because of the it´s geometry, made in the road-curve of radius of r=1000m after about 80m straight road through it, where it´s cross-section, in a shape of a "horse-shoe", has an almost constant aspect ratio of (As=1.49). In it´s length of 1067m, the explored tube of "Vranduk-2" has a denivelation between geodetically higher, eastern entrance from Zenica-side downwards to exit at the Travnik-side of 8.37m. For this CFD-based investigation, the tunnel-entrance and -exit were characterised as open (pressure) boundaries. The drop i.e. increase of 1.48Pa at the entrance and tunnel-exit, respectively, as well as overall air-movement of 1.1m/s were used for initializing the values in computational domain for the velocity and pressure, since the global temperature was set to the 300K. The fuel "pool" - the simulated fire-place, has been determined by the constant max flux rate of 0,889kg/m^2s[37] having hydraulic diameter of 1,333m for a 40MW-heptane fire[38]. The tunnel housing, tunnel road and tunnel walls as well, were presumed to be heat-transparent fences of the investigated "Vranduk-2" tunnel-tube; and this especially - because of the-type of the soil, where tunnel situated[39]. This decision was based on so-far research-experience, but also on the reality-oriented investigation on modern tunnel-construction, that obeys the thermal conductivity of a rock where through a tunnel would be built. Particularly for the "Vranduk-2" tunnel, that was built in the carstic area, in the Bosnian Highlands, the specific thermal conductivity of such rocks (λ =2,3W/mK)[39, 54].

4.3 Computational domain

The area in which the computation with applied mathematical model and subsequent performed numerical discretization, was the very volume, that can take a fluid (an air, or in case of accidental fire, the combustion gaseous products) i.e. the housing of the tunnel.

Therefore there are two computational domains to be distinguished: the "Sentvid" underground highway-facility that undergoes the Slovenian Capital of Ljubljana; and the "Vranduk-2" tunnel-tube - both of them as the traffic communications, very frequently used.

4.3.1 The Slovenian tunnel

While meshing the Slovenian tunnel, two kinds of grids (having two kind of characteristic basic cells) were applied. So, in the area of the domain where the major fluid-mechanic and thermodynamic occurrences are expected, the unstructured tetrahedral mesh was installed. In this part, where the pool with the mentioned fuel flux rate of 0,899kg/s is situated and the pool-surface temperature is 500K, a combined asymmetric mesh was employed. This mesh has different density, having characteristic cell size of 60mm up to 200mm from one side of the middle-plane of the Tunnel. Such approach was about to approve that the gained solution is mesh-independent. The other parts of the 1470m-long tunnel with alternating surface of the cross-section is meshed with unstructured penta-hedral mesh, following the explained general image of the asymmetric mesh. The entire computational domain is meshed with round 4.400.000 cells.

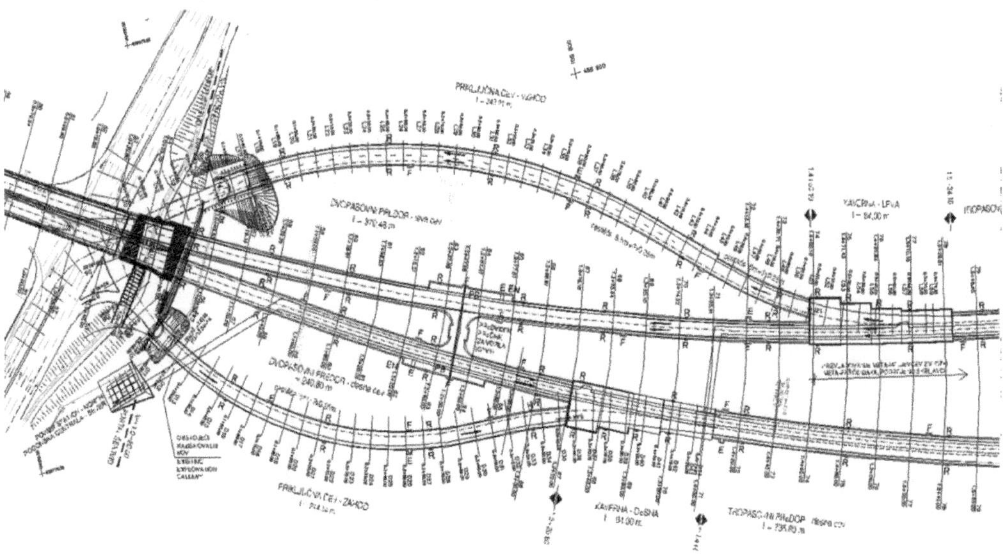

Fig. 4.2.1-1. Both north and south tube of the Sentvid tunnel

As mentioned, the tunnel-housing as well as the tunnel-road in the computational domain were defined as heat-transparent walls and fluid-domain is air with the ambient conditions and no movement. The entrance as well as the tunnel-exit, are defined as open-pressure boundaries. All of these facts were used for the initialisation.

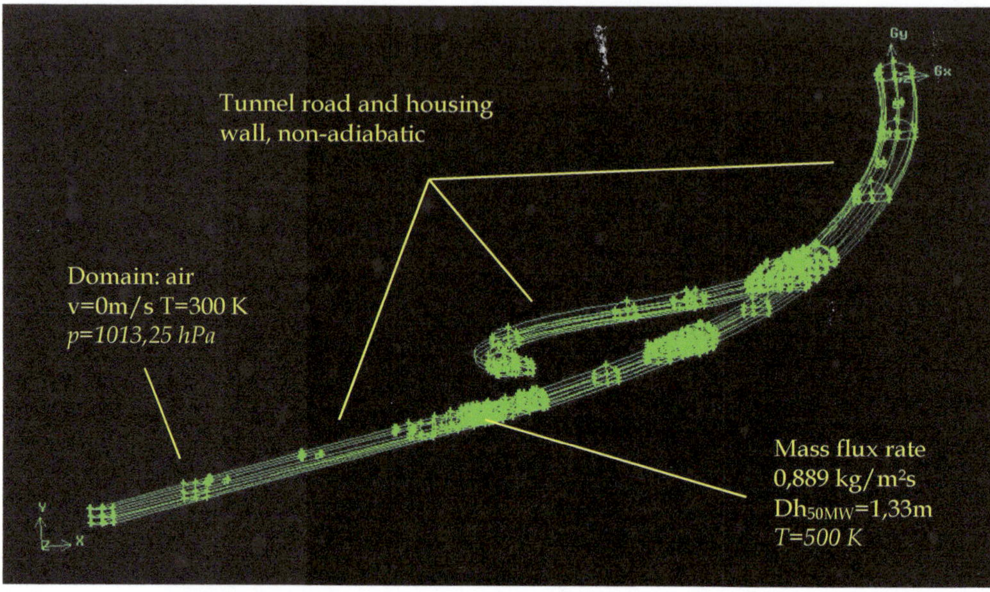

Fig. 4.2.1-2. North tube of the "Stentvid" tunnel in computational domain

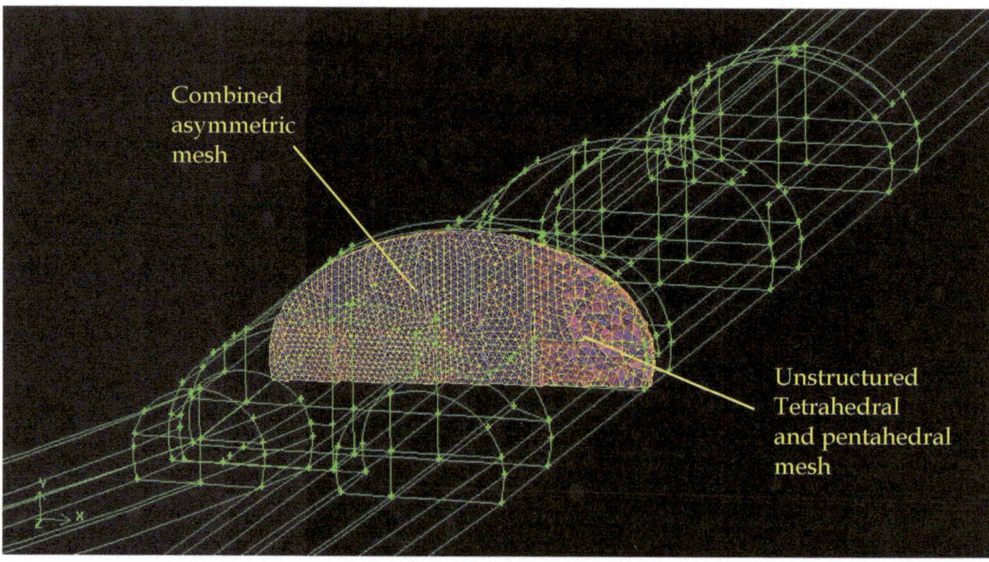

Fig. 4.2.1-3. Meshing – the near-bifurcation zone

4.3.2 The Bosnian tunnel

The mesh of this computational domain is also determined by tetrahedral-, pentahedral- and hexahedral-cells of a random structure. In this case a denser grid was also applied in the area around the fire-pool, having so more grid-points to support the major fluid-mechanic and thermodynamic occurrences. Such unstructured mesh was installed in whole computational domain. However, the subsequent parts of the 1067m long tunnel with non-alternating cross-section, are also meshed with unstructured pentahedral and hexahedral cells in the explained way, having increased cells sizes from 400mm up to 800mm - as one gains on distance from fire pool, towards the tunnel-exit and entrance. The pool with the fuel had a flux-rate of 0,899kg/s as well, and the pool-surface temperature is set also to 500K. Middle-plane of the Tunnel was going throughout the domain following the curvature of this cavity.

The tunnel-housing and the tunnel-road were in the computational domain defined as non-adiabatic walls as well; and fluid-domain is air, with the ambient conditions and fluid-movement of 1,1m/s as natural draft. The entrance from the side of the city of Zenica as well as the tunnel-exit at the Travnik-side, are in this case also defined as open-pressure boundaries with the mentioned pressure-decrease i.e. pressure surplus of 1.48Pa.

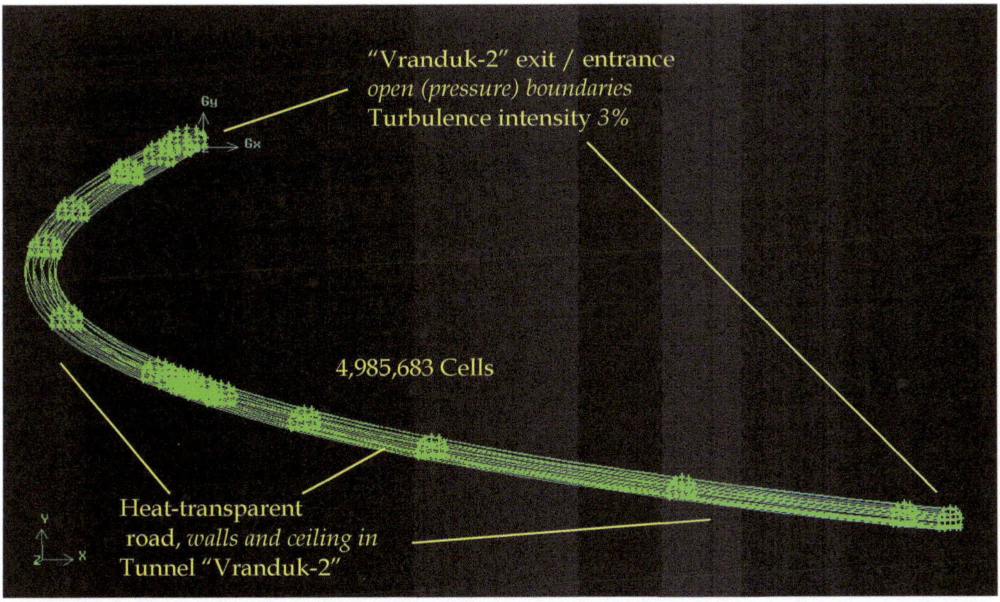

Fig. 4.2.2-1. The computational domain "Vranduk-2"

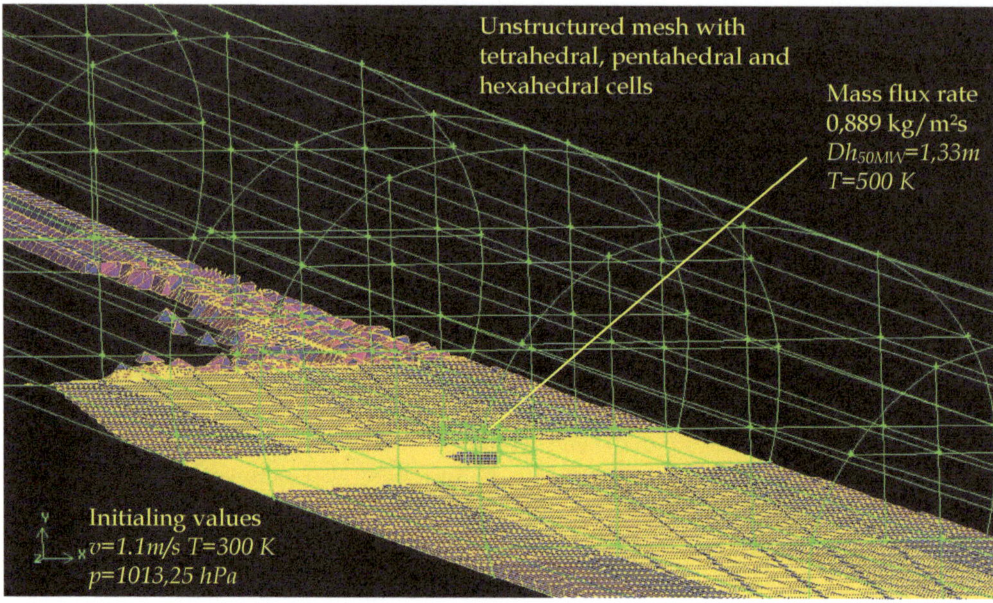

Fig. 4.2.2-2. The unstructured mesh in "Vranduk-2", at one fourth of it's length

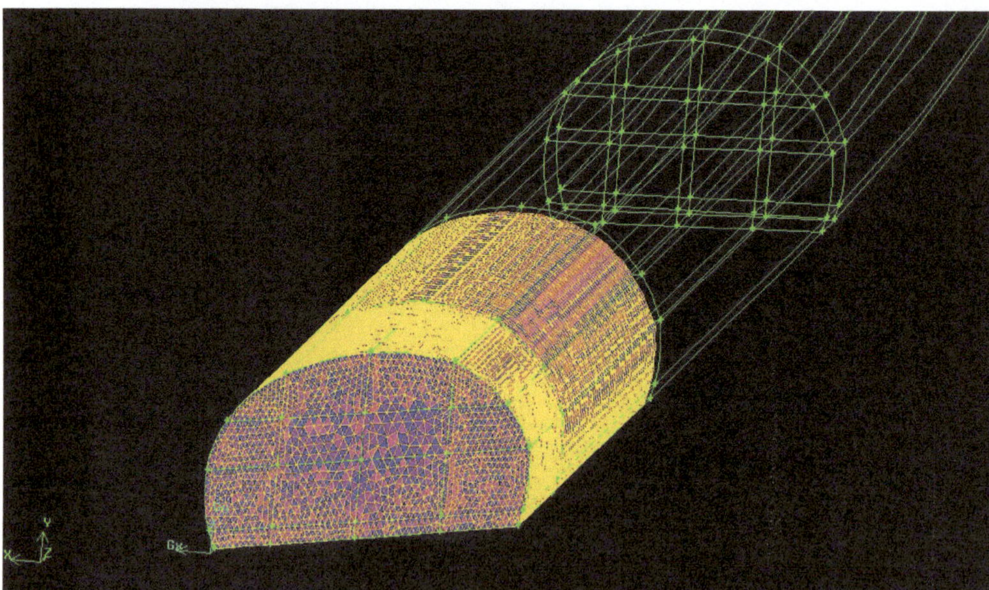

Fig. 4.2.2-3. "Vranduk-2"-exit towards city of Zenica meshed randomly - pentahedral cells

5. Discussion of the observed phenomena – Bosnian tunnel

Although applied onto the scaled model in physical experiment, a well defining the phenomena of the confined accidental thermal load (while flame was impinging up to the tunnel-ceiling), was introduced by Kurioka et. al.[41] as well as the additional characteristic phenomenon of the "rolling", when fire in enclosure[41] is "leaving" the original combustion-place. In this light it is to distinguish two major groups of the flame-shapes in enclosure: the ones, that impinge on the tunnel roof and the others that, influenced by the longitudinal air-stream, that do not. During the CFD-investigations of the large-scale flow phenomena, in both of the tunnels, one could recognise obviously the first ones, where such art of the large-scale flame could be cause for stronger damage on the Reinforced Concrete tunnel-construction.

5.1 The "Sentvid" highway-tunnel

The unwanted effect of the large-scale accidental fire can be noticed during the numerical experiment in the north tube of "Sentvid", where in first simulated scenario, the employment of the longitudinal ventilation set was not foreseen. This built-in system should react by the 2nd minute of such accidental thermal load, where until then the Reinforced-Concrete-Construction of a tunnel ought to withstand the thermal load.

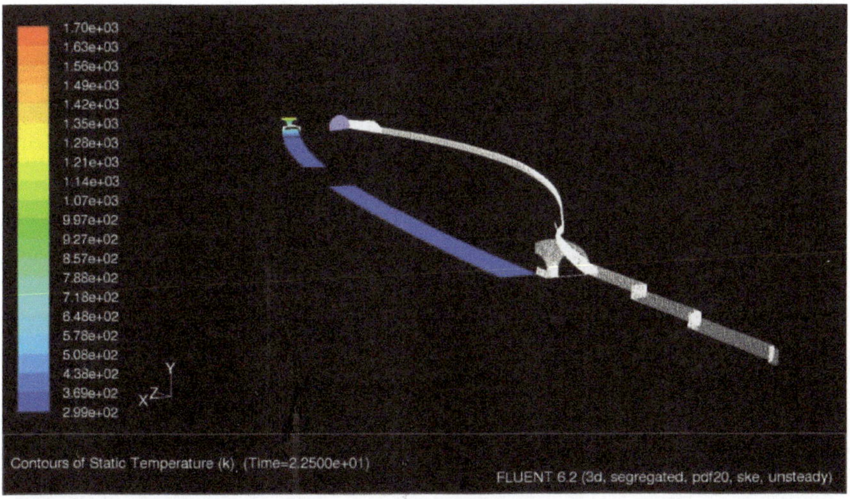

Fig. 5.1-1. A view to a part of the computational domain for the north tube of the "Sentvid"-highway-tunnel. In right part of the sketch, "coming" from the East, one can notice the begin of the bifurcation zone, marked through the part of the grid, after which highway-exit-trek is rising right up to the surface and left part is "continuing" towards the West. At the very end of this highway-section, in this part of computational domain, the modelled fire-place was set. The artificial ventilation was not employed yet.

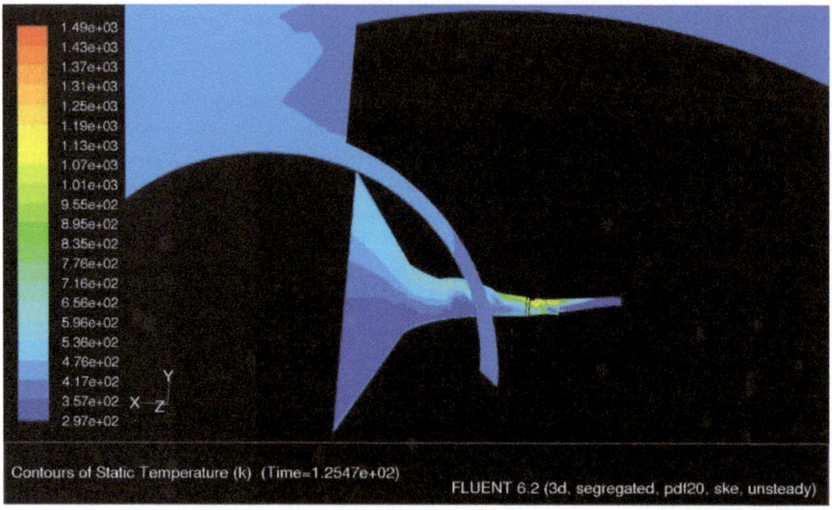

Fig. 5.1-2. "Standing" in the bifurcation zone of the north tube in "Sentvid". After the 2nd minute, , the spreading of the volatile combustion products is not confronting with longitudinal ventilation and it is rising towards the higher geodetic position at the eastern tunnel-entrance is established, while the fire-place (in the bottom of this CFD-recorded moment) is forcing the hot gases to reach the western tunnel-exit.

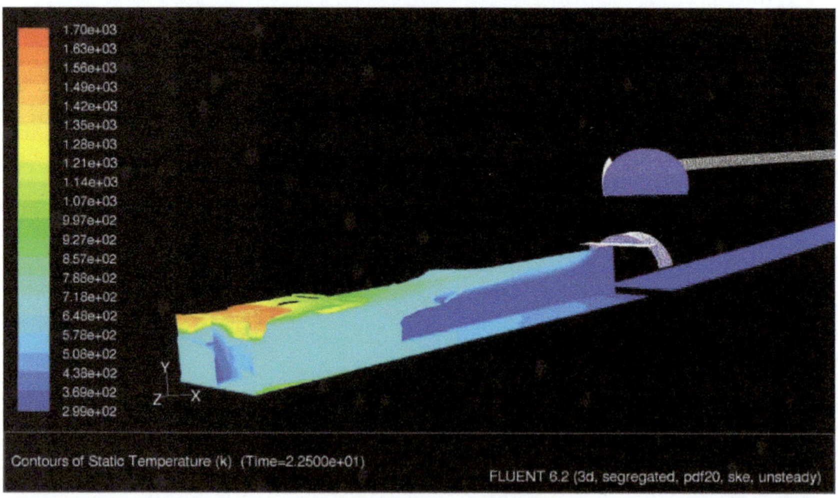

Fig. 5.1-3. Observed computational domain in the north tube of the "Sentvid-highway-tunnel: On the iso-surfaces that present draft-velocity of 4m/s due to the ventilation towards western exit, as well as on the central plane, the back-layering phenomenon is to be noticed. Since the flow of the ventilated air is still not entirely established, the movement of the hot gaseous products in 22nd second of accidental combustion is reaching the interchange between old and new tunnel-part - here made recognizable by the mesh-display of this tunnel-part (and just above it - although totally separated - is highway-tunnel-exit to the "surface").

Having firstly no artificial ventilation, the surroundings in northern cavern of the "Sentvid"-tunnel in first 22s during the numerically performed investigation (with applied ***k-ε*** turbulence model) supported the expected phenomenon of natural buoyancy towards the higher geodetic position. However, the set draft into the computational domain, caused by mentioned artificial ventilation (and according to the real-case scenario, due to the traffic-flow in opposite direction) is providing that the axis of the flame is pushed away, down the tunnel-"slope". This is because of the relative weak buoyancy, compared to the mentioned forced longitudinal air-movement[41]. As one moves further of the fire-source, towards the western tunnel-exit, the buoyancy gains on the strength due to the increased temperature, as we observe an impinging of the flame.
The aim of the to-be-employed ventilation in such accidental event is not to contribute to the extinguishing, but to limit the unwanted thermal load of a 40MW-fire and to conquer the distribution of hot gaseous products and their negative impact. This urged occurrence can be noticed in record of the 88th second of numerical modelling on establishment of the accidental fire and propagation of it´s consequences (Figure 5.1-4), where one can notice additional agreement with the approaches in some physical simulations[42,46] as well. So, the flame and/or hot gas-current do not impinge on the tunnel ceiling, but (enveloped by the ventilation stream) are heading towards "Sentvid 's" western (and geodetically lower) exit. The stream of the hot gases is losing the temperature through the convection-mechanism, the buoyant effect is becoming weaker and the damaging of the Reinforced-Concrete-Construction is limited. Although still present, back-layering along the ceiling is falling down since the gaseous combustion products are being mixed with the cooler fan-air (Fig. 5.1-4) However, the tunnel road remains under the unwanted impact of the thermal load.

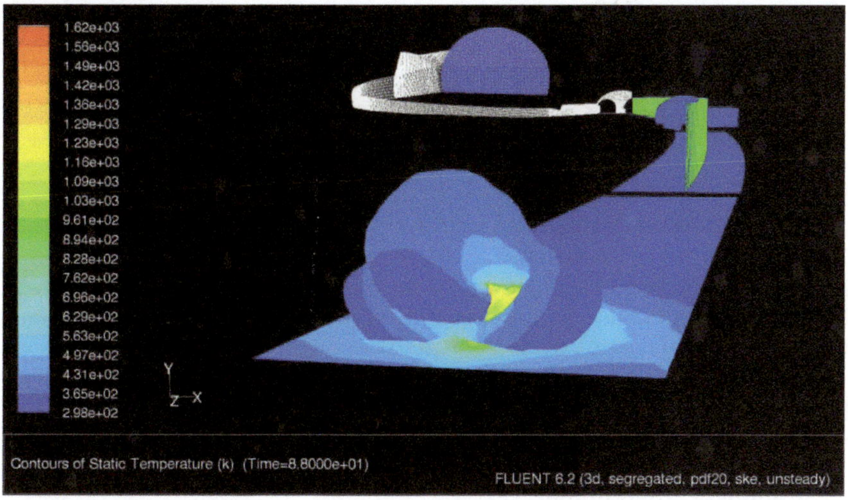

Fig. 5.1-4. The CFD-record of the 88th second: on the iso-surfaces of the 4m/s-velocity-magnitude, displayed temperatures coming from the accidental fire-event are not to pose any threat to the RC-construction of the tunnel. The thermal energy, demonstrated to the surroundings by the convection and irradiance, is being transferred. However, the accidental thermal load is further-on negatively influencing the material of tunnel-road.

Additional size of the impact of longitudinal artificial ventilation offers a view of the fire-place (Fig 5.1-5) in the 120th second of this accidental-event. In this large-scale reactive flow - the mixing processes and the chemical reactions with their physical consequences are more expressed. So, on the central plane (that intersects the fire) with the established temperature-fields, one can notice the "hotspots", occurred, due to the fluctuations in the combustion. This is to be understood as a constructive interference of the buoyancy with the longitudinal flow (towards the western lower geodetic position) - wherever the buoyant forces are stronger than convection of the fire plume and hot gas-stream, which is in the near-fire-area. These occurrences do present in this phase of the fire-development already a dangerous point for the construction of the tunnel body (Fig. 5.1-5). Further observation of the temperature-impact is holding however no proof for interaction between the cavern-curvature and the distribution of the large-scale fire consequences (Fig. 5.1-3, -4, -5) and combined plots of the temperature with velocity iso-surfaces did not point up to expected "chimney effect" due to the tunnel´s bifurcation and tube of the exit-trek. In it´s 120th second, the employed longitudinal ventilation has appear as successful enveloping of the large-scale-fire event and it's negative consequences throughout the "Sentvid´s" north-cavern.

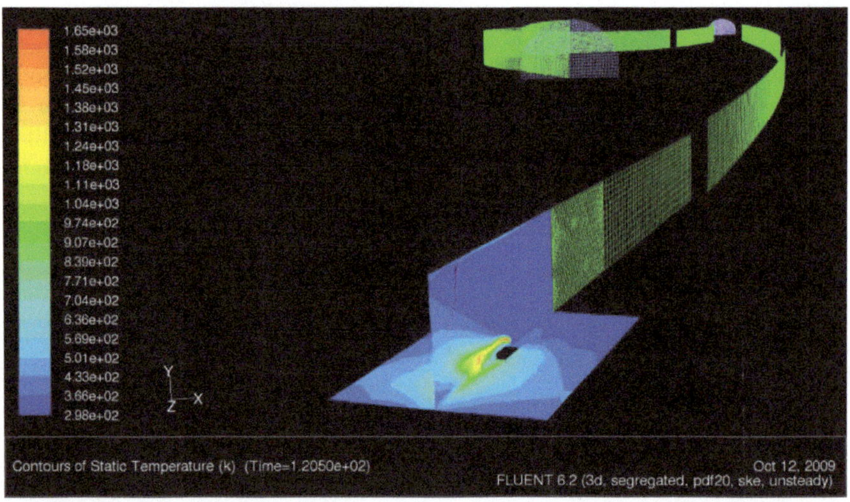

Fig. 5.1-5. The CFD-record of the 120th second: on the central plane that goes through the fire, displayed temperatures coming from the accidental-event are not to pose any threat to the RC-construction of the tunnel, since the fire-event is fully enveloped by the ventilated air-stream. The thermal energy, demonstrated to the surroundings, is being transferred into the air-flow. However, the accidental thermal load is further negatively influencing the material of the tunnel-road (here: a view towards the eastern entrance: far in the sketch in bifurcation zone. Above the fire-place, we see the tunnel exit)

5.2 The "Vranduk-2" road-tunnel

Situation in the cavern of "Vranduk-2" forecasted by the CFD-modelling, can be understood, up to the most important way, by the comparison of the simulated, developing fire-event of 40MW thermal-power, in it´s 25th and 119nd second, during which, this non-premixed combustion was going on, firstly without any ventilation in the tunnel. Employed *LES* turbulence-model was capable of demonstrating the fire shape and since we had confinement of this motorway-tunnel in both of the cases of CFD-based exploring - the temperature zones were also easily recognizable. So, the expected negative impact, we saw in "Sentvid" northern tube - was displayed in the "Vranduk-2" as well, pointing to the fact that even in the 25th second (in case of scenario with non-ventilated tunnel cavern) the temperature level, for the RC-construction to decay, was unfortunately easily reached.

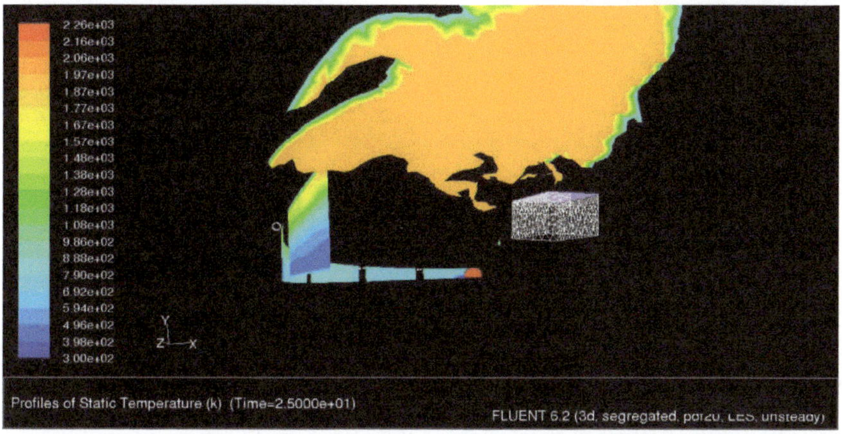

Fig. 5.2-1. In spite of the "dominance" of the iso-surface of 1800-K it is obvious that the thermal load of up to 40MW, that comes from accidental-fire in "Vranduk-2", will be negatively impacting the RC-construction (here: in the area of the fire-place, made visible by it´s mesh, looking at the tunnel-ceiling "from below", we can see the eastern tunnel exit in far-part of the sketch)

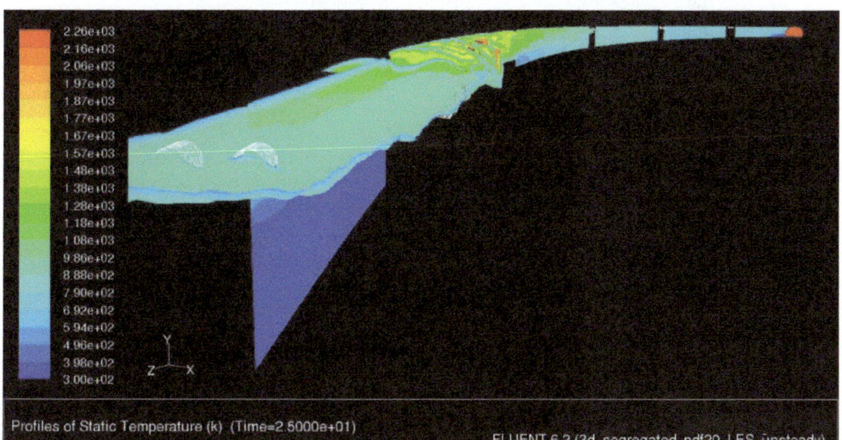

Fig. 5.2-2. The iso-surfaces of the performed temperatures due to the simulated large-scale non-premixed combustion do discover in first 25 seconds spreading of the negative thermal impact, along the "Vranduk-2" ceiling and the escorting tunnel body (the light blue-greenish temperature-area is a limit for a start of RC-decay due to the exposure to the accidental thermal load of 850 K). Far in this sketch one can notice the eastern-exit towards the city of Zenica. The canyon of the river Bosna is on the left-hand side

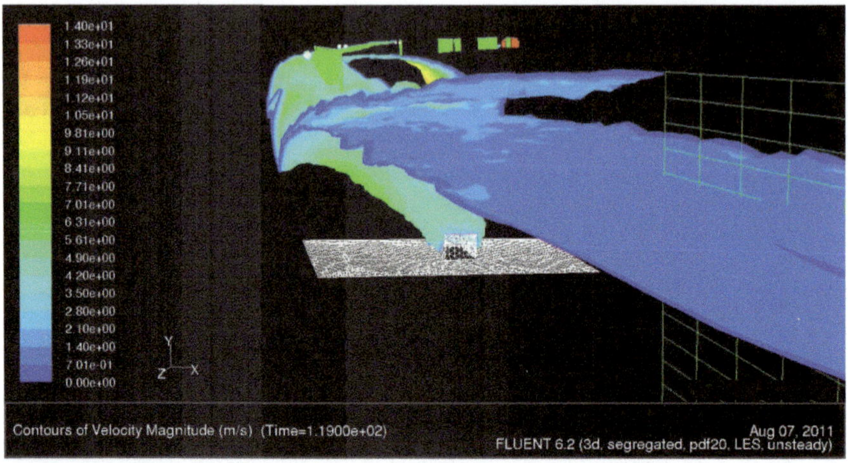

Fig. 5.2-3. The CFD-plotted iso-surface for the temperature of 600K is showing the velocities of the gaseous combustion products, as their spreading was forecasted throughout the cavern of "Vranduk-2". In spite of natural draft towards the geodetically higher positioned eastern exit (far at the bottom of the sketch) in the 119th second, the employed LES-turbulence model forecasted also the back-layering towards the western cavern-exit.

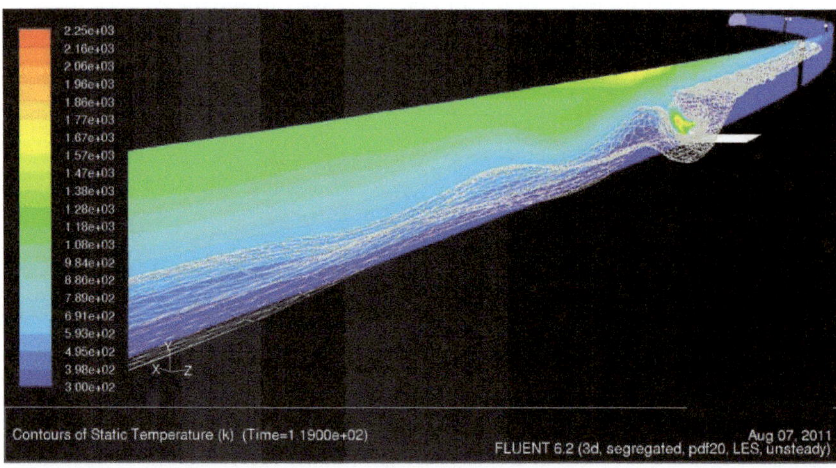

Fig. 5.2-4. The central control plane is intersecting the fire-place and pointing at the temperature zones. The displayed mesh is bordering the zone where thermal load can cause concrete destruction. Far down in the sketch one can see the western tunnel-entrance.

After 120s of modelling with CFD-tool on the scenario of a large-scale confined fire in "Vranduk-2" cavern - the artificial ventilation was employed. This forced air-flow was coming from the seven pairs of fans fixed to the ceiling along the tunnel-body, where each of fourteen of them produced a longitudinal air-velocity of 39.1 m/s. As well as in case of "Sentvid" - also here is aim of such forced fluid-flow to envelope the existing non-premixed combustion (and it´s consequences) and limit the possible damages to the RC-construction.

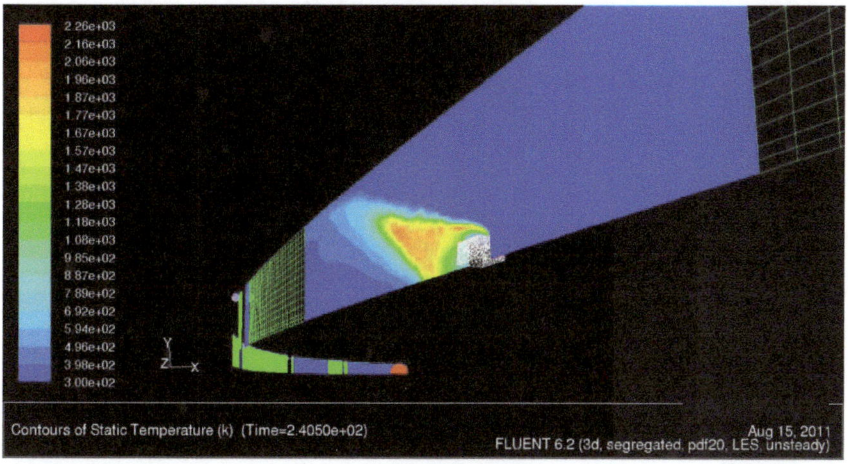

Fig. 5.2-5. The highest CFD-forecasted temperature is displayed already in the 25th second (modelled with employed LES-turbulence approach). Due to the energy transferring mechanism of convection, the heat is transmitted to the tunnel-ceiling and the iso-surfaces are pointing to the temperature decrease towards the tunnel body

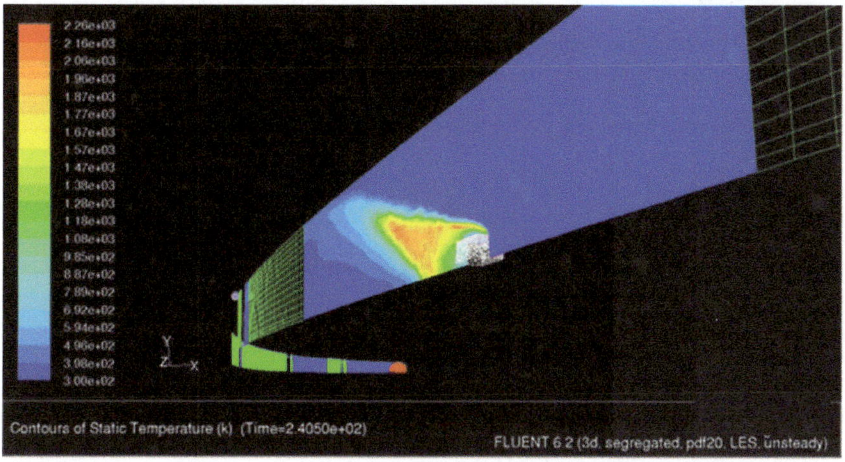

Fig. 5.2-6. In the 2nd minute of the employed ventilation (as simulated scenario was suggesting) the large scale fire and it´s unwanted thermal load are suppressed

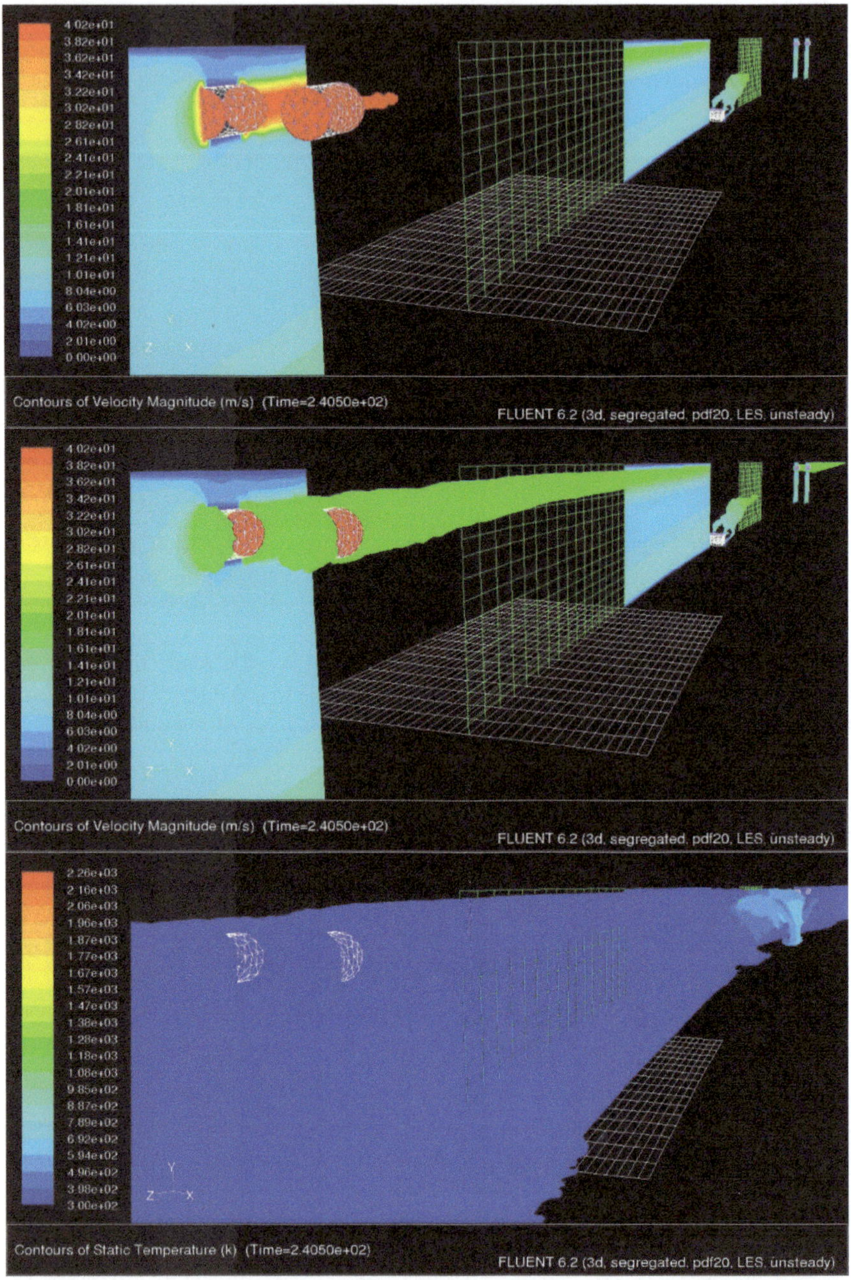

Fig. 5.2-7. The air-fans, looking towards the eastern exit of the "Vranduk-2" - And while the high temperature-zones of the fire were visible in first sketches, on the iso-surface for the 10m/s, the displayed temperature-zone is fully enveloping the near-fire-place. The accidental thermal load (deep in the sketch) is carried away by the forced air-flow

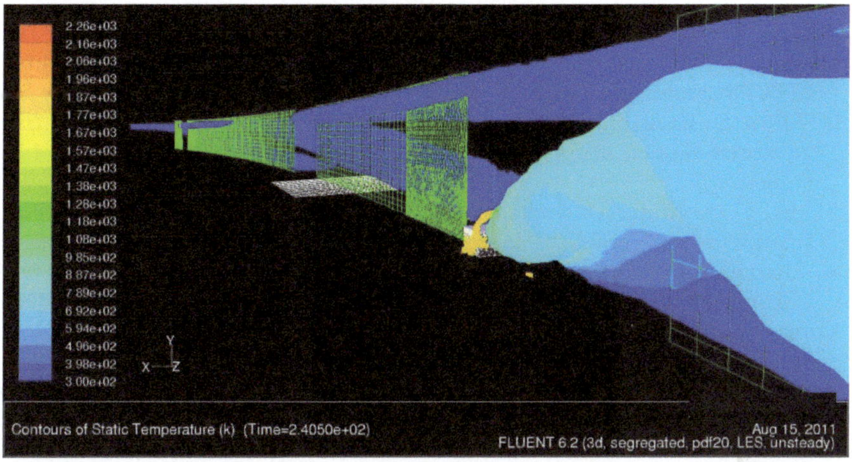

Fig. 5.2-8. Looking deeper into zones of the longitudinal ventilation - here the velocity of the forced air-flow of 13m/s is displaying the enveloping of the thermal influence in the 120th second of artificial air-flow. Looking towards the western entrance of "Vranduk-2" one can notice (deep down in the sketch) the velocity zones (blue) that are surrounding the air-fan pairs and "coming" towards the fire-place.

By performing this CFD-based study on covered traffic-objects of the given geometric characteristics within an existing road-infrastructure, the investigation on the accidental fire event was conducted according to the both planned scenarios of several experimental[33, 42, 44, 45] and computer aided[46-48] research[49, 50] research approaches. A "provocation" to conduct this research, was this specific geometry of the objects of interest, expecting new answers due to the possible impact of a reality-oriented traffic-enclosure (that was imbedded into a computational domain) onto this large-scale combustion and escorting occurrencesThe CFD-demonstration in this attempt, pointed that the geometric characteristic of explored tunnels was not "strong enough" to give major influence to the propagation of the combustion consequences in the first 120s of non-ventilated space - as well as in the case of 120s-long ventilation time of the mentioned caverns. However in the near-fire place one was able to notice the asymmetric distribution of the gaseous products in "Vranduk-2" motorway-tunnel, over the tunnels´ ceiling and temperature´s iso-surfaces.

Giving the small mosaic-stone to the entire urge in the community; which is researching on large-scale confined fires, with the results of this research-attempt, it would be the intention of the presented study to address both the civil-engineering sector[51, 52] and enlarge data-base for the medical health-protection[53].

6. Acknowledgement

Medzid Muhasilovic thanks to Prof. Dr. Michel Deville for his unselfish on-going support as well as direct and indirect advices. Additional gratitude goes to Prof. Dr. Ivica Kozar for the ideas and both administrative and scientific support. In this sense, special thanks are made to Dr. Markus Gawlowski and to Dr. Iris Vela. Last but definitely not least, is that Medzid Muhasilovic specially thaks to Dipl.-Ing. Marko Zibert and Dipl.-Ing. Drago Dolenc.

7. References

[1] G. Holmstedt, S. Bengston, H. Tuovinen, Sensitivity Calculations of Tunnel fires Using CFD. Fire Safety Journal, 1996. 1: p. 99 - 119.

[2] S. D. Miles, S. Kumar, R. D. Andrews. Validation of a CFD model for fires in the memorial tunnel. in First International Conference on Tunnel Fires. 1999. Lyon, France.

[3] T. Heins, K. Kordina, Untersuchungen über die Brand- und Rauchentwicklung in Unterirdischen Verkehrsanlagen - Katastrophenschutz in Verkehrstunneln. 1990.

[4] D. A. Charters, W. A. Gray, A. C. MacIntosh, A Computer Model to Assess Fire Hazards in Tunnels (FASIT). Fire Technology, 1994. 30: p. p. 143.

[5] S. Kumar, G. Cox. Mathematical Modelling of Fires in Tunnels. in 5th International Symposium on the Aerodynamics & Ventilation of Vehicle Tunnels. 1985.

[6] S. Kumar, G. Cox. Radiation and Surface Roughness Effects in the Numerical Modelling of Enclosure Fires. in Fire Safety Science - 2nd International Conference. 1988.

[7] Chasse, P. Sensitivity Study of Different Modelling Techniques for the Computer Simulation of Tunnel Fire: Comparison with Experimantal Measures. in First CFDS International User Conference. 1993. Oxford, UK.

[8] H. Briollay, P. Chasse, Validating and Optimazing 2D and 3D Computer Simulations for the Offenegg Tunnel Fire Test. 1994, Centre d´Etudes des Tunnels: Bron Cedex.

[9] S. Kumar, G. Cox, Mathematical Modelling of Fires in Tunnels - Validation of JASMINE. 1986, Transport and Research Laboratory Contractor: Crowthorn, UK.

[10] Tuovinen, H., Validation of Ceiling Jet Flows in a Large Corridor with Vents Using the CFD Code JASMINE. Fire Technology, 1994. 32.

[11] Kunsch, J. P., Simple model for control of fire gases in a ventilated tunnel. Fire Safety Journal, 2002. 37: p. 67-81.

[12] A. Beard, D. Drysdale, P. Holborn, S. Bishop, Cofiguration Factor for Radiation in a Tunnel ar Partial Cylinder. Fire Technology, 1993. 29.

[13] Miles, S. D., About JASMINE, D.M. Muhasilovic, Editor. 2006: Edinburgh, Essen.

[14] B. F. Magnussen, B. H. Hjertager. On mathematical modelling of turbulent combustion with special emphasis on soot formation and combustion. in 16th international symposium on combustion. 1976. Pittsburgh, USA.

[15] S. D. Miles, S. Kumar. Computer modelling to assess the benefits of road tunnel fire safety measures. in Inflam 2004. 2004. Edinburgh, Scottland, UK.

[16] K. B. McGrattan, A. Hamins, Numerical Simulation of the Howard Street Tunnel Fire. 2002, NIST: Gaithersburg, USA.

[17] http://www.tunnelfire.com.

[18] Wehlan, M., On "Memorial Tunnel Experiments" (personal communication). 2006: Washington(USA), Podgora(Croatia).

[19] M. K.-A. Neophytou, R. E. Britter, A simple model for the movement of fire smoke in a confined tunnel. Pure and Applied Geophysics, 2005. 162: p. 1941.

[20] P. Z. Gao, S. L. Liu, W. K. Chow, N. K. Fong, Large eddy simulations for studying tunnel smoke ventilation. Tunneling and Underground Space Technology, 2004. 19: p. 577.

[21] M. Peric, J. H. Ferziger, Computational Methods for Fluid Mechanics. 2001, Berlin: Springer Verlag. 423.

[22] H. R. Baum, et. al. Gravity-current transport in buildings fires. in International Conference on Fire Research and Engineering. 1995. Orlando, Florida, USA.

[23] Vladimirova, N., Model flames in the Boussinesq limit. 2006, ASC / Flash Center, Dept. of Astronomy and Astrophysics, The University of Chicago, IL 60637: Chicago, USA.

[24] T. B. Gatski, T. Jongen, Nonlinear eddy viscosity and algebraic stress models for solving complex turbulent flows. Progress in Auerospace Sciences, 2000. 36: p. 655.

[25] Leupi, C., Numerical modeling of cohesive sediment transport and bed morphology in estuaries, in La faculte sciences at techniques de língenieur. 2005, Ecole Polytechnique Federale de Lausanne: Lausanne.

[26] W. Zhang, et. al., Turbulence statistics in a fire room model by large eddy simulation. Fire Safety Journal, 2002. 37: p. 721.

[27] http://www.fluent.com.

[28] P. J. Woodburn, R. E. Britter, CFD-simulations of a tunnel fire - part one. Fire Safety Journal, 1996. 26: p. 35.

[29] N. C. Markatos, M. R. Malin, Mathematical modelling of buoyancy-induced smoke flow in enclosures. International Journal of Heat Mass Transfer, 1982. 25: p. 63.

[30] S. M. Jojo, W. K. Chow, Numerical studies on performance evaluation of tunnel ventilation safety systems. Tunneling and Underground Space Technology, 2003. 18: p. 435.

[31] C. K. Westbrook, W. J. Pitz, H. J. Curran, Chemical kinetic modelling study of the effects of Oxygenated hydrocarbons on soot emissions from diesel engines. Journal of Physical Chemistry, 2006. 110: p. 6912.

[32] O. Megret, O, Vauquelin, A model to evaluate tunnel fire characterisrics. Fire Safety Journal, 2000. 34: p. 393.

[33] H. K. Versteeg, W. Malalasekera, An Introduction to computational fluid dynamics. 1995, London: Longman Group Ltd.

[34] Hirsch, C., Numerical Computation of Internal and External Flows. Vol. I. 1988, Chichester Brisbane Toronto New York: John Wiley & Sons. 515.

[35] http://lin.epfl.ch.

[36] Muhasilovic, M., CFD-approach in investigation of consequences of accidental large-scale fires in road tunnels with natural ventilation, in STI-ISE-LIN. 2007, EPFL: Lausanne, Switzerland.

[37] Babrauskas, V., Estimating large pool fire burning rates. Fire Technology, 1983. 19: p. 251.

[38] I. Vela, et. al. Scale Adaptive Simulation (SAS) of heat radiation and soot amount in a large-scale turbulent JP-4 pool fire. in DECHEMA Tagung. 2006. Wiesbaden, Germany.

[39] http://lmr.epfl.ch.

[40] http://www.also-natursteine.de.

[41] H. Koseki, T. Yumoto, Air entrainment and thermal radiation from heptane pool-fires. Fire Technology, 1988. 24: p. 33.

[42] H. Kurioka, Y. Oka, H. Satoh, O. Sugawa, Fire Properties in Near Field of Square Fire Source with Longitudinal Ventilation in Tunnels. Fire Safety Journal, 2003. 34(4): p. 319 - 340.

[43] Modic, J., Fire Simulation in Road Tunnels. Tunneling and Underground Space Technology, 2003. 18: p. 525 - 530.

[44] H. Kurioka, Y. Oka, H. Satoh, H. Kuwanna, O. Sugawa, Properties of the Plume and Near Fire Source in horizontally long and narrow spaces. Journal of Construction Engineering, 2001(546): p. 151 - 156.

[45] O. Vauquelin, O. Megret, Smoke extraction experiments in case of fire in a tunnel. Fire Safety Journal, 2002. 37: p. 525.

[46] Ingason, H., Correlation between temperatures and oxygen measurements in a tunnel flow. Fire Safety Journal, 2007. 42: p. 75.

[47] O. Vauquelin, Y. Wu, Influence of tunnel width on longitudinal smoke control. Fire Safety Journal, 2006. 41: p. 420.

[48] P. J. Woodburn, R. E. Britter, CFD Simulation of a Tunnel Fire - part two. Fire Safety Journal, 1996. 26: p. 63.

[49] G. B. Grant, S. F. Jagger, C. J. Lea, Fires in Tunnels. Philosophical Transactions: Methematical, Physical, Engineering Sciences, 1998: p. 2873.

[50] S. R. Lee, H. S. Ryou, A numerical study on smoke movement in longitudinal ventilation tunnel fires for different aspect ratio. Building and Environment, 2006. 41: p. 719.

[51] R. O. Carvel, A. N. Beard, P. W. Jowitt, The influence of longitudinal ventilation systems on fires in tunnels. Tunneling and Underground Space Technology, 2001. 16: p. 3.

[52] V. K. R. Kodur, L. A. Bisby, M. F. Green, Experimental evaluation of the fire behaviour of insulated fibre-reinforced-polymer-strengthened reinforced concrete columns. Fire Safety Journal, 2006. 41: p. 547.

[53] F. Wald, et. al., Experimental behaviour of a steel structure under natural fire. Fire Safety Journal, 2006. 41: p. 509.

[54] F. Lestari, et. al., An alternative method for fire smoke toxicity assessment using human lung cells. Fire Safety Journal, 2006. 41: p. 605.

7

CFD and Thermography Techniques Applied in Cooling Systems Designs

Samuel Santos Borges and Cassiano Antunes Cezario
Research and Technological Innovation Department, WEG
Brazil

1. Introduction

The focus of this work consists in optimizing the water flow inside a water cooled electric motor frame, aiming at the maximization of the power/frame size ratio, the minimization of pressure drop and the avoidance of hot spots. For the development of this work computational fluid dynamics (CFD) and thermography techniques were used.

For many years water cooled electric motors have been used by industry, especially in specific applications that require features that can not be provided by conventional fan cooled motors. Among theses features are a better power/frame size relation, low noise levels, enclose applications, among others.

Cooling by means of water circulation is frequently used in large motors, typically frame sizes above IEC 315. A justification for this practice not to be widely used in smaller motors is its higher cost compared to air cooling systems. However, provided that it presents some technical and/or economical viability, water cooling can also be used in smaller frame sizes.

The technological progress have resulted in the development of new tools that facilitate the study of this motor type, such as the computational fluid dynamics and the thermography techniques, which consist, respectively, in the use of numeric models applied to fluid mechanics and the obtaining of the motor surface temperature distribution.

2. Benefits and drawbacks

The following advantages and disadvantages of this kind of motor can be mentioned, in comparison with air cooled motors.

2.1 Advantages

- Greater power/frame size ratio;
- Lower noise level;
- Higher efficiency;
- The outside waste deposit on the frame does not damage the motor cooling;
- No problems with waste deposit on the external fan blade;
- It can be used in totally enclosure environment;
- The heat removed from the motor is not directly dissipated into the environment;
- Possibility of using the same cooling fluid of the load and/or machine where it is installed;

- Better resistance to local impact.

2.2 Disadvantages

- Higher manufacturing cost;
- Auxiliary system to provide water;
- Need to control the water chemical composition;
- Risk of corrosion inside of the water circuit;
- Risk of fouling in the water circuit;
- Risk of leaks;
- More precautions on maintenance.

3. Fundamental concepts

3.1 Cooling system

The cooling systems are apparatus usually comprised of several components that interact to keep the temperature of machines and systems in general controlled in order not to exceed the limits imposed by quality, safety, performance and/or efficiency. For instance, systems such as internal combustion engines, turbines, compressors, bearings, electric machines, electronic structures, electrical conductors, chemical and welding processes, among many others, can be mentioned as some examples of systems comprising thermal restrictions. Consequently, the cooling systems are very important to industries and should therefore be designed to keep a good performance and reliability.

The more commonly used coolants are gases and liquids. Among them, due to their wide occurrence, low cost and practicality, air and water are customarily employed. They can be used in many different ways: in direct or indirect contact with the systems, by means of one or more fluid combinations, separated or mixed. And they can be forcedly or naturally moved, by physical mechanisms respectively known as forced convection and natural convection.

3.2 Heat transfer

Heat transfer concerns the exchange of thermal energy from one region of higher temperature to another of lower temperature in one or more means. This exchange occurs by mechanisms as conduction, convection, radiation and phase-change.

3.2.1 Conduction

The conduction occurs due to interactions between the particles of fluids. Thermal insulation, tanks, plates and ducts walls, fins, and others are examples of devices that exchange heat by conduction. It is described by Fourier's Law and the following equation:

$$q_k = -kA\frac{\partial T}{\partial n} \tag{1}$$

Where,

q_k: Heat flux of conduction;

K: Thermal conductivity;

A: Heat transfer area;

∂T : Temperature gradient;

n: Direction.

3.2.2 Convection

The convection is the heat transfer between a solid surface and a moving fluid. This phenomenon can be natural, induced by density difference (low fluid velocity), or forced, customarily supplied by fans or pumps (high fluid velocity).

$$q_h = hA(T_s - T_\infty) \tag{2}$$

Where,
q_h: Heat flux of convection;
h: Convection coefficient;
A: Heat transfer area;
T_s: Surface temperature;
T_∞: Fluid temperature.

3.2.3 Radiation

The thermal radiation is provided by electromagnetic waves emission. The classic example of this phenomenon is the heat transfer between sun and earth. It is given by Stefan-Boltzmann Law.

$$q_r = \varepsilon\sigma(T_1^{\,4} - T_2^{\,4}) \tag{3}$$

Where,
q_r: Heat flux of radiation;
ε: Emissivity;
σ: Stefan-Boltzmann constant;
T_1, T_2: Surface temperatures.

3.2.4 Phase – Change

It is boiling, condensation, freezing and melting. These processes are given by following equations.

$$Q = m.\Delta H \tag{4}$$

Where,
Q: Amount of heat;
m: Mass;
ΔH: Specific latent heat.

3.2.5 Heat transfer in conduits

In the heat transfer in the fluid flowing in a conduit, the temperature is neither uniform in the fluid flow direction nor in the heat flux direction. Therefore, the bulk temperature is assumed as reference temperature. The use of a temperature reference allows to execute the heat balance in steady state. In this case, the transferred heat per time unit is a direct measure of the bulk temperature difference between two conduit sections. That is,

$$q = \dot{m}.c_p.\Delta T_b \tag{5}$$

Where,
q: Amount of heat per time unit;
$\dot{m}$: Mass flow rate;
c_p: Specific heat of fluid;
ΔT_b: Bulk temperature difference of fluid among two conduit sections.

3.3 Fluid dynamics

3.3.1 Mass flow rate

The mass flow rate is given by the mass of fluid which passes through a surface per time unit and can be calculated from the following equation:

$$\dot{m} = \rho.v.A \tag{6}$$

Where,
$\dot{m}$: Mass flow rate;
ρ: Fluid density;
v: Mean velocity in the section;
A: Area of conduit section.

3.3.2 Pressure drop

The pressure drop in pipes may be occasioned by means of:

- Localized drops (valves, curves, and others);
- Friction;
- Difference of piping height.

The pressure drop is obtained as a function of Reynolds number, which depends on the fluid velocity, given by:

$$\mathrm{Re} = \frac{\rho.v.D}{\mu} \tag{7}$$

Where,
Re: Reynolds number;
ρ: Fluid density;
v: Mean velocity in the section;
D: Diameter of pipe;
μ: Viscosity.
If the Reynolds number is smaller than 2300 the flow is laminar and friction factor is calculated by:

$$f = \frac{64}{\mathrm{Re}} \tag{8}$$

Where,
f: Friction factor;
Re: Reynolds number.
But, if the Reynolds number is greater than 2300 the flow is turbulent and friction factor is calculated by:

$$\frac{1}{\sqrt{f}} = -2.\log\left(\frac{e/D}{3.7} + \frac{2.51}{\mathrm{Re}.\sqrt{f}}\right) \tag{9}$$

Where,
f: Friction factor;
Re: Reynolds number;
e: Absolute roughness;
D: Pipe diameter.
And the pressure drop is determined by:

$$\Delta p = f.\frac{L}{D}.\frac{\rho.v^2}{2} \tag{10}$$

Where,
Δp: Pressure drop;
f: Friction factor;
L: Length of pipe;
D: Pipe diameter;
ρ: Fluid density;
v: Mean velocity in the section.
Therefore, to minimize the pressure drop and consequently reduce the energy loss of flow, it is desired low fluid velocity.

3.4 Turbulence models

Among several turbulence models known in the literature, will be discussed here only those ones involved in this work: κ - ε model, κ - ω model and Shear-Stress Transport (SST).

3.4.1 The κ - ε model

The simplest complete turbulence model has a wide range of applications in industrial and engineering problems. It can be applied to mass or heat transfer, combustion, multi-phase flows simulations, and others. The κ - ε model has been incorporated in most commercial CFD codes due to its stability and numerical robustness, good accuracy and low computational cost.
This model should be avoided in complex flow applications, such as flows with boundary layer separation, sudden changes in the mean strain rate, rotating fluids, and over curved surfaces.
The standard κ - ε model is based on model transport equations for the turbulence kinetic energy κ and its dissipation rate ε. Transport equations for the standard κ - ε model are given by:

$$\frac{\partial(\rho k)}{\partial t} + \frac{\partial(\rho k \mu i)}{\partial x_i} = \frac{\partial}{\partial x_j}\left[\left(\mu + \frac{\mu_t}{\sigma_k}\right)\frac{\partial k}{\partial x_j}\right] + P_k + P_b - \rho\varepsilon - Y_M + S_k \tag{11}$$

$$\frac{\partial(\rho \varepsilon)}{\partial t} + \frac{\partial(\rho \varepsilon \mu i)}{\partial x_i} = \frac{\partial}{\partial x_j}\left[\left(\mu + \frac{\mu_t}{\sigma_\varepsilon}\right)\frac{\partial \varepsilon}{\partial x_j}\right] + C_{1\varepsilon}\frac{\varepsilon}{\kappa}(P_k + C_{3\varepsilon}.P_b) - C_{2\varepsilon}\rho\frac{\varepsilon^2}{\kappa} + S_\varepsilon \tag{12}$$

Where,
μ_t: Turbulent viscosity;
P_k: Generation of turbulence kinetic energy due to the mean velocity gradients;
P_b: Generation of turbulence kinetic energy due to buoyancy;
Y_M: Fluctuating dilatation in compressible turbulence;
ρ: Density;
μ: Viscosity;
$C_{1\varepsilon}$, $C_{2\varepsilon}$, $C_{3\varepsilon}$: Model constants;
σ_k, σ_ε: Turbulent Prandtl number for κ and ε;
S_k, S_ε: Source term.

3.4.2 The κ - ω model

The κ - ω model is widely used and is superior for boundary-layer flows both in the viscous near-wall region treatment and in the streamwise pressure gradients application. However, it is not appropriate outside the shear layer that is a free-stream boundary. The standard κ - ω model is an empirical model based on model transport equations for the turbulence kinetic energy κ and the specific dissipation rate ω, which is obtained by ε to κ ratio. Based on Wilcox, κ and ω are obtained from:

$$\frac{\partial(\rho k)}{\partial t}+\frac{\partial(\rho U_j k)}{\partial x_j}=\frac{\partial}{\partial x_j}\left[\left(\mu+\frac{\mu_t}{\sigma_k}\right)\frac{\partial k}{\partial x_j}\right]+P_k-\beta'\rho\kappa\omega+P_{kb} \tag{13}$$

$$\frac{\partial(\rho \omega)}{\partial t}+\frac{\partial(\rho U_j \omega)}{\partial x_j}=\frac{\partial}{\partial x_j}\left[\left(\mu+\frac{\mu_t}{\sigma_\omega}\right)\frac{\partial \omega}{\partial x_j}\right]+\alpha\frac{\omega}{\kappa}P_k-\beta\rho\omega^2+P_{\omega b} \tag{14}$$

$P_{\kappa b}$, $P_{\omega b}$: Turbulence buoyancy term;
U: Velocity vector;
σ_ω: Turbulent Prandtl number for ω;
β', β, a: Model constant.

3.4.3 The shear-stress transport model (SST)

The shear-stress transport κ - ω model was developed by Menter to effectively blend the robust and accurate formulation of the κ - ω model in the near-wall region with the free-stream independence of the κ - ε model in the far field. To achieve this, the κ - ε model is converted into a κ - ω formulation. The SST κ - ω model is similar to the standard κ - ω model, but the modeling constants are different in one and another and the SST model includes some changes in the standard one, such as the multiplication of both transformed κ - ε and κ - ω models by a blending function, which activates standard κ - ω model in the near-wall region and activates transformed κ - ε model away from the surface, the incorporation of a damped cross-diffusion derivative term in the ω equation, and a modification in the definition of the turbulent viscosity to account for the transport of the turbulent shear stress.

These improvements provide more accuracy and reliability for a wide range of flow classes than the standard κ - ω model. For instance, adverse pressure gradient flows, airfoils, and transonic shock waves.

The base equations are:

$$\frac{\partial(\rho k)}{\partial t} + \frac{\partial(\rho k \mu_i)}{\partial x_i} = \frac{\partial}{\partial x_j}\left(\Gamma_k \frac{\partial k}{\partial x_j}\right) + \tilde{G}_k - Y_k + S_k \tag{15}$$

$$\frac{\partial(\rho \omega)}{\partial t} + \frac{\partial(\rho \omega u_i)}{\partial x_i} = \frac{\partial}{\partial x_j}\left(\Gamma_\omega \frac{\partial \omega}{\partial x_j}\right) + G_\omega - Y_\omega + D_\omega + S_\omega \tag{16}$$

$$\Gamma_k = \mu + \frac{\mu_t}{\sigma_k}; \qquad \Gamma_\omega = \mu + \frac{\mu_t}{\sigma_\omega} \tag{17}$$

Where,
$\tilde{G}_k$: Generation of turbulence kinetic energy due to the mean velocity gradients;
G_ω: Generation of ω;
Y_k, Y_ω: Dissipation of κ and ω due to turbulence;
Γ_k, Γ_ω: Effective diffusivity of κ and ω;
G_ω: Cross-diffusion term;
σ_ω: Turbulent Prandtl number for ω;
S_ω: Source term.

3.4.4 Comparing SST and κ - ε model

In the Figures 1 and 2 two simulations of temperature and velocity of the fluid are presented, with the same boundary conditions for SST and κ - ε model. Note that the results are similar for both models, however SST models capture more flow characteristics than κ - ε model.

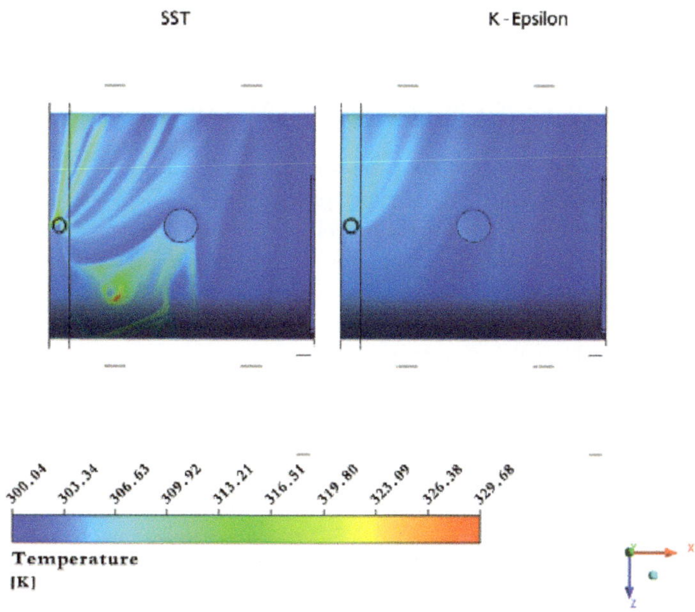

Fig. 1. Fluid temperature using SST and κ - ε model

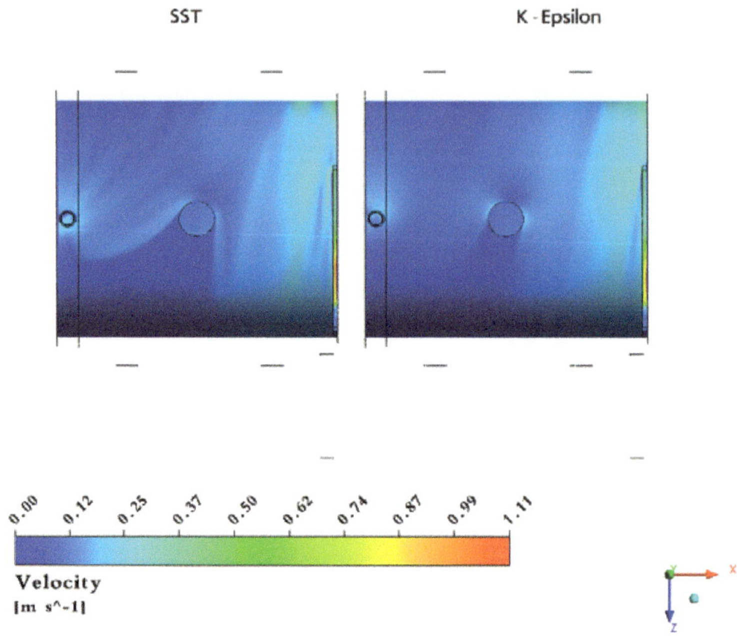

Fig. 2. Fluid velocity using SST and κ - ε model

3.5 Fouling

Fouling is a term employed to represent undesired accumulation of solid material on the surfaces of heat exchangers. This accumulation results in thermal resistance increment reducing exchanger performance. Fouling reduces the flow area and increases surface roughness, causing a reduction in the flow rate as a consequence of the pressure drop increase.

The fouling deposits may be loose particles or fix and hard layers, comprising sediments, polymers, inorganic salts, fuel or corrosion products, biological growth, and others. The maximum fouling layer can occur in few hours or may require many years to be formed. In any case, it should be avoided, because it increases capital costs, maintenance costs, production and energy losses, etc.

The main fouling mechanisms are crystallization, precipitation, sedimentation, chemical reaction, corrosion, biological and freezing fouling.

The amount of heat per time unit can be calculated by:

$$q = U.A.\Delta T \tag{18}$$

Where,

q: Amount of heat per time unit;

U: Overall heat transfer coefficient;

A: Heat transfer area;

ΔT: Bulk temperature difference between two conduit sections.

And U is obtained by the following equation:

$$U = \frac{1}{R_{hec} + R_{fou}} \tag{19}$$

Where,
U: Overall heat transfer coefficient;
R_{hec}: Resistance of cleaned heat exchanger;
R_{fou}: Resistance of fouled heat exchanger.
If the heat exchanger is clean, then R_{fou} = 0. Table 1 presents typical values of fouling resistance.

Fluid and Situation	*Fouling Resistance* R_f (m^2K/W)
Distilled water	0.0001
Seawater	0.0001 – 0.0004
Treated boiler feedwater	0.0001 – 0.0002
Clean river or lake water	0.0002 – 0.0006
About the worst waters used in heat exchangers	< 0.0020
No. 6 fuel oil	0.0001
Transformer or lubricating oil	0.0002
Most industrial liquids	0.0002
Most refinery liquids	0.0002 – 0.0009
Steam, non-oil-bearing	0.0001
Steam, oil-bearing (e.g., turbine exhaust)	0.0003
Most stable gases	0.0002 – 0.0004
Flue gases	0.0010 – 0.0020
Refrigerant vapors (oil-bearing)	0.0040

Table 1. Typical fouling resistance

Some variables must be taken into account in the design and operation of cooling systems, in order to avoid fouling. The most important are:
Flow velocity: for water, it should be kept above 2 m/s to suppress fouling and above 1 m/s to minimize fouling.
Surface temperature: for cooling towers, the water temperature must be kept below 60 °C.
Tube material: select materials to avoid corrosion, for instance.
For liquid coolants, fouling inhibitors should be used, such as: corrosion inhibitors, anti-dispersants, stabilizers, biocides, softeners, acids, and polyphosphates. However, if such fouling controls are not effective, the exchanger must be cleaned either on-line or off-line.

3.6 Corrosion

Corrosion is electrochemical destructive attack of the base material with its environment. Water in direct contact with a metal surface causes oxidation, which is a process where metal dissolved in fluid; this phenomenon generates serious problems in the worldwide. Corrosion causes:

- Efficiency reduction of machines and plants;
- Increase costs of maintenance and overdesign;
- Losses or contaminations of products;
- Plants shutdowns;
- And others.

The many aspects of corrosion problems constrain it control, which is achieved by recognizing and understanding corrosion mechanisms. The control and prevention of corrosion damage can be obtained basically by following methods:

- Employment of suitable materials;
- Use of protective coatings;
- Change to the environment;
- Application of cathodic or anodic protection and etc.

4. Water-cooled frame

The objective of this work consists in optimizing the water flow inside a water cooled frame considering the boundary conditions intrinsic to both the manufacturing process and the electric motor itself.

One of the challenges of this development is the frame manufacturing, which consists in a single cast iron piece free of welds or seals, making it difficult to obtain the water circuit.

Due to the increasing demand for smaller water cooled motors, in this work an IEC 200 frame size - 75 kW - IV Poles - 60 Hz motor was used.

5. Thermal circuit and equations

The first step of the design is the definition of the motor simplified thermal circuit as represented in Figure 3. It can be observed that the produced heat, L, is transferred through the equivalent resistance, R_{eq}, and the convection resistance, R_h, thus causing an increase in the average temperatures of the coil, T_w, the frame, T_{frame} and the fluid, T_{fluid}.

The next step is calculating the water flow that is needed to remove the heat produced by the motor. This can be obtained by means of (20) and (5), where L is the heat amount to be removed in W, $\dot{m}$ is the mass flow in kg/s, c_p is the specific heat of water in J/(kg . K), T_o is the outlet water temperature and T_i is the inlet water temperature both in K.

$$L = \dot{m}.c_p.(T_o - T_i) \tag{20}$$

Using (20) and the values shown in Table 2, it is obtained a mass flow of 0.239 kg/s.

Variable	Value
L	5.000 (W)
c_p a 20 °C	4 184.3 (J / kg / K)
T_o - T_i	5 (K)

Table 2. Value and units for variables

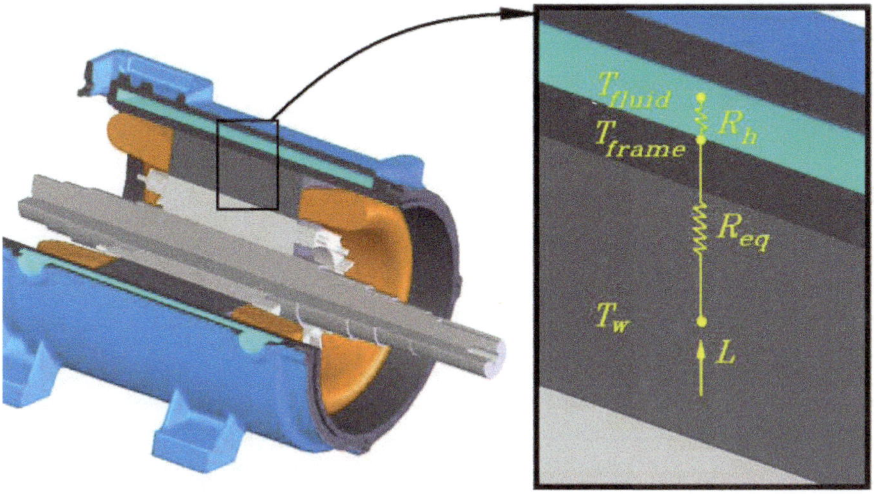

Fig. 3. Simplified thermal circuit of the motor

The value of L was considered as the total losses of the motor, that is, it is considered that all heat generated by the motor will be removed by water, which in practice does not happen, once that a portion of generated heat will be dissipated to environment through the endshields, the shaft and the terminal box. However, the heat portion removed by water is greater than by other means, so that it can be considered conservative, thereby simplifying the problem.

The next step consists in checking the circuit thermal saturation in relation to the water flow that is, quantifying how much an increase in the flow causes the coil temperature to reduce. Therefore, it is necessary to verify the behavior of the coil average temperature fluctuation, ΔT_w, with the water flow. For this verification equations obtained by heat transfer laws will be used, as described next.

The equations (21) and (22) relate L in W with T_w, T_{frame} and T_{fluid} in K and with R_{eq} and R_h in K/W.

$$L = \frac{\left(T_w - T_{frame}\right)}{R_{eq}} \tag{21}$$

$$L = \frac{\left(T_{frame} - T_{fluid}\right)}{R_h} \tag{22}$$

And to obtain R_h it is used (23), where the convection coefficient, h, is in W / (m2 . K) and the thermal exchange surface between the frame and the fluid, A, is in m^2.

$$R_h = \frac{1}{h.A} \tag{23}$$

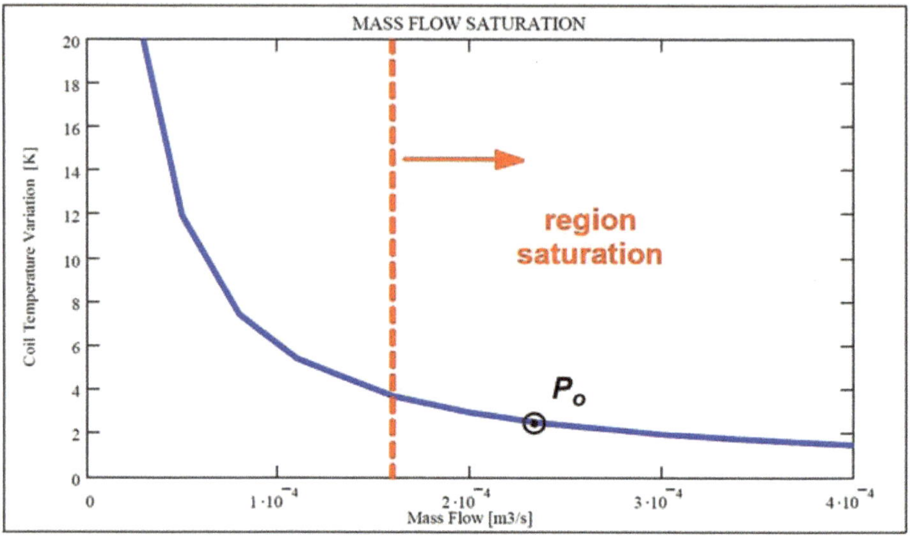

Fig. 4. Mass flow saturation

Knowing that h is a variable dependent on the fluid properties, the channel geometry and the flow, then the fluctuation temperature, ΔT_w, can be related to the mass flow $\dot{m}$, according to (21), (22), (23), the fluid properties and the channel geometry, obtaining the curve presented in Figure 4. It can be observed that the motor operation point, P_o, is located on the curve saturation region, what means that when increasing the water flow, the coil temperature reduction can be despised, because it is necessary a great increase the flow for the temperature to be slightly reduced. Therefore, this flow is correct for the design.

For the obtainment of this curve the following considerations were made:

a. R_{eq} is constant and, due to the difficulty to theoretically calculate its value, it was obtained through (21) and (22) using experimental data of preliminary tests;
b. Fluid properties are constant;
c. Dimensions of channel: 0.400 x 0.015 m;
d. Average diameter of the channel: 0,350 m.

6. Numerical analyses

Once analytically defined the flow value to be used in the design, the numerical analyses are initiated to check the water flow behavior inside the frame. For better use of the computational capacity, some considerations and simplifications were made concerning both the physical phenomenon and the geometry.

6.1 Heat transfer

The possible hot spots inside the water circuit can be generated basically by three causes, local heat generation, cooling fluid shortage in determined regions, and/or fluid recirculation. For simulation purposes local heat generation is neglected, because the heat flow from the stator to the frame is uniformly distributed throughout the surface. As for fluid shortage, it can be considered that the water circuit will be totally filled and therefore there will be no shortage. So the possible hot spots will be generated only by water flow recirculation, thus allowing the elimination of the heat transfer calculation on the numerical simulations, making possible to identify the hot spots by means of association with the recirculation. It is emphasized that with this consideration the hot spots can be only identified, but not thermally quantified.

6.2 Geometry

The water circuit geometry has as basic shape a cylinder with an inlet and an outlet of water as presented in Figure 5, where the difference between the external radius, *re*, and the internal radius, *ri*, is 0.015 m and the length, *l*, is 0.400 m. These values are limited by both the manufacturing process and the motor geometry. Flow directional guides will be inserted if needed, after analyzing the simulation results.

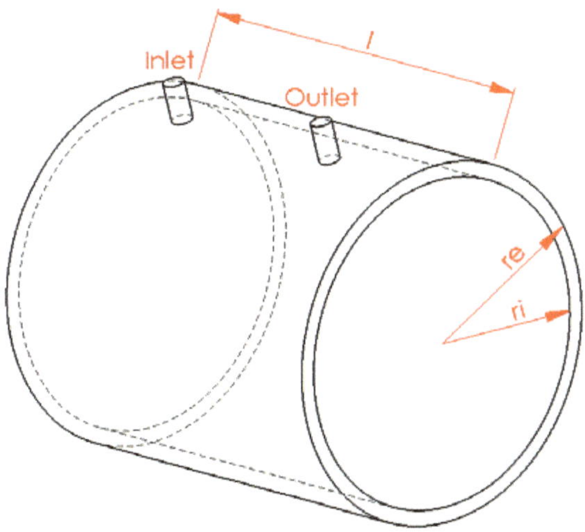

Fig. 5. Basic geometry of water circuit

To make easy the study of the flow inside the frame, the cylinder was transformed into a planned shape with a separation guide between inlet and outlet (configuration I) as presented in Figures 6 and 7. In Figure 6 it can be observed the flow inside the cylinder and in Figure 7 the flow inside configuration I. It is possible to observe the similar behavior of the flow in both of them, allowing for the use of configuration I as the base for the circuit optimization. This similarity happened because the gravitational force is considered null and, as the channel average speed is approximately 0.04 m/s, the centrifugal force caused by the cylinder curvature can be despised.

6.3 Inlet and outlet channels

The frame inlet and outlet channels are the main responsible for the pressure drop; therefore, an optimization in these regions can significantly reduce the total pressure drop.

In the analysis of the inlet and outlet channels it was used the planned cylinder technique and nine simulations were made. The three more significant ones from these simulations will be presented.

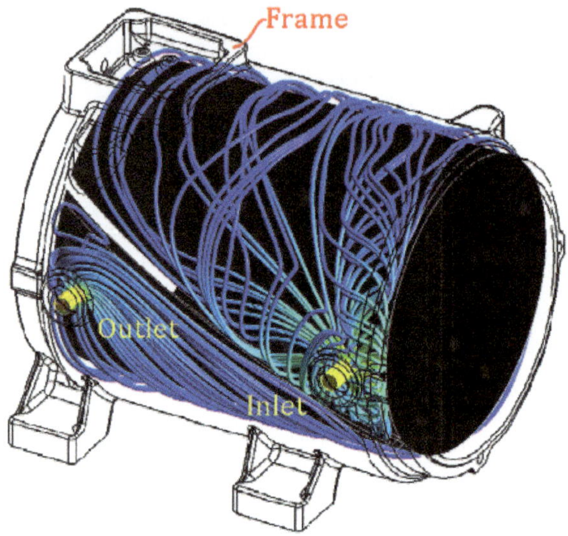

Fig. 6. Original cylinder

Fig. 7. Configuration I

The first simulation presented in Figure 8(a) consists in an inlet channel with circular pipe format located in the left side of the water main circuit. In this case, a water flow concentration is making tangent to the circuit walls occasioning a low flow in the middle of the circuit, which would cause an heterogeneous heat exchange.

In the second case, presented in Figure 8(b), the flow was improved if compared with the first case. In this case a diffuser on the inlet, instead of a pipe, was used. The flow further improved when the inlet diffuser was moved to the middle of the main circuit that can be seen in the Figure 8(c).

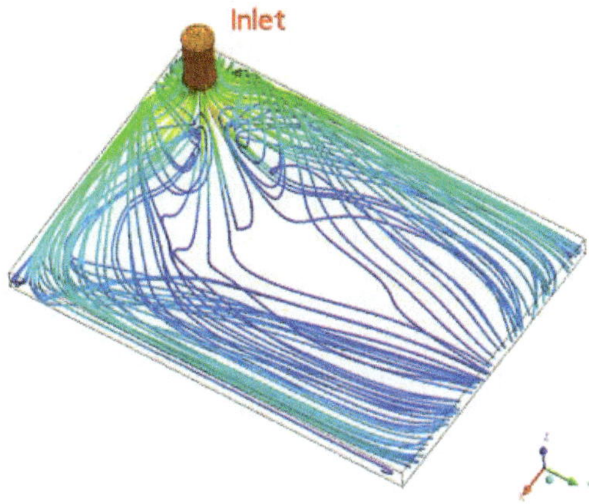

Fig. 8(a). Inlet and outlet channel

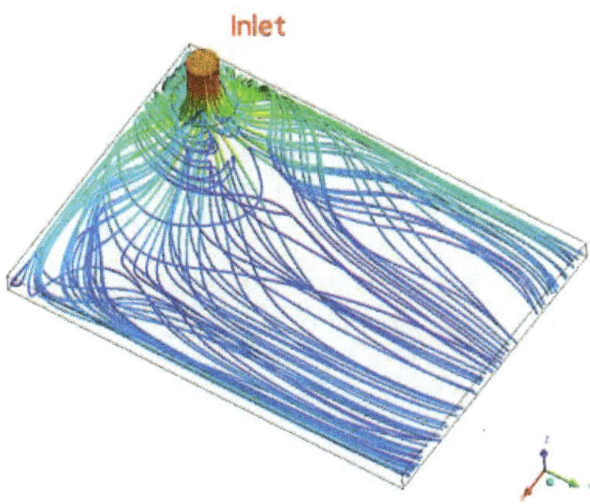

Fig. 8(b). Inlet and outlet channel

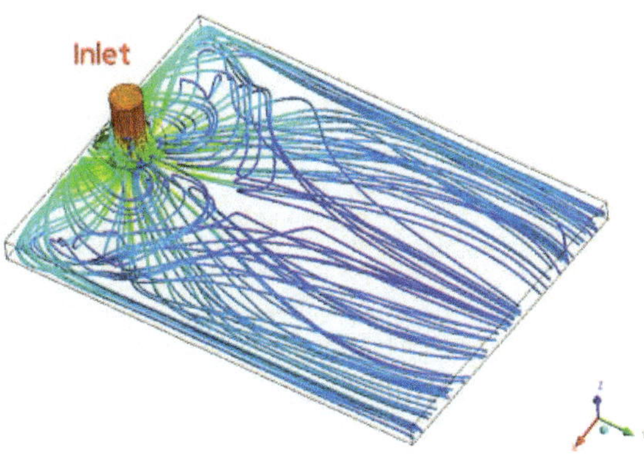

Fig. 8(c). Inlet and outlet channel

Due to the similarities observed during the work between the flow behavior in the inlet and in the outlet channels, the improvement of the inlet channel also could be applied to the outlet channel.

6.4 Water circuit

When the source of heat generation is constant in a surface, the cooling system needs to keep a constant fluid speed, keeping the heat transfer coefficient also constant in order to avoid undesired speed variation. Then, according to (24), where Q is the volumetric flow in m3/s, V the average speed in m/s and S the area of the transversal section of the flow in m^2, in order to keep the speed constant, the ratio Q/S must be kept constant. In this case, as the flow is constant the area can also be kept constant throughout the circuit. Thus the whole design of the cooling circuit will be based on this premise.

$$Q = V.S \tag{24}$$

In accordance with the Figure 7 the configuration I presented a good flow, but without guides inside the circuit. The absence of these guides could compromise the frame stiffness; therefore, other configurations with guides inside the circuit were simulated. Among the 16 simulated configurations, it was initially expected that the configuration II would present a good result, what did not happen, because such configuration caused low speeds in the back of the guides, as shown in Figure 9.

Due to the result obtained with the configuration II, the configuration III, presented in Figure 10, is proposed. In this configuration it was obtained the double water speed of the initial configuration I, besides presenting a good flow and low pressure drop.

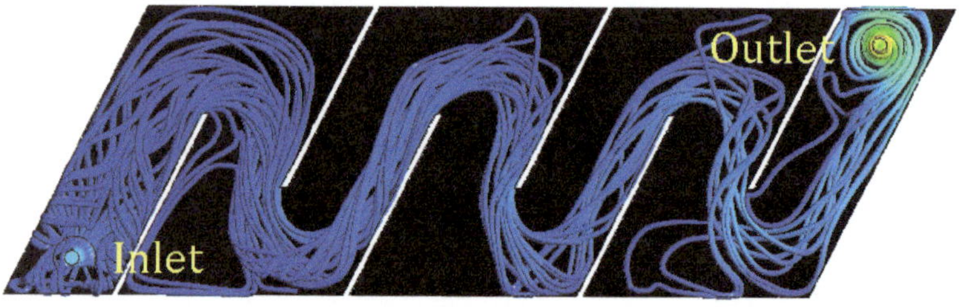

Fig. 9. Configuration II - intercalated opposite guides

Fig. 10. Configuration III - central guide with curved guide

6.5 Simulation data

The simulation meshes were made with the software *ICEM CFD 11.0.1*. The used elements were prisms on the walls and tetras inside the mesh, always respecting the quality criterion *Smooth Elements Globally - Quality* with value above 0.2.

The simulations were made with the software *CFX 11.0*, using the following simulation parameters:

- Heat transfer: *None*;
- Buoyance: Non buoyancy;
- Turbulence: Shear Stress Transport;
- Convergence criteria: 10^{-5} *(RMS)*.

7. Experiments and thermography

As yet only one motor with water circuit similar to the configuration I was tested. Due to the manufacturing process the dividing guide had been interrupted, therefore, the water inlet and outlet were not separated. This fact was confirmed using thermography techniques.

7.1 Thermal visualization techniques

To check the guide arrangement inside the channel the following procedure was accomplished: the motor was turned on and kept running for 30 minutes without water circulation. The water inlet valve was then opened, while keeping the motor working. The phenomenon of water filling in the frame channel was recorded by the thermographic camera. The sequence captured on the video can be visualized in Figure 11. In this case, the outlet is located on the top and not on the lateral of the frame, thus assuring the removal of air from the frame inside, as presented in Figures 11(d) and 11(e).

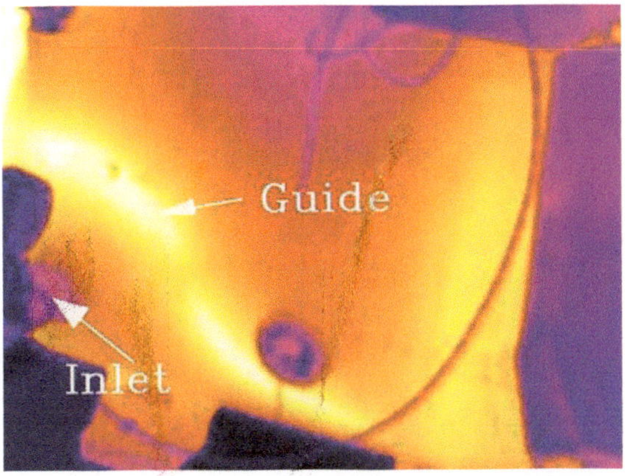

Fig. 11(a). Water filling frame channel - Closed valve

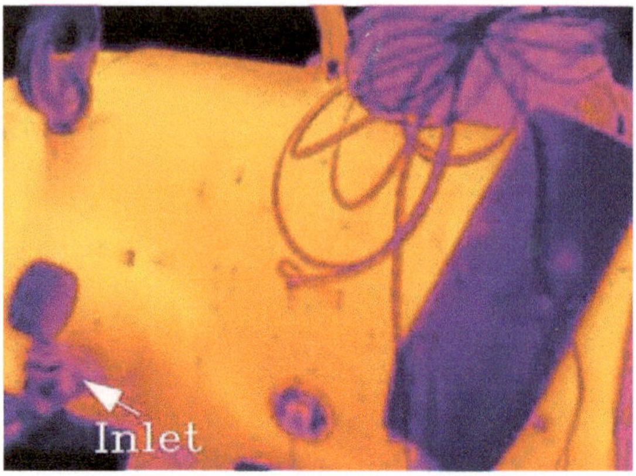

Fig. 11(b). Water filling frame channel - time: 10 s

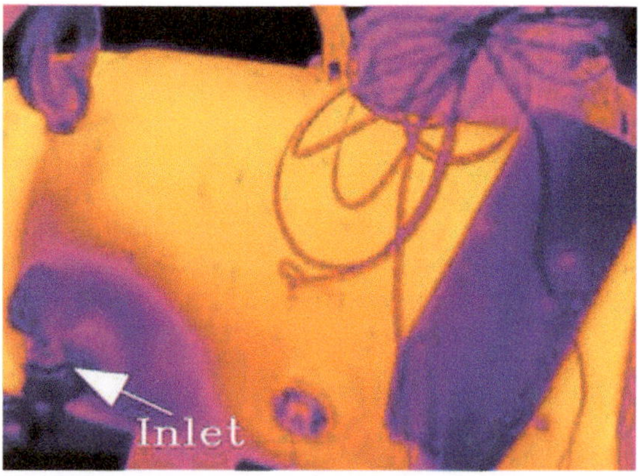

Fig. 11(c). Water filling frame channel - time: 1 min and 10 s

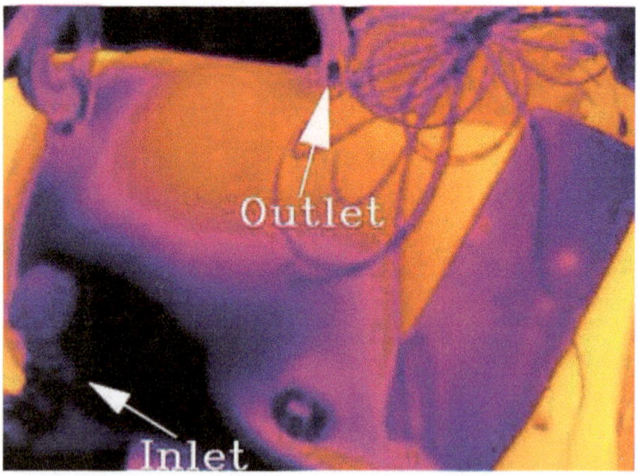

Fig. 11(d). Water filling frame channel - time: 3 min and 10 s

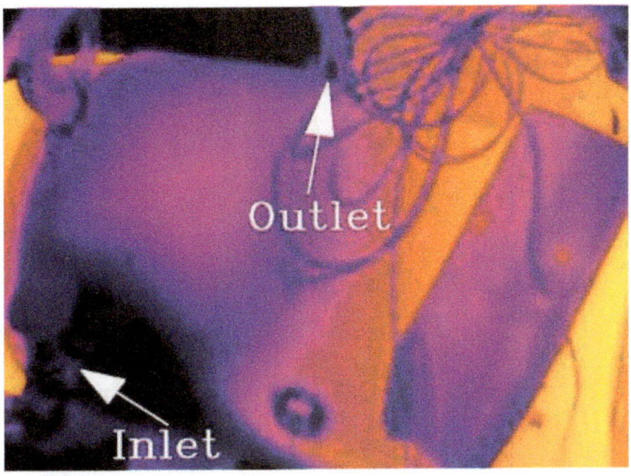

Fig. 11(e). Water filling frame channel - time: 10 min

Through this test it can be observed that two dark bands appear between the inlet and the outlet, allowing to affirm that two openings exist on the dividing guide. Theses bands can be visualized in Figures. 11(d) and 11(e). Therefore, when the outlet is located on the frame lateral, as proposed in the configuration I, the flow passes directly from the inlet to the outlet.

7.2 Results

Although the dividing guide between inlet and outlet to have been interrupted, thus causing recirculation, this motor presented a temperature rise of 73 °C on the windings, below the higher limit of its thermal class (80 °C).

7.3 Final considerations

The next steps of this work consist in testing motors of higher power rates in the same frame size using the configuration III previously presented.

When increasing the power in these frames, the bearing temperature can be a limiting factor for the motor integrity. Therefore, it is also intended to design motors with water cooled bearings.

Note that in this design step the fouling and corrosion were not considered, however, due to its meaning should be contemplated before product execution.

8. Conclusions

With the support of the CFD technique it was possible to foresee the water flow behavior inside the frame channel without visualizing directly the water flow. Through the use of the CFD it was also possible to double the water speed inside the frame channel, thus reducing the pressure drop in 18 % when compared with the initial design.

Another advantages provided by CFD were the reduction obtained both in the development time and in the number of prototypes. These advantages have been proven in this work: in spite of the 22 frame configurations that have been initially proposed, after the simulation results were obtained only the three best prototypes have been actually manufactured.

The use of thermography techniques in the field of electric motors helps to improve not only the development of products, but also the quality control and the maintenance of operating machines in the field. In this case, specifically, it was possible to visualize the guides configuration and the water flow inside the frame channel preventing the need to destroy it.

9. Acknowledgement

The authors would like to thank all Weg colleagues that directly or indirectly participated in this project.

10. References

Fox, R. W. & Mcdonald, A. T., *Introdução à mecânica dos Fluidos*, fifth edition, 2001.

Incropera, F. P. & Witt, D. P., *Introduction to Heat transfer*, second edition by John Wiley & Sons, 1985.

John H. Lienhard IV & V, *A Heat Transfer Textbook*, third edition, published by Phlogiston Press, 2004.

Jones, W. P. & Launder, B. E. (1972), *The Prediction of Laminarization with a Two-Equation Model of Turbulence*, International Journal of Heat and Mass Transfer, vol. 15, 1972, pp. 301-314.

Kreith, F., *Princípio da transmissão de Calor*, translation of the 3rd american edition, 1977.

Kreith, F., *The CRC Handbook of Thermal Engineering*, CRC Press LLC, 2000.

Launder, B. E. & Sharma, B. I. (1974), *Application of the Energy Dissipation Model of Turbulence to the Calculation of Flow Near a Spinning Disc*, Letters in Heat and Mass Transfer, vol. 1, no. 2, pp. 131-138.

Launder, B. E. & Spalding, D. D., *Mathematical Models of Turbulence*, Academic Press, London, 1972.

Menter, F. R. (1993), *Zonal Two Equation k-ω Turbulence Models for Aerodynamic Flows*, AIAA Paper 93-2906.

Menter, F. R. (1994), *Two-Equation Eddy-Viscosity Turbulence Models for Engineering Applications*, AIAA Journal, vol. 32, no 8. pp. 1598-1605.

Nailen, R. L., *Understanding the TEWAC Motor*, IEEE Transactions on Industry Applications, vol. IA-11, nº. 4, July / august, 1975.

Roberge, P. R., *Handbook of Corrosion Engineering*, McGraw-Hill, 2000.

Wilcox, D.C., *Multiscale model for turbulent flows*, In AIAA 24th Aerospace Sciences Meeting. American Institute of Aeronautics and Astronautics, 1986.

8

Aerodynamic Design of the Vertical Takeoff Hopper Concept of Future Launchers Preparatory Programme

Giuseppe Pezzella
Italian Aerospace Research Centre - CIRA, via Maiorise, Capua Italy

1. Introduction

In the frame of the Future Launchers Preparatory Program, carried out by the European Space Agency, the vertical take-off (VTO) Hopper - reusable launch vehicle concept is investigated (Guédron, 2003) (Kauffmann, 2006).

The VTO Hopper is a reusable, winged sub-orbital single stage vehicle, featuring a circular cross-section, designed for vertical take-off. The launch vehicle is composed of a reusable booster (the Hopper) and an expendable upper stage, mounted on top of it (Pezzella et al., 2009).

After the staging at suborbital altitude (> 130 km), the reusable booster will follow a ballistic arc trajectory, will re-enter the Earth's atmosphere, and will then perform a downrange landing (Pezzella et al., 2009).

The upper stage will deliver a payload up to 8 Mg in geostationary transfer orbit.

In this chapter the current aerodynamic analysis related to the launcher design is described. The goal was to define the aerodynamic data-base of the vehicle in order to provide the necessary input for the Flight Mechanics analysis.

Different design approaches were adopted in this work. In fact, aerodynamic analysis has been performed starting from engineering based methods, in order to rapidly accomplish the preliminary aerodynamic database to generate a number of possible re-entry trajectories, able to fulfill the program requirements. To do these analyses, a three dimensional (3-D) panel methods code, based on the simplified Newtonian approach and local inclination methods typical of hypersonics, was employed (Bertin, 1994).

Increasing the order of complexity, a number of detailed computational fluid dynamics (CFD) analyses have been carried out to more deeply characterize the hypersonic re-entry environment of the vehicle.

To this end, 3-D Euler and Navier-Stokes numerical flowfield computations were performed at different Mach numbers and angles of attack, at the most critical flight conditions of the vehicle descent trajectory.

In the chapter, qualitative summaries of the results are given for each aerodynamic force and moment coefficient as function of Mach number, Reynolds number and angle of attack.

2. The VTO-Hopper concept

The VTO-Hopper is one of the possible future launch systems which are investigated within the Future Launchers Preparatory Program (FLPP) (Tomatis et al., 2006). The vehicle architecture is shown in Fig. 1 (Pezzella et al., 2010).

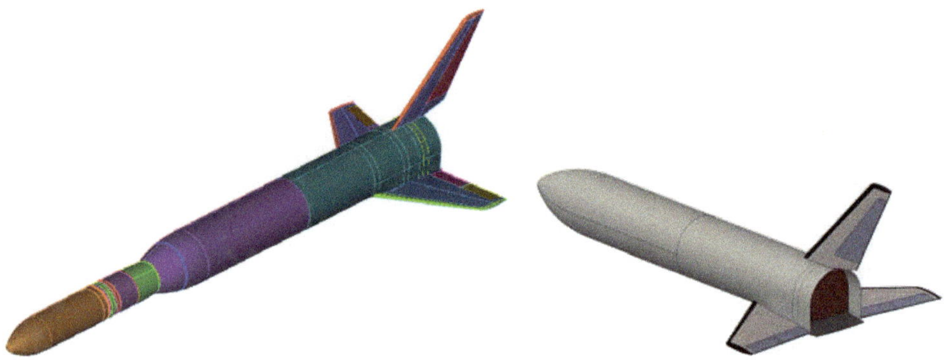

Fig. 1. The VTO-Hopper concept in ascent (left) and descent Configuration (right)

The vehicle concept is a two stage space transportation system comprising a fully reusable first stage, which starts vertically and performs a parabolic suborbital trajectory, and a dispensable second stage, carrying the payload.
The first stage, designed as a winged reentry body, will return to Earth to a point downrange of the launch site and land horizontally.
The current shape features a circular cross section of the body fuselage with wings in low position and a vertical fin. The circular shape has been adapted in order to introduce body flap.

2.1 The mission scenario

The design trajectory of the VTO-Hopper on which the aerodynamic design activities were performed is depicted in Fig. 2.
It shows, in the upper side, the altitude versus time spent to reentry starting at entry interface (e.g. 120 Km), while in lower part the Mach and angle of attack (AoA) profiles versus time to reentry are reported.

3. Description of design approach and used tools

A summary review of the aerodynamic characteristics of the VTO-Hopper concept is performed. These evaluations were aimed only to carry out a preliminary aerodynamic database (AEDB) for such configuration, compliant with a phase-A design level (Pezzella et al., 2008). The range between Mach 2 and Mach 20 was analyzed, with the goal to provide aerodynamic database for the flight mechanics analyses. The aerodynamic coefficients have been provided as a function of Mach number, and angle of attack (zero sideslip angle and no active control surface deflections) according to the space-based design approach (Pezzella et al., 2009).

This approach dictates the generation of a complete data set as function of a number of independent parameters (i.e. M_∞, Re_∞, α, β). Then, by using engineering based design, one can rapidly develop aerodynamic database as a function of the freestream parameters in a matter of hour (Bertin, 2004).

In the present analysis only continuum regime (supersonic and hypersonic speed ranges) with the air modeled as perfect gas has been studied.

It is worth to underline, however, that at high altitudes the rarefaction and real gas effects should be taken into account when the vehicle is flying at high Mach number, being the AEDB strongly affected by both these aspects (Bertin, 2004). Therefore, for the prosecution of vehicle design (i.e. phase B and C) more reliable design methodologies are mandatory.

In the following paragraphs the tools used for the analysis are described.

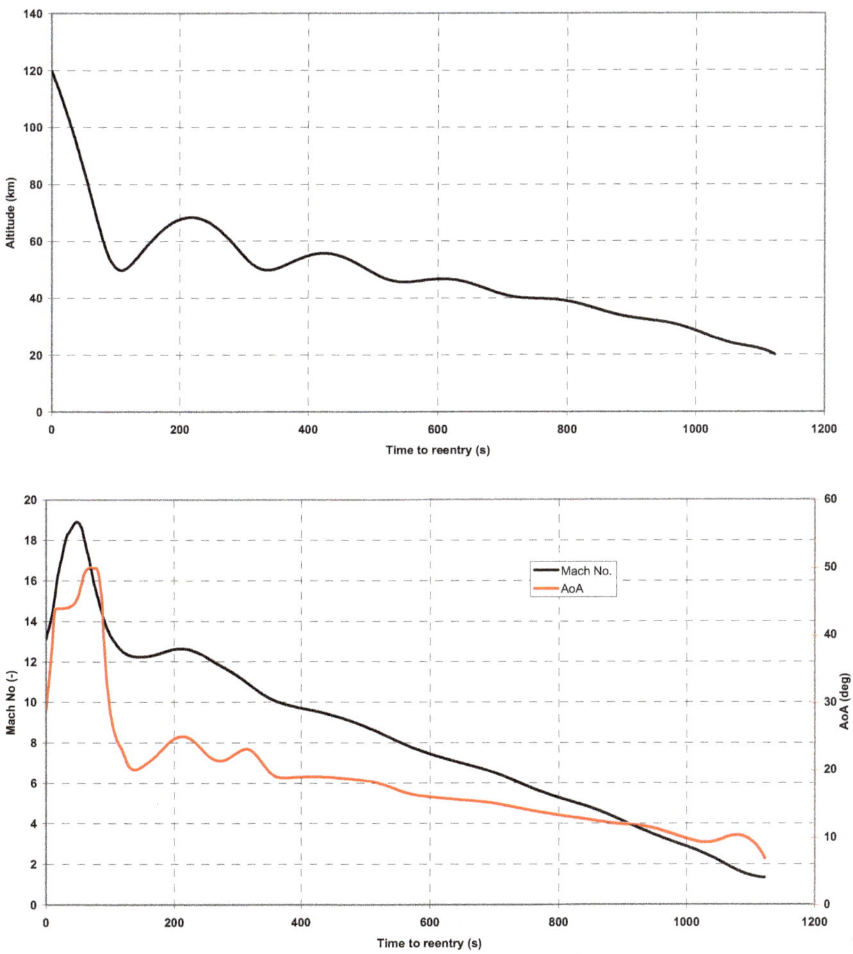

Fig. 2. The VTO-Hopper design trajectory in Altitude-time map (up) and Mach no/AoA vs time to reentry (down)

3.1 Aerodynamic analysis tools

The VTO-Hopper has a number of extreme loading flight conditions for which analyses are required. It must return from orbit, fly trimmed throughout hypersonic and supersonic regimes until landing is gained, and withstand severe aeroheating. An accurate aerodynamic analysis of all these flight conditions is very complex and time consuming, and is not compatible with a Phase A design study, in which fast predicting methods are mandatory (Pezzella, 2011). Therefore, the evaluations of the vehicle AEDB and of its reentry aerothermal environment were mainly performed by means of engineering tools, while a limited number of more advanced CFD computations were performed in order to verify the attained accuracy and to focus on some critical design aspects not predictable with simplified tools. Engineering based aerodynamic and aerothermodynamic analyses were extensively performed by using a 3D Panel Methods code developed by CIRA (SIM-Surface Impact Method) in the frame of its research activities on preliminary design of reentry vehicles (Pezzella, 2011). This tool at high supersonic and hypersonic speeds is able to accomplish the aerodynamic and aerothermodynamic analyses of a complex reentry vehicle configuration by using simplified approaches as local surface inclination methods and approximate boundary-layer methods, respectively. Surface Impact Methods (SIM), typical of hypersonics, are: Newtonian, Modified Newtonian, Tangent cone and Tangent Wedge theories (Bertin, 2004).

In Fig. 3 is shown a typical mesh surface of VTO-Hopper that has been used for the engineering level computations (Pezzella, 2011).

On the other hand the numerical code used to carry out the CFD analyses of the VTO vehicle concept is the CIRA code H3NS (Pezzella et al., 2009). It solves the flowfield governing equations, including chemical and vibrational non-equilibrium, with a finite volume approach; a flux difference splitting upwind scheme is used for the convective terms, with a 2nd order ENO-like reconstruction of cell interface values. The viscous fluxes are calculated by central differencing, i.e. computing the gradients of flow variables at cell interfaces by means of Gauss theorem. Time integration is performed by employing an Euler Forward scheme coupled with a point implicit treatment of the species and vibration energies source terms. Also a parallel version of the code is currently available. Several boundary conditions are available for the viscous computations, including different catalycity models and the possibility to assign at the wall a fixed temperature or a radiative equilibrium condition (Pezzella et al., 2009).

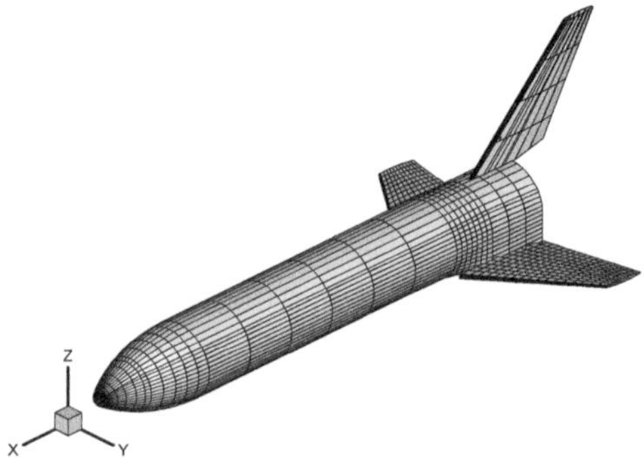

Fig. 3. Example of surface mesh used for engineering analysis

CFD computations have been carried out on a multiblock structured grid similar to that shown in Fig. 4 that is generated with the commercial tool ICEM-CFD. The grid used for Euler calculations has consisted of 62 blocks for an overall number of 829000 cells (half body).

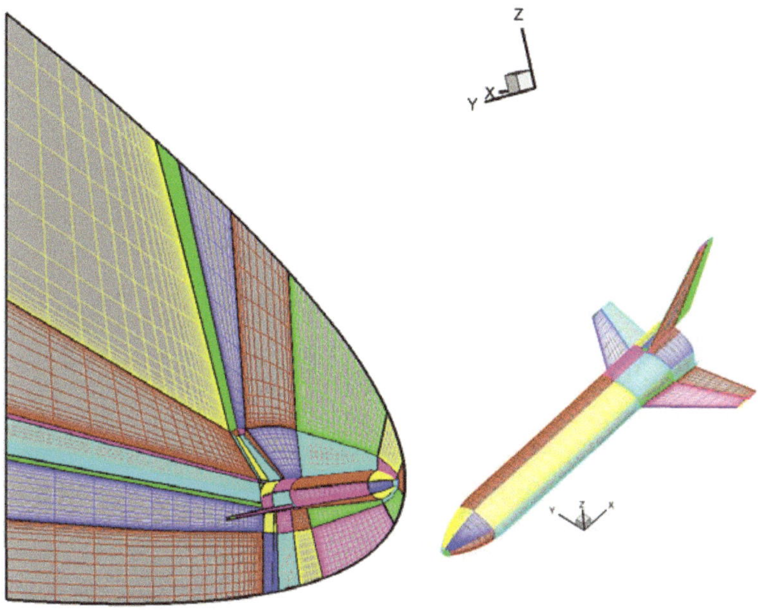

Fig. 4. Example of multiblock CFD domain. Mesh on symmetry plane and vehicle surface (left). Mesh on vehicle surface (right)

The grid is tailored for the freestream conditions of the trajectory check points, that are summarized in Table 1.
The distribution of surface grid points was dictated by the level of resolution desired in various areas of vehicle such as stagnation region and base fillet, according to the computational scopes. Grid refinement in strong gradient regions of flowfield was made through a solution adaptive approach.

4. Aerodynamic analysis and features of the VTO-Hopper vehicle in ascent and descent configuration

The VTO Hopper launcher consists of a rather conventional slender missile-like vehicle with a small delta planform wing (37.2° leading edge sweep) at very rear position and a central vertical stabilizer, as basic shape. The concept shows a circular cross section with a loft fillet on the belly side to accommodate the wing (blended wing body interface). The Wing geometry data are: root/tip chord: 11.70 m / 4.914 m; half span b/2: 11.63 m; wing leading / trailing edge angle: 37.23° / 10°; apex longitudinal position (from the base of the fuselage): 11.70 m; angle of incidence (setting) of the wing: 3°; dihedral: 3°.
With these data the wing surface is equal to 193.23 m^2.
The VTO-Hopper is characterized by a clean aerodynamic configuration, (i.e. controls in neutral position), that is depicted in Fig. 5 for ascent and descent flight.

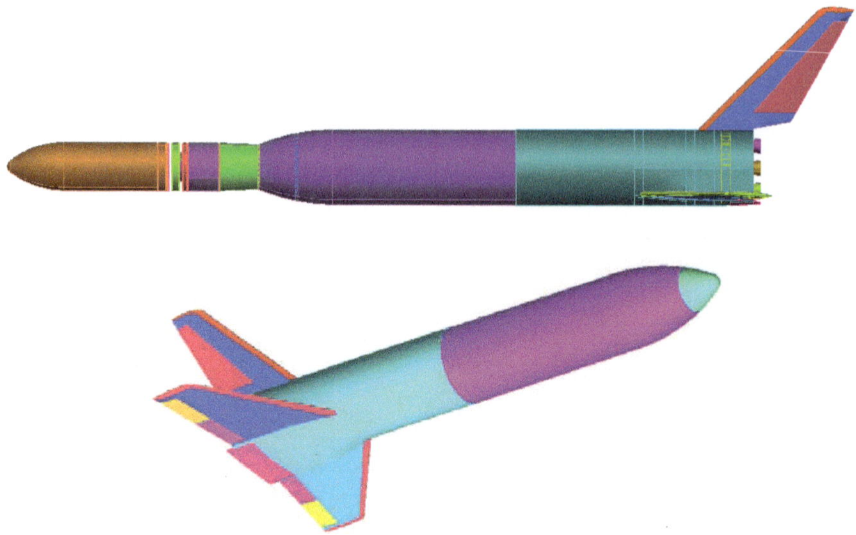

Fig. 5. Vehicle clean aerodynamic configuration for ascent (up) and descent flight (down)

The aerodynamic controls comprise rudders on the vertical tail, elevons and ailerons on the wings, and a body flap underneath the main engines in order to provide maneuverability and longitudinal stability during atmospheric descent. At hypersonic speeds a surface behind the vehicle centre of gravity (CoG) balances the nose up pitching moment typical of such kind of vehicle configuration at hypersonic speeds (Bertin, 2004). Normalizing the vehicle overall dimensions by fuselage length (L_{ref}=58.8 m), the VTO-Hopper is characterized by the following normalized reference data:

- B′ (wing span)=0.54;
- S′ (reference surface)=0.056;
- X'_{MRP}=0.69; Y'_{MRP}=0; Z'_{MRP}=0;

In the following, the aerodynamic analysis is shown in term of lift (C_L), drag (C_D) and pitching moment (C_{My}) coefficients which are calculated according to Eq. 1 and Eq. 2, respectively.

$$C_i = \frac{F_i}{\frac{1}{2}\rho_\infty v_\infty^2 S_{ref}} \qquad i = L, D \tag{1}$$

$$C_{Mj} = \frac{M_j}{\frac{1}{2}\rho_\infty v_\infty^2 L_{ref} S_{ref}} \qquad j = Y \tag{2}$$

4.1 The VTO-Hopper aerodynamic for ascent flight

The aerodynamic performance of VTO Hopper in ascent flight has been carried out starting from the results summarized in (Guédron et al., 2003) for the French concept RFS. The geometry of RFS launcher configuration with booster, expendable upper stage and fairing is shown in Fig. 6.

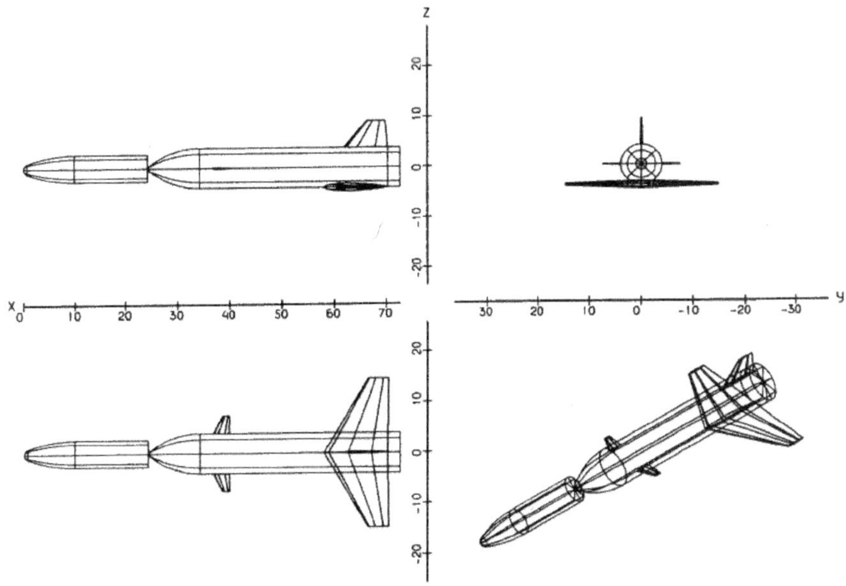

Fig. 6. RFS Configuration. Dimensions in [m] (Guédron et al., 2003)

As one can see this vehicle concept is characterized by an aerodynamic configuration very close to the VTO one, except for canard flight control surfaces mounted in front of launcher. Therefore, the preliminary ascent AEDB of the VTO-Hopper has been built by properly scaling the RFS's AEDB on the base of VTO aerodynamic configuration features.
For the RFS launcher configuration the aerodynamic coefficients are summarized in Fig. 7 (Guédron et al., 2003).

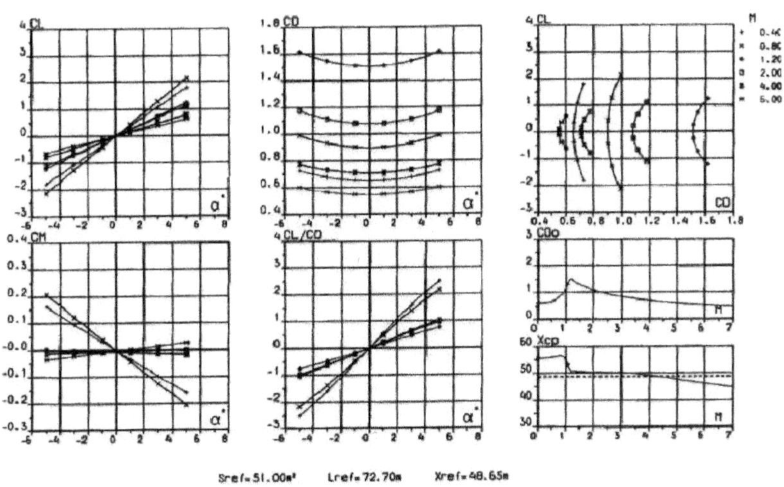

Fig. 7. Aerodynamic coefficients for ascent flight of RFS launcher (Guédron et al., 2003)

Starting from those data the preliminary AEDB of the VTO-Hopper is synthesized in the following Fig. 8 through Fig. 10.

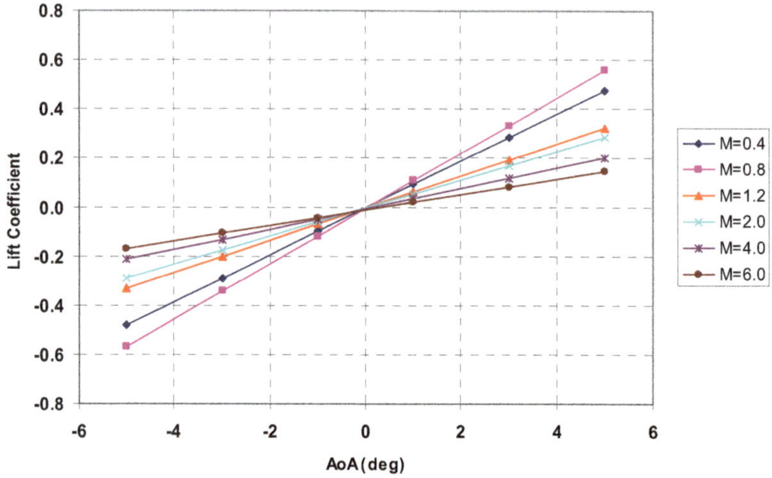

Fig. 8. Lift coefficients vs AoA for ascent flight

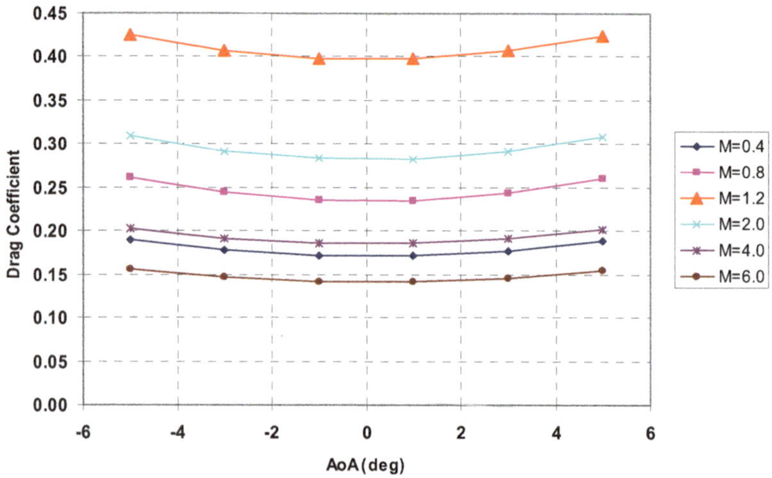

Fig. 9. Drag coefficients vs AoA for ascent flight

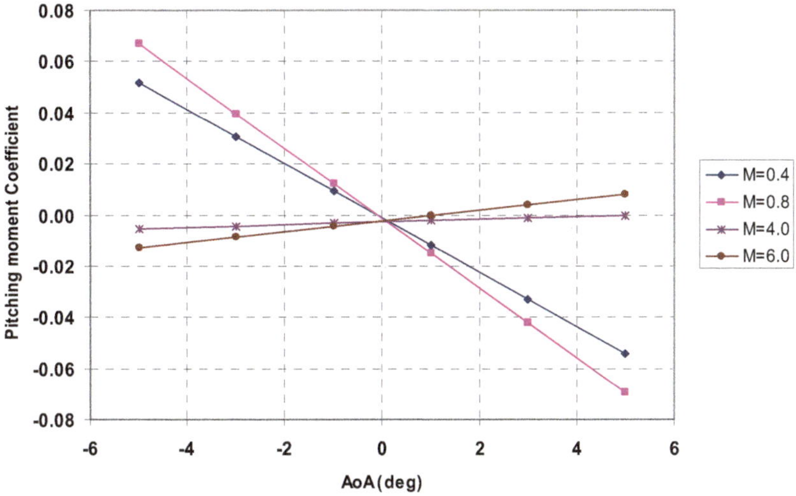

Fig. 10. Pitching moment coefficients vs AoA for ascent flight

4.2 The VTO-Hopper aerodynamic for descent flight in clean configuration

The re-entry scenario of the VTO-Hopper launcher is summarized in the following Fig. 11 where the flight profile refers to the altitude-velocity map. Moreover, the Mach and Reynolds numbers grid is also reported in order to characterize the aerodynamic flight scenario of the booster.

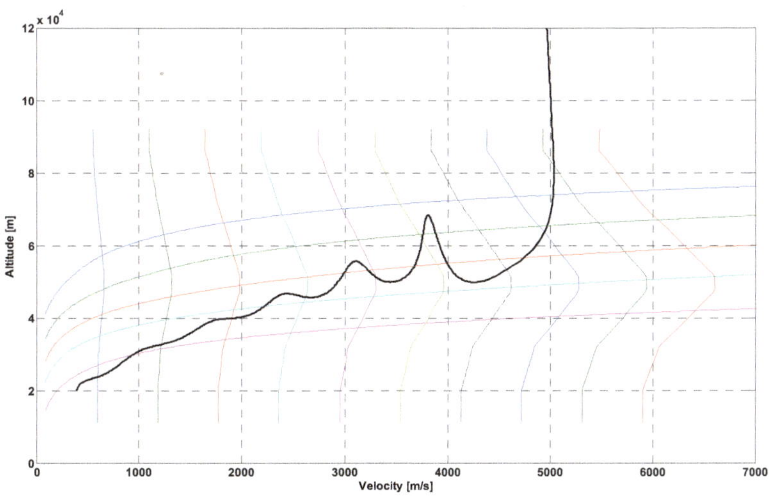

Fig. 11. The VTO-Hopper re-entry trajectory in the altitude-velocity map

Mach number ranges from 2 to 20 while five Reynolds numbers (i.e. [1, 3, 8, 20, 70]x10^6) with respect to L_{ref} are displayed. It must be noted that the ranges of Mach and Reynolds numbers were chosen in such a way to cover a wide part of the reentry flight, especially the most critical one from the aeroheating point of view (i.e. M_∞=13.4) (Pezzella et al., 2009).
For the VTO-Hopper booster, the following reference parameters have been chosen in order to make aerodynamic forces and moments non-dimensional coefficients (see Fig. 12):

- L_{ref} =58.8 [m] (Fuselage length - longitudinal reference length);
- S_{ref} =193.23 [m^2] (wing wetted area - black in Fig. 12 -)
- Pole coordinates are (0,0,0) [m] (e.g., vehicle nose);

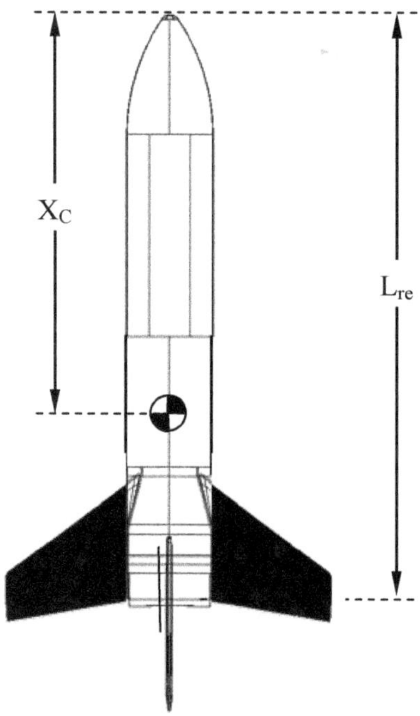

Fig. 12. The VTO-Hopper booster reference parameters

Based on reentry flight scenario summarized in Fig. 11 the aerodynamic data set was generated for the following ranges:

- 2 $< M < 20$ [2, 3, 5, 7, 10, 15, 20];
- 0° $< \alpha < 50°$ [0, 5, 10, 15, 20, 25, 30, 35, 40, 45, 50];
- $10^6 < Re < 70 \times 10^6$ [1, 3, 8, 20, 70] x 10^6;
- β =0°

Neither lateral directional analysis nor wing and body flap effects have been taken into account in this report.
As an example, results for M_∞=5 are summarized in Fig. 13, where the static pressure distribution over the wetted vehicle surface for two AoAs (i.e. 10 and 30 deg) is reported.

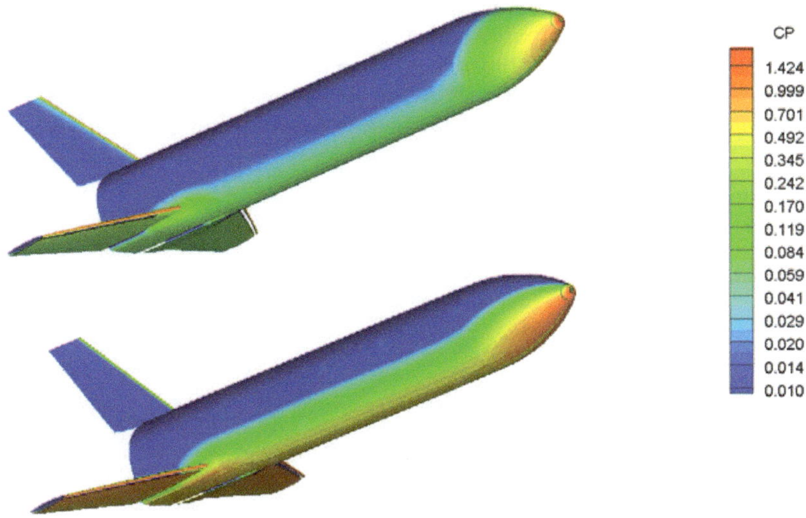

Fig. 13. Pressure coefficient contours on the vehicle surface at M_{∞}=5, AoA=10 deg (up) and 30 deg (down)

Ranging AoA from 10 to 30 deg, as highlighted by pressure coefficient contours, the Hopper configuration must exhibit very different aerodynamic performance. For example, Fig. 14 shows that starting from AoA=0 deg the booster aerodynamic efficiency straightly increases reaching the peak of about 1.6 at AoA=15 deg, and then it decreases up to about 0.7 at AoA=30 deg.

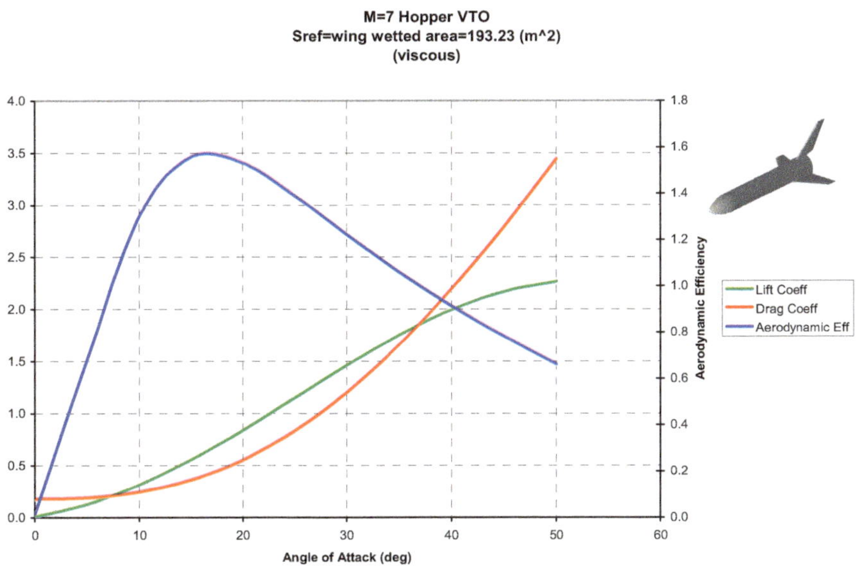

Fig. 14. Booster aerodynamic characteristics at M_{∞}=7

By conclusion, looking at Fig. 15 one can appreciate an overview of booster's aerodynamic efficiency for the whole Mach number and AoA regimes.

Note that lift-to-drag ratio (L/D) is the most important feature of vehicle aerodynamic performance, which has a direct impact on cross-range capability of a re-entry vehicle that reaches its nominal landing site (DLRS) at the end of space mission by unpowered flight.

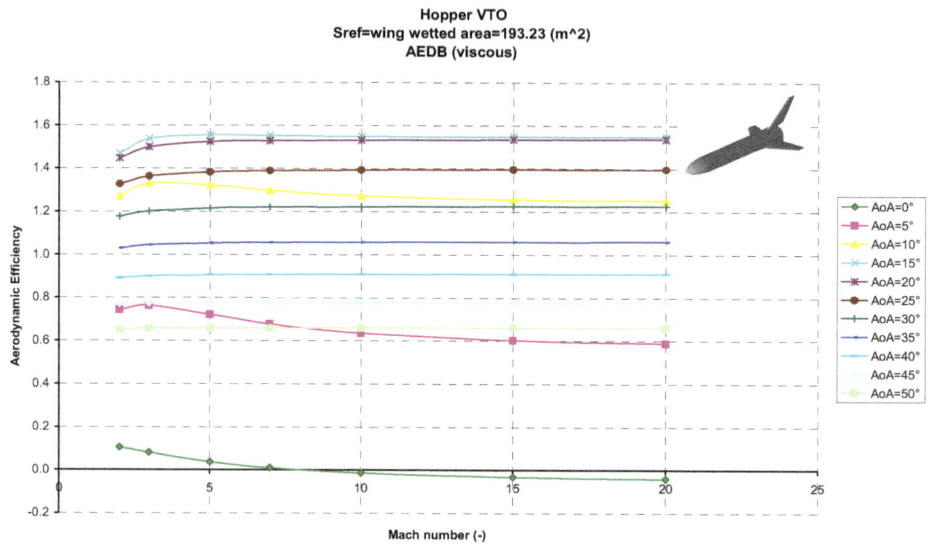

Fig. 15. Booster aerodynamic efficiency vs Mach number, for different AoA

As clearly shown, the Oswatich principle (independence of aerodynamic coefficients to M_∞) is satisfied starting already from Mach=7 (Bertin, 2004).

4.2.1 Preliminary evaluation of wing planform shape on the booster aerodynamic performance

Generally speaking the wing shape may influence markedly the aerodynamic and aerothermodynamic performance of a launcher (Viviani and Pezzella, 2009).

In hypersonic flow conditions, for example, the presence of a wing sweep angle (Λ) influences both the aerodynamic efficiency (E_{ff}) and wing leading edge aeroheating.

Fig. 16 shows qualitatively that the more wing sweep angle (Λ) the more the aerodynamic efficiency. This is true only up to Λ^* where a reversal occurs and E_{ff} begins to decrease. Typically Λ^* ranges between 65 deg and 75 deg.

Detailed trade-off analyses on wing planform shape are beyond the scopes of this chapter. However, the effect of five different wing planform shapes on whole launcher's aerodynamic efficiency has carried out in a very preliminary analysis.

A simplified approach has been used to build the wing. Each of them has been derived by scaling up the baseline geometry, in x and y directions, by different factors.
An overview of these wing planforms are shown in Fig. 17. The first wing, identified as step #1, has been obtained scaling up the baseline one in x-direction by 1.5, taking constant y. The second one (e.g., step #2) is characterized by a x scaling of a factor of two.

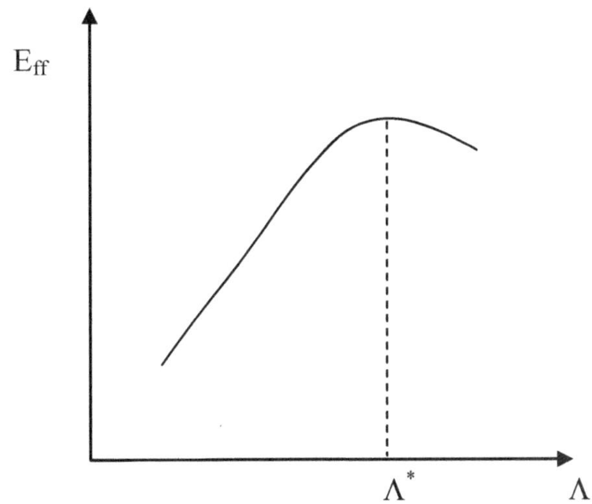

Fig. 16. Qualitatively behavior of Eff vs wing swept.

The wing number three (step #3) differs from the baseline one for a y scaling by a factor of 1.3. The step #4 identifies a wing obtained by x and y scaling up the baseline by a factor of 1.5 and 1.3, respectively.
Finally, the wing step #5 is obtained by x and y scaling the baseline wing by a factor of 2 and 1.3, respectively.
Fig. 18 shows the comparison between the baseline configuration on the left side of the booster and the modified wing (e.g., step #i) on the right side. Therefore, one can appreciate at same time the wing differences.
The effect of each wing shape is summarized in Fig. 19 where is shown the ΔE_{ff} (e.g., aerodynamic efficiency with respect the baseline at the same AoA) versus AoA at M_{∞}=7. Note that this Mach number was chosen as representative of whole hypersonic regime on the base of Oswatich principle. As one can see, a strong effect can be appreciated only around AoA=10 deg. Therefore, if the vehicle re-enters flying at an AoA higher than 20 deg no matter is wing planform shape on aerodynamic performance for a configuration as such that of VTO Hopper.

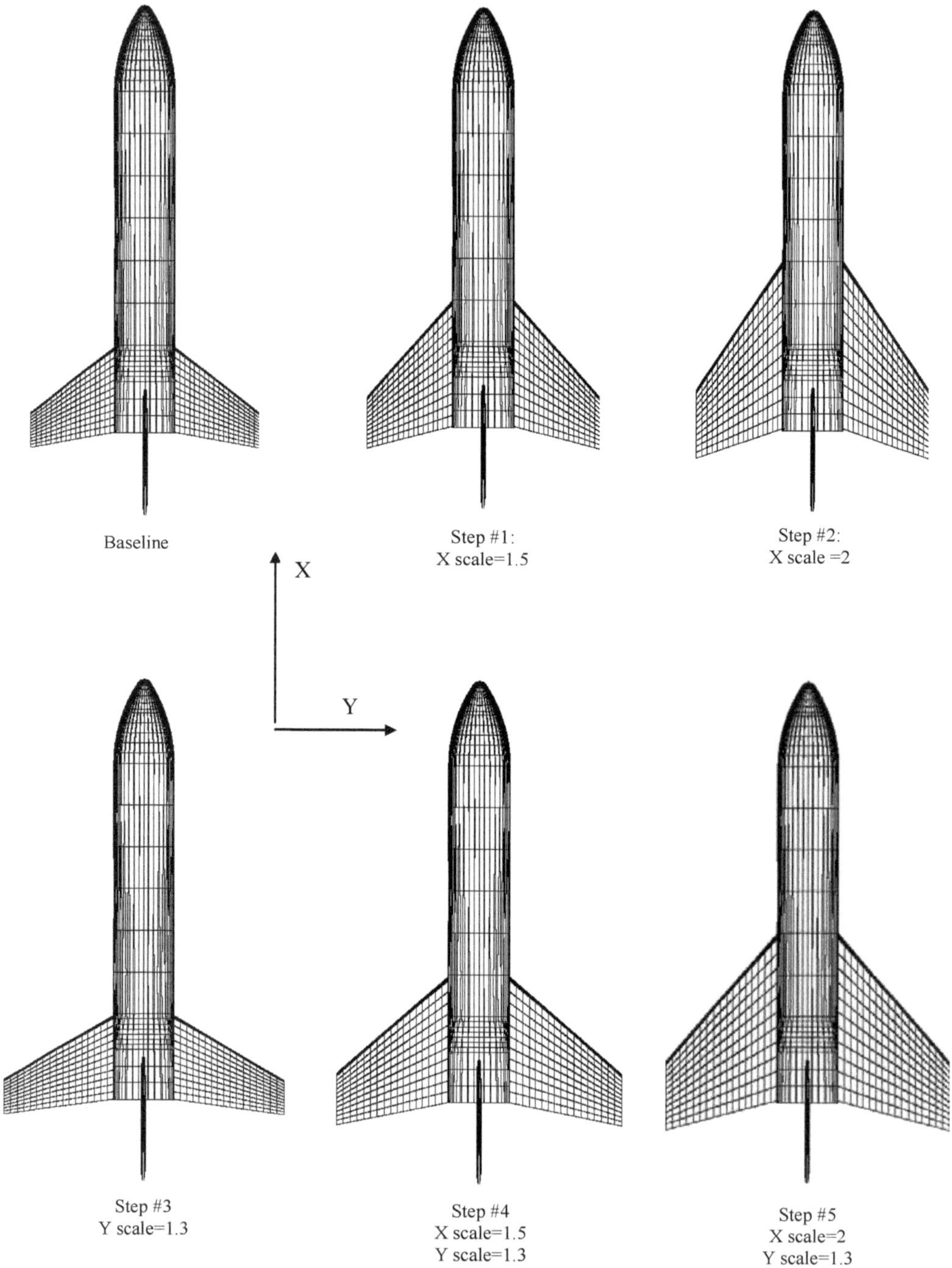

Fig. 17. The VTO-Hopper booster with different wing planform shapes

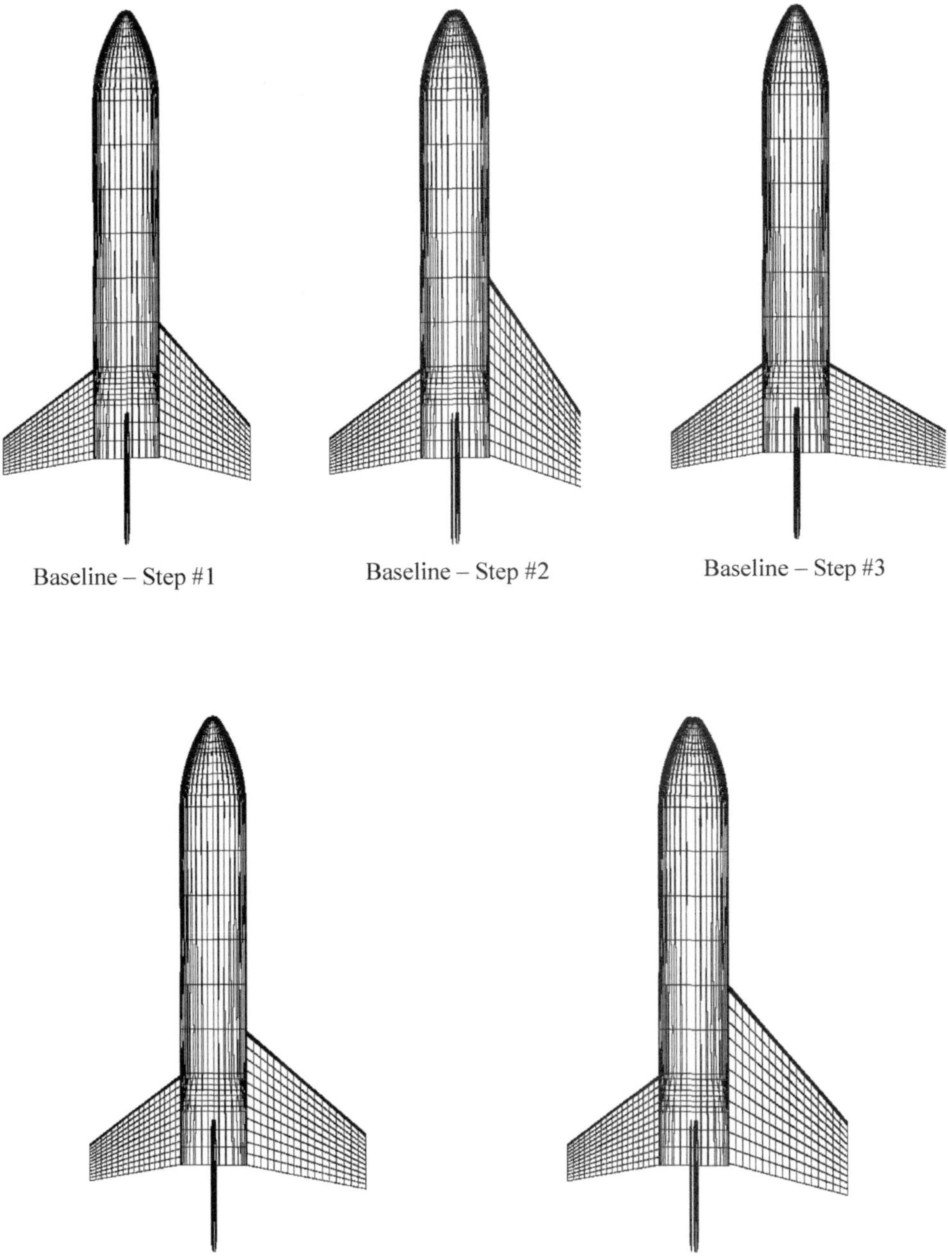

Fig. 18. The VTO-Hopper booster with base line wing (left side) and different wing planform shapes (right side)

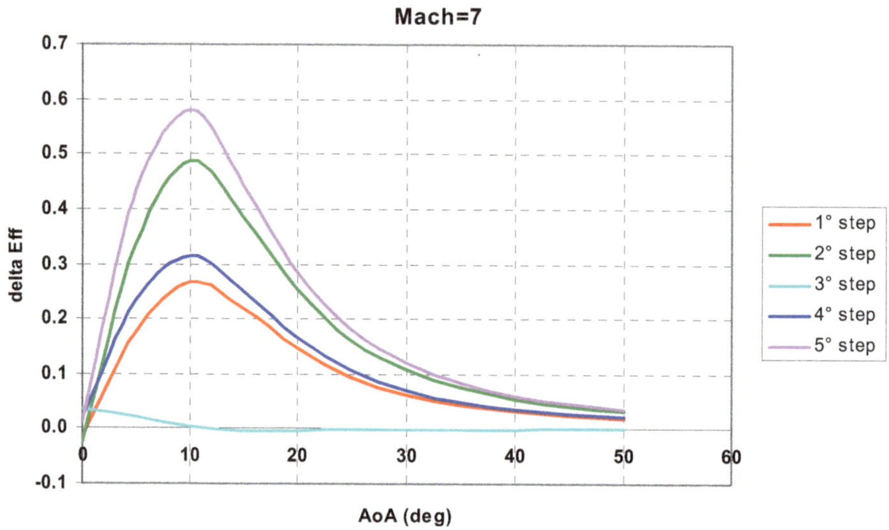

Fig. 19. ΔE_{ff} vs AoA for each different wing planform shape

5. Reliability of design activities

By conclusion several comparisons are provided between results of present analyses and data provided by more reliable CFD computations.
For instance, since aerodynamic analyses are based on empirical correlations and approximate theories, it is important that they are calibrated against the more accurate CFD results. Therefore, on the base of the reentry scenario of VTO Hopper (see Fig. 11) a number of flight conditions have been chosen to carry out CFD computations. Those numerical computations will allow anchoring engineering analyses in order to verify the attained accuracy of these simplified analyses and to focus on some critical design aspects not predictable with engineering tools. Those control points are reported in Fig. 20 while the CFD test matrix is summarized in Table 1.
As one can see, each check point lies within the flight scenario foreseen for a typical mission profile of the booster. Note that each red box identifies a CFD run (e.g., check point), therefore the results of twelve CFD Euler computations were considered within the vehicle design.
In the following figures some of the main results obtained for booster configuration are shown.

For example, Fig. 21 shows the normalized temperature contours field both on the vehicle symmetry plane and booster outer surface when the vehicle is flying at M_∞=5 and AoA=10 deg.

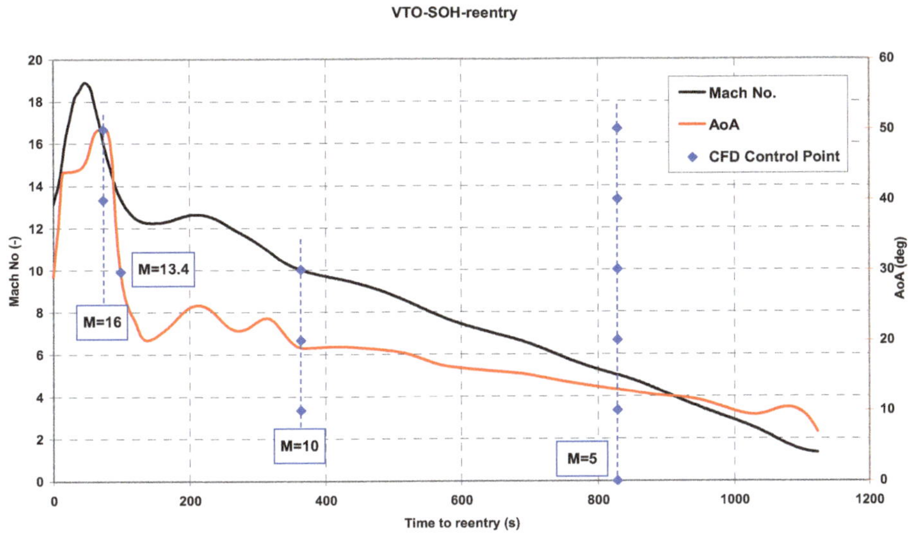

Fig. 20. The VTO-Hopper booster re-entry scenario with control points for CFD analyses

AoA [deg]	Mach [-]			
	5	10	13.4	16
0				
10				
20				
30				
40				
50				

Table 1. The CFD Test Matrix

Looking at contour field on the vehicle symmetry plane one can appreciate the strong bow shock that occurs ahead of vehicle during descent. This shock surface envelopes the vehicle and may impinge on wing leading edge (e.g., shock-shock interaction - SSI) thus increasing locally the heat flux (overheating) that the vehicle thermal shield has to withstand, as shown in Fig. 22 where the Mach number contour field in the cross plane where bow shock impinges over the wing for M_∞=10, AoA=20 deg is displayed. Therefore, analyses of SSI with overloads (pressure and heat flux) at impingement are mandatory for a reliable vehicle design (Viviani and Pezzella, 2009).

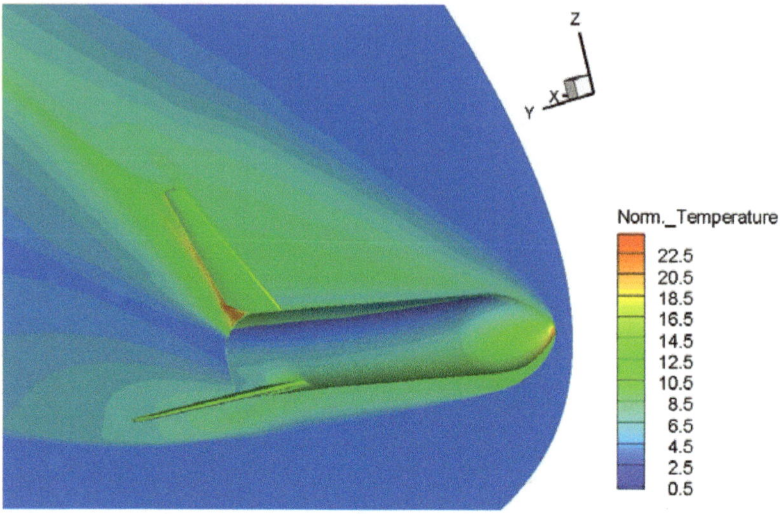

Fig. 21. Normalized temperature contours on symmetry plane and vehicle surface at M_∞=5, AoA=10 deg. Euler computation.

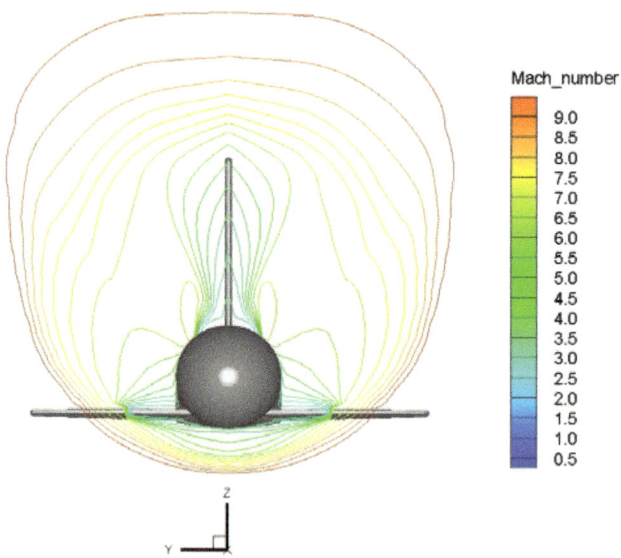

Fig. 22. Mach number contours in the cross plane where bow shock impinges over the wing (SSI) for M_∞=10 and AoA=20 deg. Euler CFD computation

Further flowfield features can be recognized in Fig. 23, where the normalized pressure contour field is shown for M_∞=5 and AoA=20 deg.

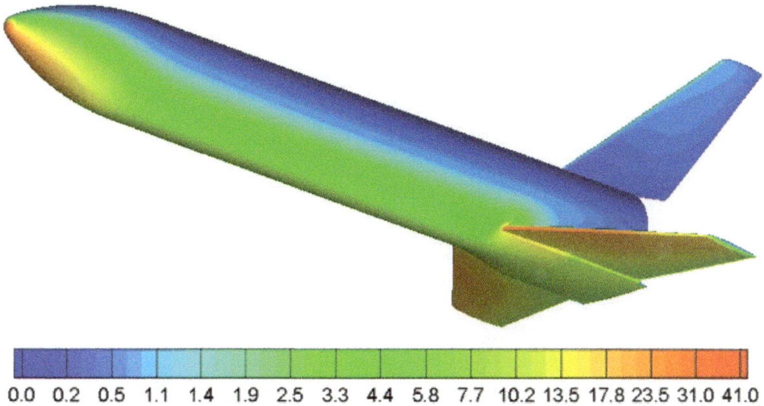

Fig. 23. Normalized pressure contours on vehicle surface at M_∞=5, AoA=20 deg. Euler computation

For what concerns the reliability of aerodynamic coefficient from Fig. 24 to Fig. 29 it can be clearly noted the good agreement between the engineering-based aerodynamics and the CFD data.

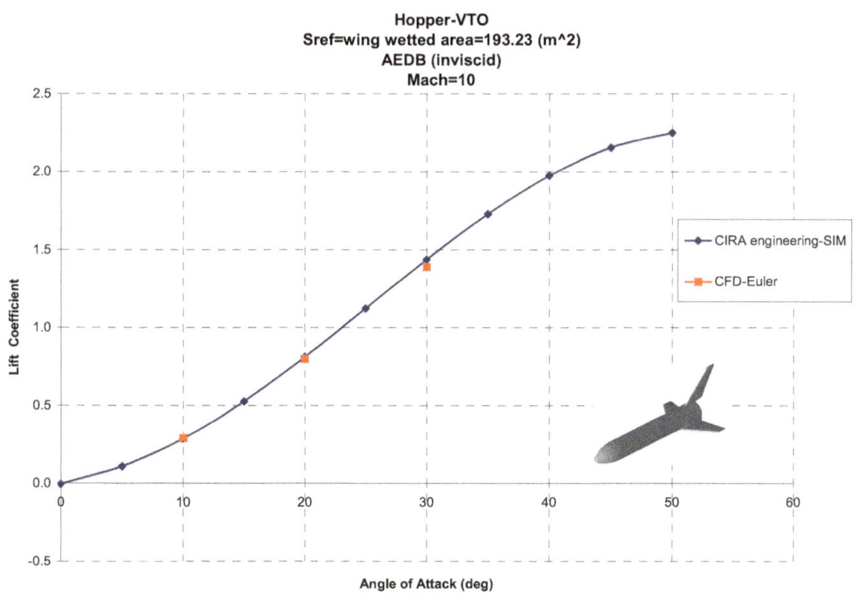

Fig. 24. Lift coefficient vs AoA. Comparison between SIM and CFD Euler at M_∞=10.

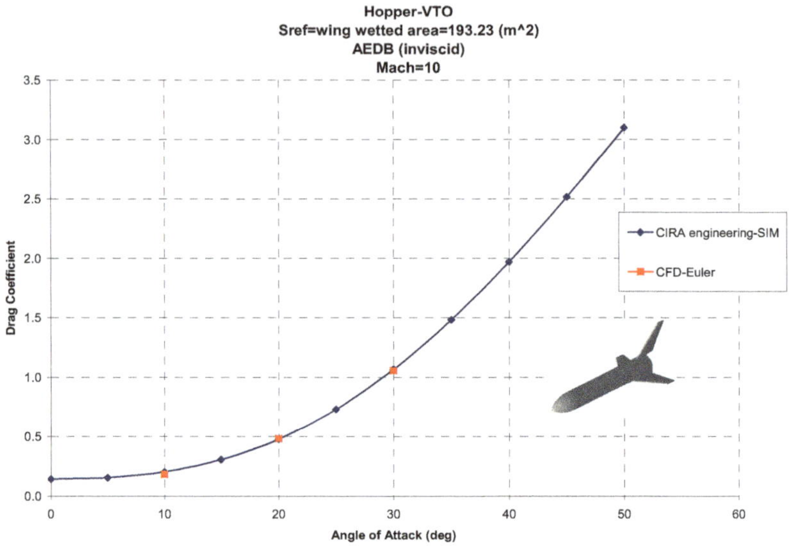

Fig. 25. Drag coefficient vs AoA. Comparison between SIM and CFD Euler at M_∞=10.

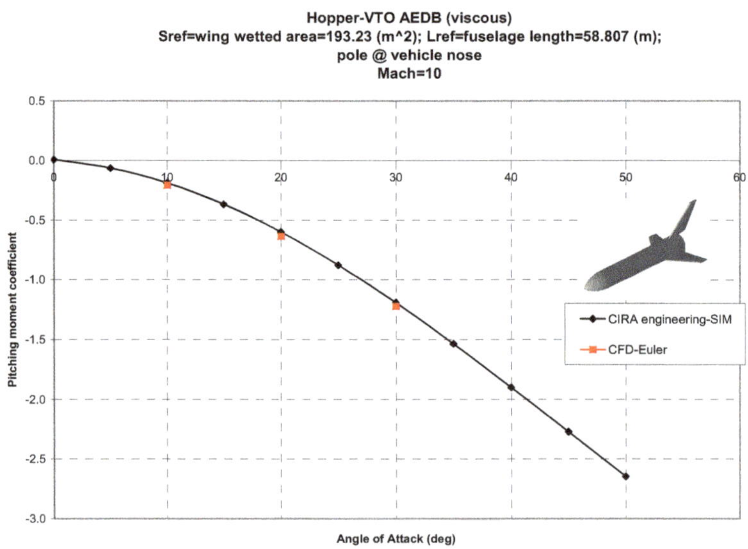

Fig. 26. Pitching moment coefficient vs AoA. Comparison between SIM and CFD Euler at M_∞=10.

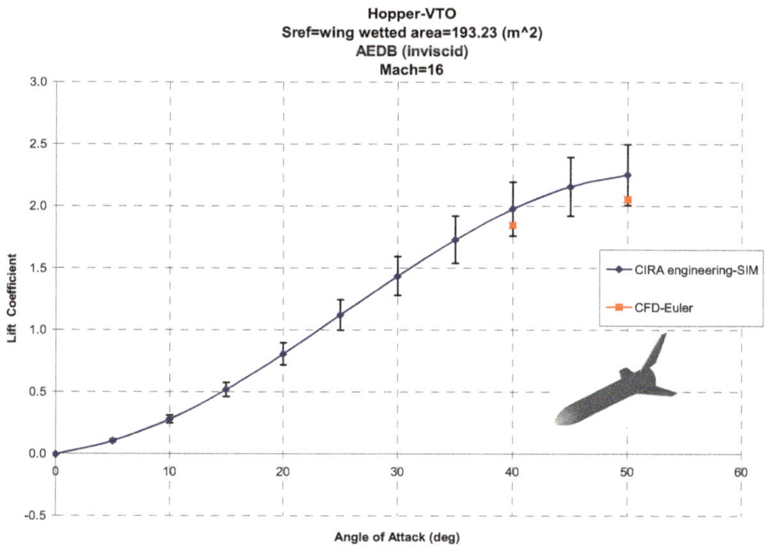

Fig. 27. Lift coefficient vs AoA. Comparison between SIM and CFD Euler at M_∞=16 (error bar for 10%).

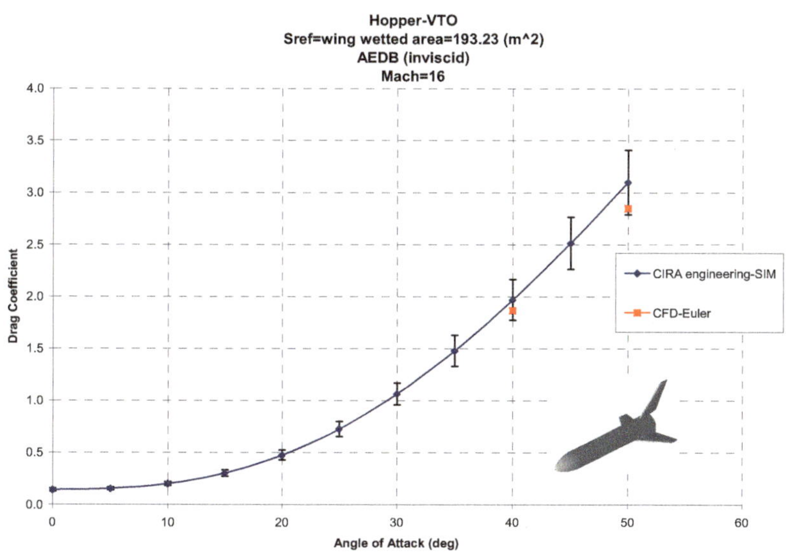

Fig. 28. Drag coefficient vs AoA. Comparison between SIM and CFD Euler at M_∞=16 (error bar for 10%).

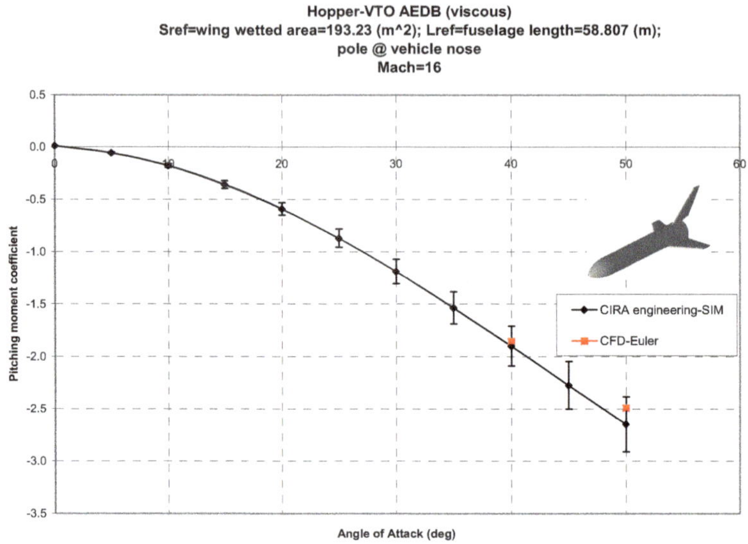

Fig. 29. Pitching moment coefficient vs AoA. Comparison between SIM and CFD Euler at M_∞=16 (error bar for 10%).

As clearly evident, results of engineering based approach and CFD computations compare very well at each Mach number and AoA considered. The maximum difference was found at Mach 16 and AoA=50 deg even if it is comprised within an error bar of 10%.

6. Concluding remarks

Preliminary hypersonic aerodynamics dataset for the VTO-Hopper, under development as launcher system concept within the *Future Launchers Preparatory Program* of *European Space Agency*, have been carried out in the present chapter.
The Hopper configuration is a missile-like re-entry body designed for performing a suborbital, parabolic trajectory with a horizontal landing.
A possible re-entry mission profile has been considered as design trajectory with respect to the aerodynamics of the vehicle concept. Hopper aerodynamics review analyses refer to Mach number ranging from 2 to 20 and AoA from 0 to 50 deg, which are conditions covering the whole reentry scenario of the vehicle concept (**space-based design approach**).
It must be underlined, however, that present analysis has not been obtained by means of accurate and more reliable CFD computations. Therefore, a proper margin should be adopted in recognizing vehicle aerodynamics.
Overall available results highlight that the difference between eulerian CFD and engineering based design is smaller than 10%, thus confirming that surface impact method (SIM) represent a reasonable preliminary design approach in order to accomplish the aerodynamic characterization of a re-entry vehicle across the hypersonic regime.
Note, in conclusion, that in the next phases of VTO-Hopper design, further analyses have to be performed on specific topics in order to increase the reliability of the aerodynamic

database and to reduce the design margins. In particular, the attention has to be focused on real gas and rarefaction effects, as well as shock-shock interactions and laminar-to-turbulent transition. The real gas effects can be important because, during atmospheric re-entry, dissociation and ionization processes take place in the shock layer, which can have an influence on the aerodynamic coefficients. Real gas effects are expected to affect stability and control derivatives of vehicle, in particular its pitching moment, as highlighted by first U.S Space Shuttle reentry (STS-1) where an unexpected higher nose-up pitching moment required a body-flap deflection twice the one predicted by the pre-flight analyses to trim the Orbiter. Moreover, real gas effects cause a shock that lies closer to the vehicle with respect to the position that would characterize a perfect gas case. These effects obviously occur only at higher Mach numbers.

Further, regarding to the aerodynamic coefficients, it is well known that at very high altitude, when the Reynolds number decreases and rarefaction effects are present, there is a strong increase of the drag coefficient and a consequent reduction of the maximum efficiency. Another aspect that has to be considered is the interaction of the bow shock with the shock generated by the wing. It is an important phenomenon that has to be investigated because it can have a significant impact on the local pressure and heat flux distribution.

7. References

Bertin, John J., *Hypersonic Aerothermodynamics*, AIAA Education Series, 2004.

Kauffmann, J. *Future European Launch Systems in the FLPP Overview and Objectives*. 57th International Astronautical Congress, 2-6 Oct. 2006, Valencia, Spain. IAC-06-D2.4.05.

Guédron, S., Prel, Y., Bonnal, C., Rojo, I., *RLV Concepts and Experimental Vehicle System Studies: Current Statuse*, 54th International Astronautical Congress, 29 Sept.-3 Oct. 2003, Bremen, Germany, IAC-03-V.6.05.

Pezzella, G., Marini, M., Roncioni, P., Kauffmann, J., Tomatis, C., *Preliminary Design of Vertical Takeoff Hopper Concept of Future Launchers Preparatory Program*, Journal of Spacecraft and Rockets 2009. ISSN 0022-4650 vol.46 no.4 (788-799) doi: 10.2514/1.39193

Pezzella, G., Marini, M., De Matteis, P., Kauffmann, J., Dapra, A. and Tomatis, C., *Aerothermodynamic Analyses of Four Reusable Launchers in the Framework of ESA Future Launchers Preparatory Programme*. Aerotecnica Missili & Spazio (The Journal of Aerospace Science, Technologies & Systems). Vol. 89 No. 1 January 2010.

Pezzella, G., *Aerodynamics and Aerothermodynamics Analysis of Three Winged Re-Entry Vehicle Concepts*. International Journal of Engineering and Allied Sciences (IJEAS), Volume (1): Issue (1). 2011. ISSN 2248-8568.

Pezzella, G., *Aerodynamic and Aerothermodynamic Trade-off Analysis of a Small Hypersonic Flying Test Bed*. Acta Astronautica. doi:10.1016/j.actaastro.2011.03.004. Volume (69): Issue (3-4). 2011. ISSN (0094-5765): pag.: 209-222.

Pezzella, G., *Preliminary Aerodynamic and Aerothermodynamic Assessment of the VTO Hopper Booster*. ISRN Mechanical Engineering. Volume 2011, Article ID 215785, 15 pages. doi:10.5402/2011/215785.

Pezzella, G., Marini, M., Roncioni, P., Kauffmann, J., Tomatis, C., *Aerodynamic and Aerothermodynamic Evaluation of the VTO Hopper Concept in the Frame of ESA Future Launchers Preparatory Program*, 15th AIAA International Space Planes and

Hypersonic Systems and Technologies Conference. Apr. 28-1, 2008. Dayton, Ohio (USA), paper AIAA-2008-2639.

Pezzella, G., Kauffmann, J., Dapra, A., Tomatis, C., *An Italian Contribution to the Next Generation Launcher in the Framework of ESA Future Launchers Preparatory Programme*, XX AIDAA Congress, Milan, Italy, June 29-July 3, 2009.

Tomatis, C., Bouaziz, L., Franck, T., Kauffmann, J., *RLV Candidates for European Future Launchers Preparatory Programme*. 57th International Astronautical Congress, 2-6 Oct. 2006, Valencia, Spain. IAC-06-D2.2.07.

Viviani, A., Pezzella, G., *Heat Transfer Analysis for a Winged Reentry Flight Test Bed*, International Journal of Engineering (IJE), Volume (3): Issue (3). 2009. ISSN: 1985-2312.

9

Computational Fluid Dynamics (CFD) Modeling of Photochemical Reactors

Masroor Mohajerani, Mehrab Mehrvar and Farhad Ein-Mozaffari
Department of Chemical Engineering, Ryerson University, Toronto, Ontario Canada

1. Introduction

Advanced oxidation processes (AOPs) play an important role in the degradation or the production of a wide range of organic materials. Many organic compounds such as pharmaceuticals, dyes, herbicides, and pesticides have been subjected to degradation and remediation purposes in water and wastewater treatment systems using AOPs. Some of the organic compounds such as drugs, vitamins, or fragrances could be also produced by oxidation processes.

As the standard of living increases, many chemicals such as pharmaceuticals, pesticides, herbicides, and dyes are extensively consumed. Each of these products may cause health issues by their accumulation in aquatic environment. Pharmaceuticals such as antibiotics are partially metabolized and excreted by humans and animals. Improper disposal, dumping, and accidental discharge of drugs lead to the increase of the concentration of compounds such as analgesics, antibiotics, steroids, and hormones in aquatic environment, which cause environmental and health problems. Residual pesticides and herbicides originate from the direct pollutant in production plant, disposal of empty containers, equipment washing, and surface runoff. High levels of these compounds are toxic, mutagenic, carcinogenic, and tumorigenic. Some other wastes such as landfill leachate are subjected to advanced treatment methods. Old landfill leachates (>10 years) are nonbiodegradable in nature due to the existence of organic compounds with high molecular weights. Although the composition of landfill leachates varies widely with respect to the age of the landfill, type of wastes, and climate conditions, they can be categorized into four groups of dissolved organic matter, inorganic macro components, heavy materials, and xenobiotic organic substances. Another type of toxic chemicals which cannot be removed using conventional treatment methods is endocrine disrupting compounds (EDCs). EDCs, especially the steroidal hormones, are well recognized exogenous agents that interfere with the synthesis, action, and/or elimination of natural hormones in the body. Conventional processes are not effective in destruction of these types of organic compounds; therefore, powerful advanced treatment processes are required in order to mineralize them. There are several options for choosing an oxidation process: wet air oxidation, supercritical water oxidation, incineration, and AOPs.

AOPs have been promising in the treatment of contaminated soils and waters. The AOPs could be employed to fully or partially oxidize organic pollutants usually using a combination of different oxidants. In contrary to the conventional physical and chemical treatment processes, AOPs do not transfer pollutants from one phase to another, i.e., organic pollutants are completely destroyed. Most AOPs are able to generate hydroxyl radicals. These hydroxyl radicals are active and powerful, capable of reacting with almost all types of organics, including non-biodegradable and recalcitrant compounds. These oxidizing hydroxyl radicals are initiated by photolysis, photocatalysis, sonolysis, radiolysis, and other AOPs alone or in combination in the presence of some reagents such as hydrogen peroxide (H_2O_2), ozone, and homogeneous or heterogeneous catalysts. Hydroxyl radicals are extremely powerful, short-lived oxidizing agent, and non-selective in nature and could react with a wide range of organic chemicals with reaction constant of several orders of magnitude higher than the reaction with molecular ozone under the same conditions. AOPs are used to convert toxic, non-biodegradable, inhibitory, and recalcitrant pollutants into simpler and less harmful intermediates, so they could be treated subsequently in biological processes. Due to the high operating and capital costs of AOPs, the complete mineralization of organic compounds using AOPs is practically impossible; therefore they are mainly combined with other processes such as biological systems to be cost-effective. With sufficient contact time and proper operating conditions and design, the mineralization efficiency of the AOPs is maximized by optimization of the processes in such a way that the total costs of the processes are minimized while the removal rates of organic pollutants are maximized.

Since conducting experiments is time-consuming and expensive or sometimes impossible to do, the modeling of the photoreactors is necessary. Computational fluid dynamics (CFD) is one of the numerical methods to solve and analyze the transport equations. CFD has been used recently in order to find the velocity, radiation intensity, pollutant concentration distribution inside the photoreacotrs (Mohajerani *et al.*, 2010,2011; Qi *et al.*, 2011; Duran *et al.*, 2011a,b, 2010; Vincent *et al.*, 2011; Denny *et al.*, 2009,2010; Elyasi and Taghipour, 2006; Taghipour and Mohseni, 2005; Mohseni and Taghipour, 2004). In this chapter, the velocity distribution in ten different configurations of single lamp and multi-lamp photoreactors in both laminar and turbulent regimes are provided. Moreover, the light intensity distribution inside the photoreactors is also presented.

2. Photoreactor design

The design and analysis of photochemical reactors have attracted researchers' interests for the past four decades. The main advantages of the photoreactors over the conventional thermal excitation reactors are their selectivity and low operating temperature. In spite of these advantages, photoreactors are not widely used in industrial scale. The use of a photoreactor results in higher product costs; therefore, the photochemical process is used when there are no other available feasible alternative conventional (thermal or catalytic) processes. Some other disadvantages such as size limitations, design and construction difficulties, and fouling on lamp walls have restricted industrial applications of these types of reactors.

Radiation field is the main characteristic of photoreactors determining the kinetics and the photoreactor performance. The photochemical reaction rate is proportional to the local volumetric rate of energy absorption (LVREA), an important parameter playing a crucial role in the photoreactor design. The LVREA depends on the radiation field presents within the photoreactor. The LVREA is a complex function of the lamp intensity, the concentration of absorbing species, the geometrical characteristics of the photoreactor system, and some other physicochemical properties. The light attenuation caused by the absorbing species makes the LVREA non-uniform. A Photochemical reactor design requires the solution of momentum, energy, and mass balances. The expression for the conservation of momentum is similar to that of the conventional reactor. It is believed that light irradiation does not affect the photochemically reactive fluid flow. The mass balance or continuity equations for each compound in the system should be solved simultaneously to find the concentration profile of each compound inside the photoreactor.

The mass and energy conservation equations are directly related to the light irradiation. The energy balance can be divided into two parts: thermal and radiant energy balances. Although for isothermal photochemical reactions, thermal balances are negligible, a complete equation of change for energy balance is required for exothermic and endothermic reactions. The first step in the LVREA evaluation is to state a radiant energy balance at steady state condition in a homogeneous system. A radiation source model is also needed to be considered. Different source models have been developed during the past years to describe the radiant energy field and to calculate the LVREA.

Gaertner and Kent (1958) were among the first group that modeled photochemical reactions. They studied the photolysis of aqueous uranyl oxalate in an annular tubular reactor. The results of their experiments were correlated with a remarkable accuracy by assuming that the rate of photolysis is proportional to the residence time multiplied by the intensity of radiation at a particular radial position. They also proposed a mathematical model without considering the reactant concentration. Therefore, their model was applicable when the conversion was small (less than 12%). The effect of diffusion was also neglected in their model.

The modeling of radiation field is classified into two categories: incidence and emission models. In the former approach, a specific radiation distribution exists in the photoreactor, while in the latter approach, the photon emission rate is employed to derive an incidence model. Although incidence models are mathematically simple to implement, there is no way of using this approach without experimentally adjustable parameters. Therefore, this approach cannot be employed as a reasonable systematic method to design photoreactors. Different incidence models have been developed and used for photoreactor modeling including radial incidence model (Gaertner and Kent, 1958; Schechter and Wissler, 1960; Cassano and Smith, 1966, 1967; Cassano *et al.*, 1968; Matsuura and Smith, 1970b, 1971; Santeralli and Smith, 1974; Dolan *et al.*, 1965; Jacob and Dranoff 1969; Williams and Ragonese, 1970), partially diffuse model (Matsuura and Smith, 1970a), and diffuse incidence model (Jacob and Dranoff, 1969; Williams, 1978; Matsuura *et al.*, 1969; Matsuura and Smith, 1970c; Harada *et al.*, 1971; Roger and Villermaux, 1983).

In emission model, a lamp is considered as an emitting line, a surface, or a volume source. Different emission models have been developed including line source with parallel plane emission model (Harris and Dranoff, 1965), line source with spherical emission model (Jacob and Dranoff, 1966, 1968, 1970; Jain *et al.*, 1971; Dworking and Dranoff, 1978; Magelli and Santarelli, 1978; Pasquali and Santarelli, 1978), and line source with diffuse emission model (Akehata and Shirai, 1972). Line and surface models are classified into specular and diffuse emission models. In specular emission, the magnitude of the light intensity vector is not a function of the emission angle. Photoreactors equipped with mercury arc and neon lamps belong to this class, while fluorescent lamps, in which the magnitude of the light intensity vectors depend on the angle of the emission, fall into the diffuse emission. There are some great reviews in which developed models describing the light distribution in photoreactors are presented in details in the open literature (Alfano *et al.*, 1986a,b; Cassano *et al.*, 1995; Alfano and Cassano, 2009).

Photoreactors are generally grouped into homogeneous and heterogeneous classes. Homogeneous reactors operate in a single phase, gas or liquid, in air and water purification systems. In heterogeneous photoreactors, a photocatalyst is added to the system in order to increase the process efficiency. Heterogeneous photoreactors are subcategorized into attached (immobilized) or suspended (slurry) modes. Most of the early photoreactors have employed a titanium dioxide (TiO_2) suspension because it offers a high surface area for the reactions and almost no mass transfer limitation exists in the system. In slurry photoreactors, a recovery step is necessary to separate, regenerate, and reuse the photocatalyst. The efficiency of an immobilized system is less than that of the slurry photoreactor but the photocatalyst is continuously used for a longer period of time.

The disadvantages of the slurry photocatalysis include 1) difficulty and time consuming process of separation or filtration of the photocatalyst after the photocatalytic process; 2) particle aggregation and agglomeration at high photocatalyst concentration; and 3) difficulty of using the suspended photocatalyst in continuous processes (Sopyan *et al.*, 1996). To overcome these drawbacks, immobilized photocatalysts are usually recommended. Photocatalysts could be immobilized on various supports such as glasses (Sabate *et al.*, 1992; Kim *et al.*, 1995; Fernandez *et al.*, 1995; Wang *et al.*, 1998; Piscopo *et al.*, 2000; Sakthivel *et al.*, 2002; Neti *et al.*, 2010), tellerette packings (Mehrvar *et al.*, 2002), silica (Van Grieken *et al.*, 2002; Ding *et al.*, 2001; Xu *et al.*, 1999; Alemany *et al.*, 1997; López-Muñoz *et al.*, 2005; Marugán *et al.*, 2006), polymers (Kasanen *et al.*, 2009), and clays (An *et al.*, 2008).

The analysis and design of photochemical reactors have received increasing interests in chemical engineering recently. The photochemical reactors could be used to either produce and synthesize chemicals or destroy water and wastewater contaminants. The photochemical reactors are one of the least well known industrial reactors. As mentioned earlier, the main advantages of the photochemical reactors over the conventional ones are their selectivity and low reaction temperature. These advantages have become the subject of several reviews (Doede and Walker, 1955; Marcus *et al.*, 1962; Cassano *et al.*, 1967; Shirotsuka and Nishiumi, 1971; Roger and Villermaux, 1979).

3. Radiation field properties

A bundle of rays with a specific energy carrying photons is called spectral specific intensity and radiance. This parameter plays a detrimental role in radiation field inside a photoreactor. The spectral specific intensity is defined as the total amount of radiative energy passing through a unit area perpendicular to the direction of propagation, per unit solid angle about the direction, per unit frequency, and per unit time. Therefore, the spectral specific intensity unit is Einstein per square meter per stradian, per unit frequency, and per second (Cassano *et al.*, 1995):

$$I_v(x,\Omega,t,v) = \lim_{dA,d\Omega,dt,dv\to 0}\left(\frac{dE_v}{dA\cos\theta d\Omega dt dv}\right) \tag{1}$$

Another important photochemical property is the spectral incident radiation that could be calculated by integrating the spectral specific intensity over the entire photoreactor volume as follows:

$$G_v = \int_{\Omega} I_v d\Omega \tag{2}$$

In spherical coordinates, the spectral incident radiation could be written as follows (Cassano *et al.*, 1995):

$$G_v = \int_{\theta_1}^{\theta_2} \int_{\phi_1}^{\phi_2} I_v \sin\theta d\phi d\theta \tag{3}$$

Equation (3) is valid for the monochromatic radiation while for the polychromatic radiation, an extra integration is required as follows (Cassano *et al.*, 1995):

$$G_v = \int_{v_1}^{v_2} \int_{\theta_1}^{\theta_2} \int_{\phi_1}^{\phi_2} I_v \sin\theta d\phi d\theta dv \tag{4}$$

The local volumetric rate of energy absorption (LVREA), the absorbed intensity, is the product of the spectral specific intensity and the absorption coefficient. The rate of a photochemical reaction is the product of the LVREA and the quantum yield. The quantum yield indicates the number of moles of chemicals reacted per moles of photons absorbed.

4. Photon transport equation

The photon balance in a photochemical reactor can be written as follows (Ozisik, 1973; Cassano *et al.*, 1995; Alfano and Cassano, 2008):

$$\begin{bmatrix} \text{time rate of change} \\ \text{of } \Omega, v \text{ photons in} \\ \text{the volume} \end{bmatrix} + \begin{bmatrix} \text{net flux of } \Omega, v \text{ photons} \\ \text{leaving the volume} \\ \text{across the surface } A \end{bmatrix} = \begin{bmatrix} \text{net gain of } \Omega, v \text{ photons owning} \\ \text{to emission, absorption, in-scattering,} \\ \text{and out-scattering in the volume} \end{bmatrix} \tag{5}$$

The right hand side of Equation (5) could be divided into two sources (emission and in-scattering) and two sink (absorption and out-scattering) terms, therefore, the photon balance equation could be written as follows (Cassano *et al.*, 1995):

$$\frac{1}{c}\frac{\partial I_{\Omega,v}}{\partial t} + \nabla.\left(I_{\Omega,v}\Omega\right) = \left(W_{\Omega,v}^{em} + W_{\Omega,v}^{in-s}\right) - \left(W_{\Omega,v}^{ab} + W_{\Omega,v}^{out-s}\right) \tag{6}$$

The absorption and emission terms are useful for both homogeneous and heterogeneous photochemical reactors. The loss of photons due to the absorption could be calculated using the following expression (Cassano *et al.*, 1995):

$$W_{\Omega,v}^{ab} = K_v(x,t) I_v(x,\Omega,t) \tag{7}$$

where K_v is the absorption coefficient, which shows the fraction of the incident radiation that is absorbed by the molecule per unit length along the path of the beam. The emission term highly depends on temperature as shown in Equation (8). Photolytic and photocatalytic reactors usually operate at low temperatures, therefore, this term can be neglected for photoreactor modeling:

$$W_{\Omega,v}^{em} = K_v(x,t) I_{v,b}\left[T(x,t)\right] \tag{8}$$

The following two equations account for the gain and loss of photons due to in-scattering and out-scattering, respectively (Cassano *et al.*, 1995). In homogeneous photoreactors, these two terms (in-scattering and out-scattering) are neglected due to the absence of semiconductors:

$$W_{\Omega,v}^{s-in} = \frac{1}{4\pi} \int_{\Omega'=4\pi} \int_{v'=0}^{\infty} \sigma_v(x,t) p(v' \to v, \Omega' \to \Omega) I_{v'}(x,\Omega',t) dv' d\Omega' \qquad (9)$$

$$W_{\Omega,v}^{s-out} = \sigma_v(x,t) I_v(x,\Omega,t) \qquad (10)$$

5. Principles of photocatalysis

The photocatalytic process in a heterogeneous photoreactor takes place in the presence of a semiconductor photocatalyst, in which the whole process is divided into five stages as follows (Herrmann, 1999; Fogler, 1998; Qi *et al.*, 2011):

- Mass transfer of the organic compounds in the liquid phase to the photocatalyst surface;
- Adsorption of the pollutants onto the photoactivated semiconductor surface;
- Heterogeneous photocatalytic reaction;
- Desorption of the product(s) from the photocatalyst surface; and
- Mass transfer of the product(s) from the interface region to the bulk solution.

The most common photocatalyst is titanium dioxide (TiO_2) which catalyzes the reaction as a result of the interaction of the electrons and holes generated during the photocatalytic process. Titanium dioxide has a diverse range of applications especially in paint and coatings, plastics, sunscreens, ointments, and toothpaste. The whiteness, brightness, and opacity of titanium dioxide make the powder as a good option for a wide range of applications. Titanium dioxide exists in three different crystalline polymorphs: rutile, anatase, and brookite. TiO_2 powders could be employed in solar and UV irradiation systems (photoreactors) because this semiconductor has a high transparency to visible light, high refractive index, and low absorption coefficient.

Wide range of metal oxides and sulfides have been used as photocatalysts including ZnO (Daneshvar *et al.*, 2004; Sakthivel *et al.*, 2003; Kormann *et al.*, 1988; Khodj *et al.*, 2001; Gouvêa *et al.*, 2000; Lizama *et al.*, 2002; Kansal *et al.*, 2007; Chakrabarti and Dutta, 2004), WO_3 (Waldner *et al.*, 2007; Sayama *et al.*, 2010; Saepurahmanet *et al.*, 2010; Cao *et al.*, 2011), WS_2 (Jing and Guo, 2007), Fe_2O_3 (Chen *et al.*, 2001; Bandara *et al.*, 2001; Pal and Sharon, 1998), V_2O_5 (Akbarzadeh *et al.*, 2010; Teramura *et al.*, 2004a,b), CeO_2 (Lin and Yu, 1998; Coronado *et al.*, 2002; Ji *et al.*, 2009; Song *et al.*, 2007), CdS (Bessekhouad *et al.*, 2004; Reutergårdh and Langphasuk, 1997; Tang and Huang, 1995), ZnS (Torres-Martínez *et al.*, 2001), and CuO (Lim and Kim, 2004; Sathishkumar *et al.*, 2011; Nezamzadeh-Ejheieg and Hushmandrad, 2010).

The light intensity in homogeneous and heterogeneous photoreactors determines the LVREA which is proportional to the rate of photochemical reaction. The light distribution in homogeneous photoreactors is well established. However, the heterogeneous photoreactor containing semiconductor particles makes the photoreactor modeling extremely difficult. In homogeneous photoreactors, the main solution parameter is the absorption coefficient, while in heterogeneous photoreactors, the absorption and scattering coefficients must be determined simultaneously.

6. Momentum balance and computational fluid dynamics

The momentum balance equations are applied for two flow regimes: the laminar and turbulent flow for these two single and multi-lamp photoreactors. The fluid flow in laminar regime is described by Navier-Stokes equation. For the case of the turbulent flow, different known models have been developed. Among them, the *k*-ε model has shown a great attraction to researchers for modeling turbulent flows. The *k*-ε model is a two-equation model in which fluctuating velocities and Reynolds stresses have been related to the properties of the turbulent flow itself such as *k* and ε. These two fluid flow properties are the turbulent kinetic energy per unit mass of the fluctuating components (*k*) and the turbulent dissipation rate of the kinetic energy (ε), respectively. The continuity equation and steady state momentum balances for an incompressible fluid are described by following equations (Bird *et al.*, 2002):

$$\nabla \cdot V = 0 \tag{11}$$

$$\rho V \cdot \nabla V = -\nabla P + \nabla \cdot (\mu + \rho \eta_T)\left(\nabla V + (\nabla V)^T\right) + F \tag{12}$$

In these equations, ρ, V, and P, represent the density, time-averaged turbulent velocity, and time-averaged pressure, respectively. The μ and η_T are the dynamic and kinematic turbulent viscosities, respectively. F is also the external force on the control volume.
The *k*-ε model depicts that the kinematic turbulent viscosity (η_T) at any point should depend only on *k* and ε at that point according to the following expression (Bird *et al.*, 2002):

$$\eta_T = C_\mu \frac{k^2}{\varepsilon} \tag{13}$$

where C_μ is an adjustable model constant. The turbulent kinetic energy (*k*) is the average kinetic energy per unit mass of eddies in the turbulent flow that is produced by the buoyant thermal and mechanically generated eddies based on the following equation (Bird *et al.*, 2002):

$$k = \frac{1}{2}\left(\overline{u'^2} + \overline{v'^2} + \overline{w'^2}\right) \tag{14}$$

Parameters u', v', and w' are the fluctuating velocities in x, y, and z directions, respectively. The steady state equations for the turbulent kinetic energy (*k*) and the turbulent dissipation rate (ε) are as follows (Wilkes, 2005):

$$\rho V \cdot \nabla k = \nabla \cdot \left[\left(\mu + \rho \frac{C_\mu k^2}{\sigma_k \varepsilon}\right)\nabla k\right] + \rho C_\mu \frac{k^2}{\varepsilon}\left(\nabla V + (\nabla V)^T\right)^2 - \rho\varepsilon \tag{15}$$

$$\rho V \cdot \nabla \varepsilon = \nabla \cdot \left[\left(\mu + \rho \frac{C_\mu k^2}{\sigma_\varepsilon \varepsilon}\right)\nabla \varepsilon\right] + \rho C_{\varepsilon_1} C_\mu k\left(\nabla V + (\nabla V)^T\right)^2 - \rho C_{\varepsilon_2} \frac{\varepsilon^2}{k} \tag{16}$$

All five model constants have been selected as follows such that the *k*-ε model gives an estimation that fits reasonably well with the experimental data (Wilkes, 2005):

$$C_\mu = 0.09; \quad C_{\varepsilon 1} = 1.44; \quad C_{\varepsilon 2} = 1.92; \quad \sigma_k = 1.0; \text{ and } \quad \sigma_\varepsilon = 1.3 \tag{17}$$

Equations (15) and (16) along with continuity and momentum Equations (11) and (12) should be solved simultaneously. Boundary conditions for the momentum balance are as follows:
Inlet of the photoreactor: The inflow velocity is specified by the ratio of the volumetric flow rate to the surface area. The k and ε are also calculated as follows:

$$k=\left(\frac{3I_T^{\,2}}{2}\right)\left(\overline{V}^2\right);\quad \varepsilon=C_\mu^{\,0.75}\frac{\left[\frac{3I_T^{\,2}}{2}\left(\overline{V}^2\right)\right]^{1.5}}{L_T}; \tag{18}$$

where I_T and L_T are the turbulent intensity scale (initial turbulence intensity) and turbulent length scale (eddy length scale). Their magnitudes could be calculated by Equations (19) and (20), respectively (Wilkes, 2005):

$$I_T = 0.16Re^{-1/8} \tag{19}$$

$$L_T = 0.07L \tag{20}$$

Outlet of the photoreactor: A normal flow is assumed so that the normal stress in the outlet is zero. Therefore, the gradient of k and ε are also calculated as follows (Wilkes, 2005):

$$\nabla k = 0 \quad ; \qquad \nabla \varepsilon = 0 \tag{21}$$

Walls of the photoreactor: No slip boundary condition could be assumed for the photoreactors and all the walls of the lamps. However, the logarithmic wall function is used for walls as a modification of the k-ε model. The boundary conditions for the k-ε model at a no-slip wall are obvious, but the near-wall behavior of the model, especially the ε-equation, is not appropriate. Therefore, the model produces poor results when integrated to the wall without modification. In fact, the integration of the k-ε model through the near-wall region and application of the no-slip conditions yield unsatisfactory results. Therefore, the logarithmic wall function is used as the boundary condition on the walls of the photoreactor and lamps. In the near-wall region, therefore, the equation is solved for the first grid node away from the wall. The logarithmic wall function for a smooth pipe, therefore, is as follows (Wilkes, 2005):

$$u_z^{\,+} = 5.5 + 2.5\ln y^+ \quad \text{where} \quad y^+ = \frac{y\sqrt{\tau_w/\rho}}{\mu/\rho} \quad \text{and} \quad u_z^{\,+} = \frac{\overline{u_z}}{\sqrt{\tau_w/\rho}} \tag{22}$$

in which y^+, u_z^+, and τ_w are the dimensionless distance from the wall, the dimensionless velocity, and the wall shear stress, respectively. The y and $\overline{u_z}$ are the distance from the wall and time-averaged axial velocity, respectively.
Simultaneous solution of Equations (11), (12), (15), and (16) as well as applying boundary conditions give various flow characteristics such as velocity distribution, minimum and maximum velocities, turbulent kinetic viscosity, turbulent kinetic energy, turbulent dissipation rate distribution, and vorticity. Equations (10) to (21) are applicable to both single and multi-lamp photoreactors.

7. Case study

Annular photoreactors are the most common continuous photoreactors. These types of photoreactors contain one or more UV lamps with quartz sleeve around them. In this study, 10 different photoreactor configurations were considered for modeling purposes as listed below:

a. Single lamp photoreactor, the UV lamp is located at the center of the photoreactor;
b. Two lamp photoreactor, each of them located at 0.5*R* from the center line of the photoreactor;
c. Two lamp photoreactor, each of them located at 0.25*R* from the center line of the photoreactor;
d. Two lamp photoreactor, each of them located at 0.75*R* from the center line of the photoreactor;
e. Four lamp photoreactor, each of them located 0.5*R* from the center line of the photoreactor;
f. Four lamp photoreactor, each of them located 0.25*R* from the center line of the photoreactor;
g. Four lamp photoreactor, each of them located 0.75*R* from the center line of the photoreactor;
h. Four lamp photoreactor, two of them at 0.5*R* and two at 0.25*R* from the center line of the photoreactor;
i. Four lamp photoreactor, two of them at 0.5*R* and two at 0.75*R* from the center line of the photoreactor; and
j. Four lamp photoreactor, two of them at 0.25*R* and two at 0.75*R* from the center line of the photoreactor.

In all cases, the diameter of the quartz sleeve is assumed to be 15 mm with the photoreactor radius (*R*) of 200 mm. The cross sections of these photoreactors (a to j) are shown in Figures 1, 2, and 3.

The flow domains inside these photoreactors were simulated both in laminar and turbulent regimes. The inlet velocities for laminar and turbulent regimes were assumed to be 0.01 and 1 ms^{-1}, respectively. In laminar flow, the 3D Navier-Stokes equations were solved in cylindrical coordinates for single lamp and Cartesian coordinates for multilamp photoreactors. In turbulent regime, the k-ε turbulent model was used. The velocity field determination is important in the photoreactor design. The local velocity determines the residence time of the solution at a specific point. As mentioned earlier, the radiation intensity inside the photoreactor is highly non-uniform; therefore, there must be a trade-off between the residence time and the radiation intensity received at particular location.

Figures 1 and 2 depict the velocity distributions at the outlet of the photoreactors in turbulent and laminar regimes, respectively. From these two figures, three distinct regions could be considered: the region around the UV lamp(s), the region around the photoreactor wall, and the remaining parts. A lower velocity is observed around the UV lamps due to the no-slip boundary condition; therefore, a higher residence time is obtained near the UV lamps. On the other hand, around the UV lamp, a higher LVREA exist. Both higher residence time and greater radiation intensity result in a higher degradation of pollutants in water. The region near the photoreactor wall also has lower velocity and higher residence time, but the LVREA in this region is the lowest compared to those of the other regions. The light intensity was also simulated for these ten photoreactors as shown in Figure 3. In multilamp photoreactors, the light intensity of UV lamps could provide a synergetic effect depending on their location. Although closer UV lamps (in four lamp photoreactor) provide higher light intensity in the photoreactor center, the shadowing effect of the lamps reduces the light intensity in the region near the photoreactor wall. A comparison between Figures 3e and 3f show that the photoreactor in Figure 3f performed a higher light intensity at the center, however, the shadowing effect is high, therefore, Figure 3e shows a good trade-off between the synergetic effect and the shadowing effect.

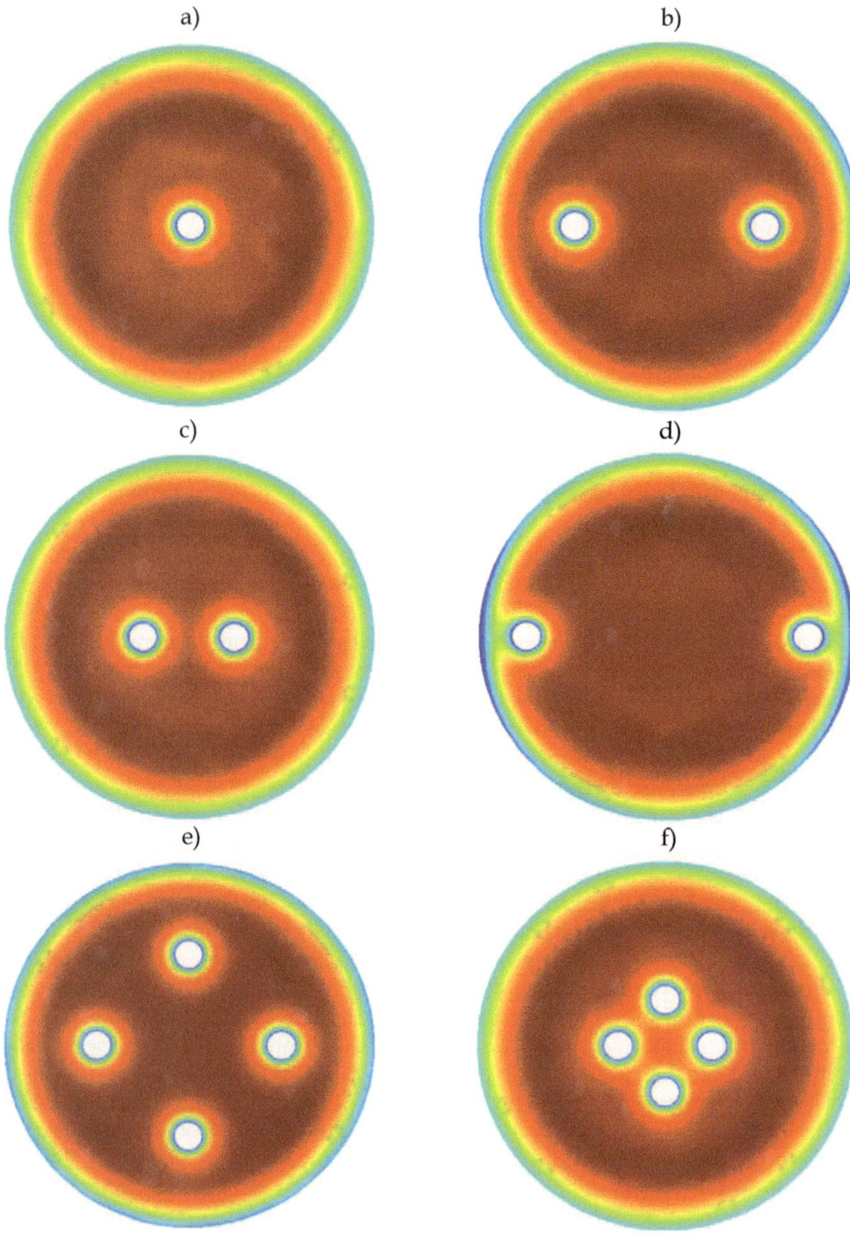

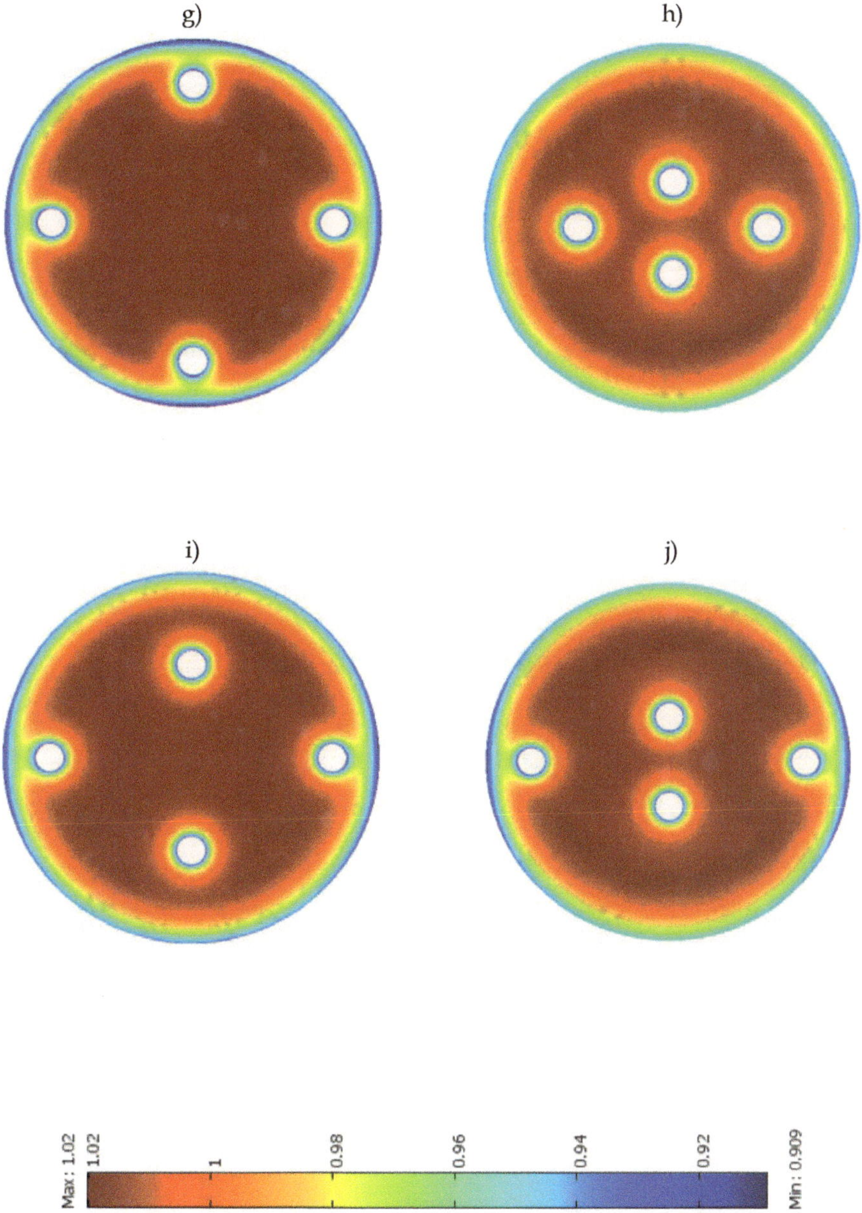

Fig. 1. Velocity distribution at the outlet of the photoreactors (turbulent regime). All geometries are defined in Section "Case study".

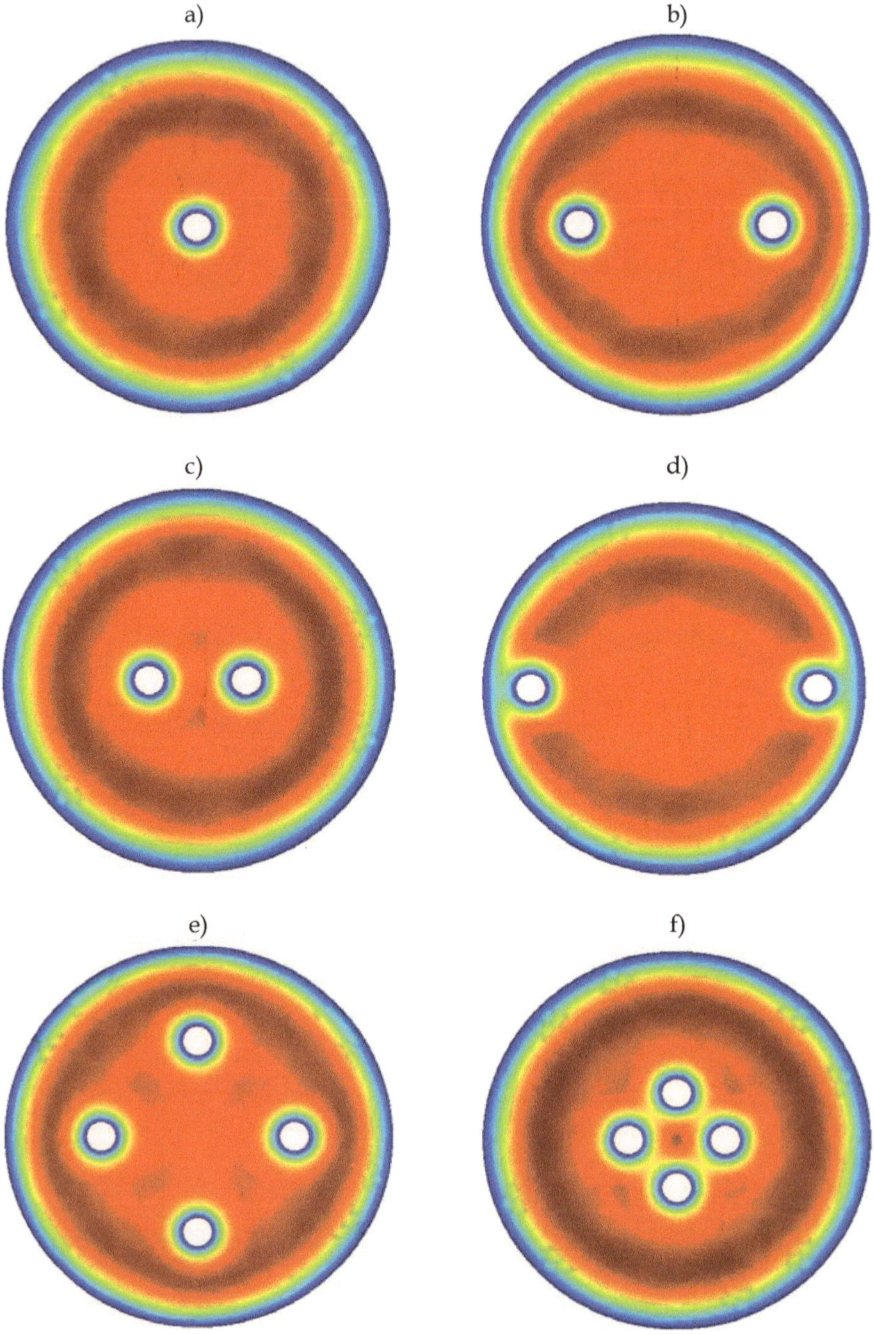
a)
b)
c)
d)
e)
f)

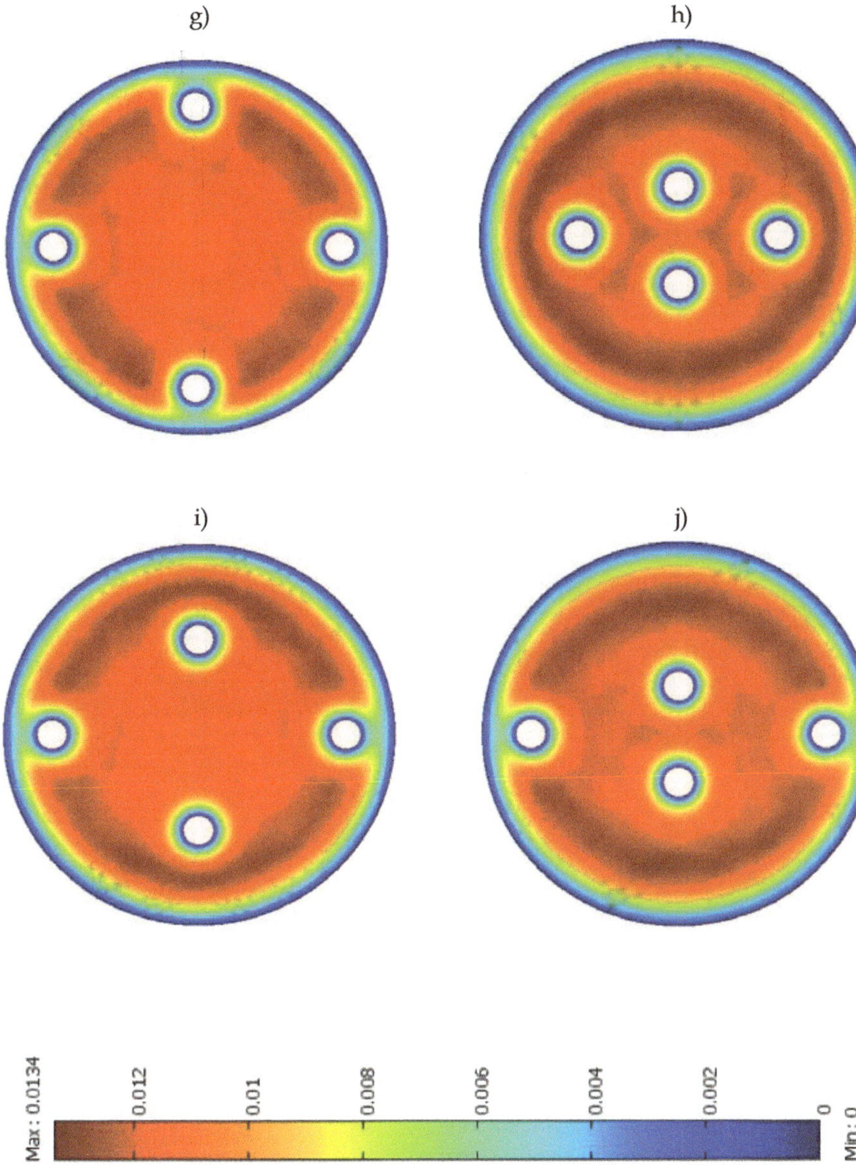

Fig. 2. Velocity distribution at the outlet of the photoreactors (laminar regime). All geometries are defined in Section "Case study".

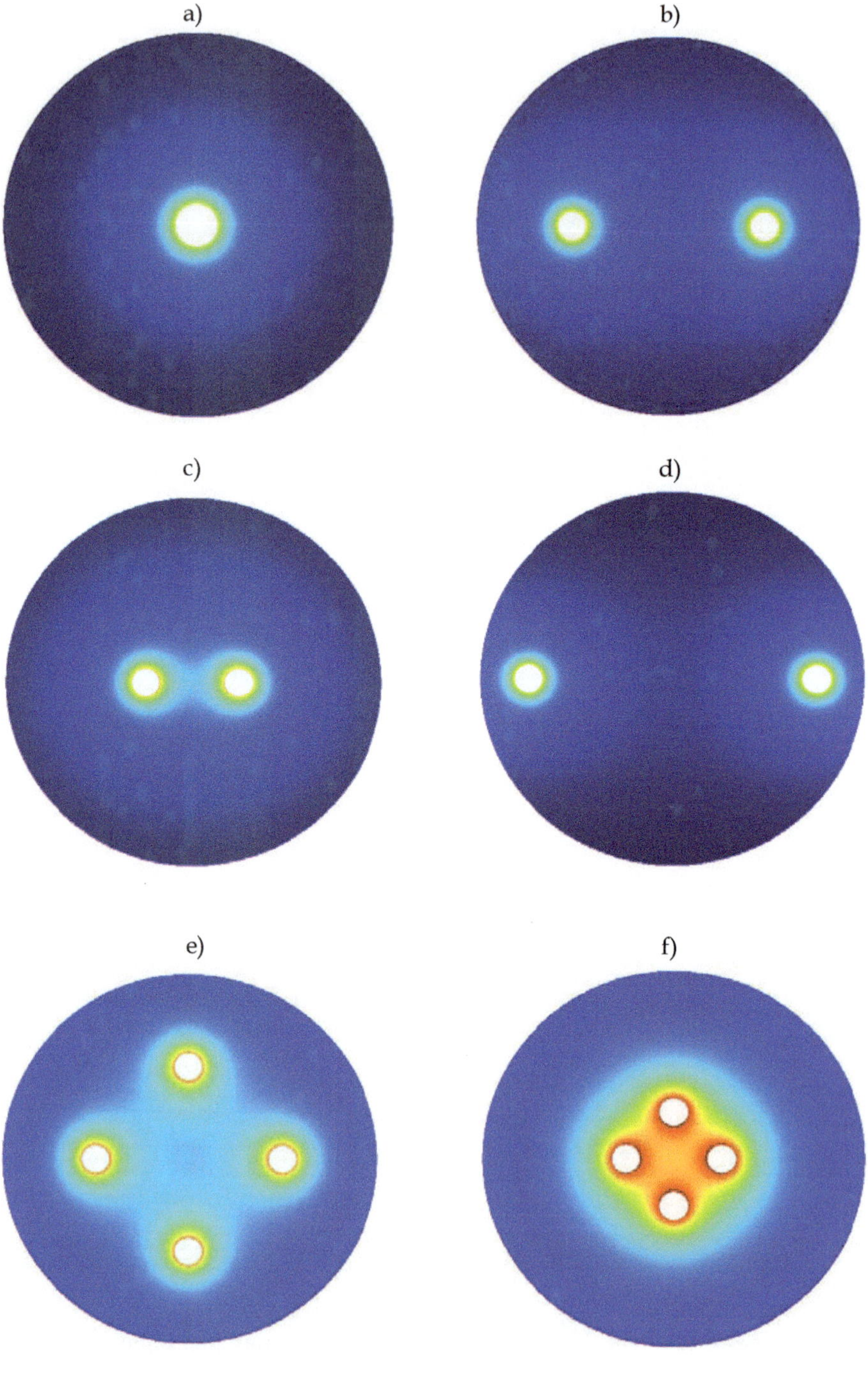
a)
b)
c)
d)
e)
f)

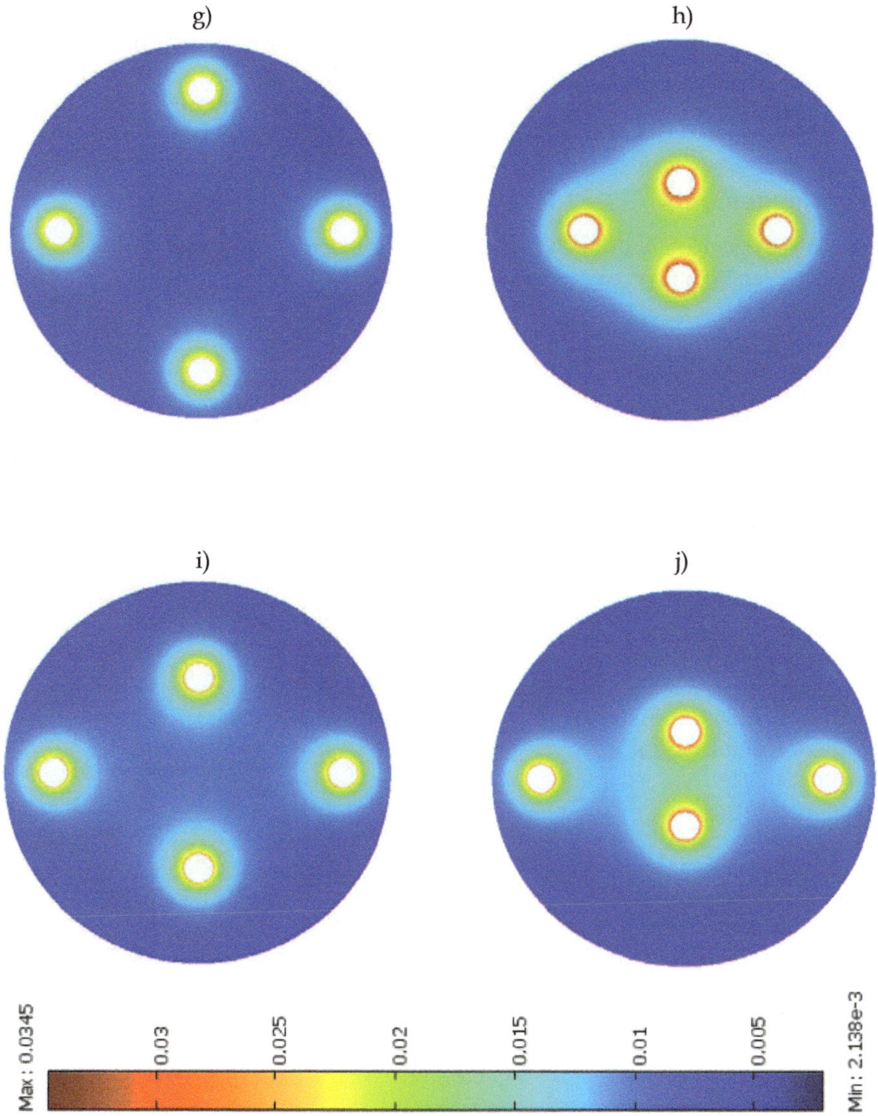

Fig. 3. Light intensity distribution at the outlet of the photoreactors for both laminar and turbulent regimes. All geometries are defined in Section "Case study".

8. Conclusions

The degradation and mineralization of polluted waters containing non-biodegradable compounds have become increasingly important. CFD modeling helps to design, optimize, and scale-up photochemical reactors. Velocity, light, and pollutant(s) concentration distribution would give better understanding of photoreactors operation for further decisions. Velocity and light intensity distribution in ten different annular photoreactors were described. Velocity profiles were sketched for both laminar and turbulent flows. The effect of number and location of UV lamps was studied to find the enhancement in light distribution in multi-lamp photoreactors. Closer UV lamps performed a higher light intensity in the center of the photoreactor but the closer lamps provided a higher shadowing effect.

9. Acknowledgements

The financial support of Natural Sciences and Engineering Research Council of Canada (NSERC) and Ryerson University is greatly appreciated.

10. Nomenclature

A	area, m^2
c	light speed, $m.s^{-1}$
E_v	activation energy for frequency v, $J.mol^{-1}$
F	external force, $kg.m.s^{-2}$
G_v	incident radiation, $Einstein.s^{-1}.m^{-2}$
I_v	Specific intensity for frequency v, $Einstein.s^{-1}.m^{-2}.sr^{-1}$
I_T	turbulent intensity scale, dimensionless
k	turbulent kinetic energy, $m^2.s^{-2}$
K	reaction rate constant, $m^3.mol.^{-1}s^{-1}$
L_T	turbulent length scale, m
L	equivalent diameter of the photoreactor, m
P	time-averaged pressure, pa
R	photoreactor radius, m
t	time, s
T	temperature, K
u'	fluctuating velocity in x direction, $m.s^{-1}$
u_z^+	dimensionless velocity
V	time-averaged turbulent velocity, $m.s^{-1}$
v'	fluctuating velocity in y direction, $m.s^{-1}$
w'	fluctuating velocity in z direction, $m.s^{-1}$
W	gain or loss of energy, $Einstein.s^{-1}.m^{-3}.sr^{-1}$
x	position vector, m
y	position on y-axis (distance from the wall), m
y^+	dimensionless distance from the wall

Greek letters

ε	turbulent dissipation rate, $m^2.s^{-3}$
η_T	turbulent kinematic viscosity, $m^2.s^{-1}$
θ	spherical coordinate, rad
μ	dynamic viscosity, $kg.m^{-1}.s^{-1}$
v	frequency, s^{-1}
ρ	density, $kg.m^{-3}$
σ	volumetric scattering coefficient, m^{-1}
τ_w	wall shear stress, $kg.m^{-1}.s^{-2}$

ϕ spherical coordinate, rad
Ω unit vector in the direction of propagation, dimensionless

11. References

Akbarzadeh, R.; Umbarkar, S.B.; Sonawane, R.S.; Takle, S. & Dongare, M.K. (2010) Vanadia-titania thin films for photocatalytic degradation of formaldehyde in sunlight. *Appl. Catal. A: General*, Vol. 374 No. 1-2 pp. 103-109.

Akehata, T. & Shirai, T. (1972) Effect of light-source characteristics on the performance of circular annular photochemical reactor. *J. Chem. Eng. Japan* Vol. 5 pp.385-391.

Alemany, L.J.; Bañares, M.A.; Pardo, E.; Martin, F.; Galán-Fereres, M. & Blasco, J.M. (1997) Photodegradation of phenol in water using silica-supported titania catalysts. *Appl. Catal. B: Environ.*, Vol. 13, No. 3-4, pp. 289-297.

Alfano, O.M.; Romero, R.L. & Cassano, A.E. (1986a) Radiation field modelling in photoreactors-II. Heterogeneous media. *Chem. Eng. Sci.*, Vol. 41 No. 5 pp. 1137-1153.

Alfano, O.M.; Romero, R.L. & Cassano, A.E. (1986b) Radiation field modelling in photoreactors-I. Homogeneous media. *Chem. Eng. Sci.*, Vol. 41 No. 5 pp. 421-444.

Alfano, O.M. & Cassano, A.E. (2008) Photoreactor modeling: applications to advanced oxidation processes. *Int. J. Chem. Reactor Eng.*, Vol. 6, P2, pp.1-18.

Alfano, O.M. & Cassano, A.E. (2009) Scalling-up of photoreactors. Applications to advanced oxidation process. *Adv. Chem. Eng.*, Vol. 36 pp. 229-287.

An, T.; Chen, J.; Li, G.; Ding, X.; Sheng, G.; Fu, J.; Mai, B. & O'Shea, K.E. (2008) Characterization and the photocatalytic activity of TiO_2 immobilized hydrophobic montmorillonite photocatalysts. Degradation o decabromodiphenyl ether (BDE 209). *Catal. Today* Vol. 139 No. 1-2, pp. 69-76.

Bandara, J.; Mielczarski, J.A.; Lopez, A. & Kiwi, J. (2001) Sensitized degradation of chlorophenols on iron oxides induced by visible light comparison with titanium oxide. *Appl. Catal. B: Environ*, Vol. 34 No. 4 pp. 321-333.

Bessekhouad, Y.; Robert, D. & Weber, J.V. (2004) Bi_2S_3/TiO_2 and CdS/TiO_2 heterojunctions as an available configuration for photocatalytic degradation of organic pollutant. *J. Photochem. Photobiol. A: Chemistry*, Vol. 163 No. 3 pp. 569-580.

Bird, R.B.; Steward, W.E. & Lightfood, E.N. Transport Phenomena, 2nd ed. John Wiley & Sons, Inc.: NY, 2002

Cao, J.; Luo, B.; Lin, H. & Chen, S. (2011) Photocatalytic activity of novel $AgBr/WO_3$ composite photocatalyst under visible light irradiation for methyl orange degradation. *J. Hazard.s Mater.*, Vol. 190 No. 1-3 pp. 700-706.

Cassano, A.E. & Smith, J.M. (1966) Photochlorination in a tubular reactor. *AIChE J.*, Vol. 12 pp. 1124-1133.

Cassano, A.E. & Smith, j.M. (1967) Photochlorination of propane. *AIChE J.*, Vol. 13, pp. 915-925.

Cassano, A.E.; Silverston, P.L. & Smith, J.M. (1967) Photochemical reaction engineering. *Ind. Eng. Chem.*, Vol. 59, pp. 18-38.

Cassano, A. E.; Matsuura, T. & Smith, J.M. (1968) Batch recycle reactor for slow photochemical reactions. *Ind. Eng. Chem. Fundam.*, Vol. 7 pp. 655-660.

Cassano, A.E.; Martín, C.A.; Brandi, R.J. & Alfano, O.M. (1995) Photoreactor analysis and design: Fundamentals and applications. *Ind. Eng. Chem. Res.*, Vol. 34, No. 7, pp. 2155-2201.

Chakrabarti, S. & Dutta, B.K. (2004) Photocatalytic degradation of model textile dyes in wastewater using ZnO as semiconductor catalyst. *J. Hazard. Mater.*, Vol. 112 No. 3 pp. 269-278

Chen, F.; Xie, Y.; Zhao, J. & Lu, G. (2001) Photocatalytic degradation of dyes on a magnetically separated photocatalyst under visible and UV irradiation. *Chemosphere*, Vol. 44 No. 5 pp. 1159-1168.

Coronado, J.M.; Javier Maira, A.; Martínez-Arias, A.; Conesa, J.C. & Soria, J. (2002) EPR study of the radicals formed upon UV irradiation of ceria-based photocatalysts. *J. Photochem. Photobiol. A: Chemistry*, Vol. 150 No. 1-3 pp. 213-221.

Daneshvar, N.; Salari, D. & Khataee, A.R. (2004) Photocatalytic degradation of azo dye acid red 14 in water on ZnO as an alternative catalyst to TiO_2 *J. Photochem. Photobiol. A: Chemistry*, Vol. 162; No. 2-3; pp.317-322.

Denny, F.; Scott, J.; Pareek, V.; Peng, G.-D. & Amal, R. (2010) Computational fluid dynamics modelling and optimal configuring of a channeled optical fibre photoreactor. *Chem. Eng. Sci.*, Vol. 65 No. 17 pp. 5029-5040.

Ding, Z.; Hu, X.; Yue, P.L.; Lu, G.Q. & Greenfield, P.F. (2001) Synthesis of anatase TiO_2 supported on porous solids by chemical vapor deposition. *Catal. Today*, Vol. 68, No.: 1-3; pp. 173-182.

Doede, C.M. & Walker, C.A. (1955). Photochemical engineering. *Chemical Engineering*, pp. 159-178.

Dolan, W.J.; Dimon, C.A. & Dranoff, J.S. (1965) Dimensional analysis in photochemical reactor design. *AIChE J.*, Vol. 11 pp. 1000-1005.

Duran, J.E.; Taghipour, F. & Mohseni, M. (2011a) Evaluation of model parameters for simulating TiO_2 coated UV reactors. *Water Sci. Technol.*, Vol. 63 No. 7 pp. 1366-1372.

Duran, J.E.; Mohseni, M. & Taghipour, F. (2011b) Design improvement of immobilized photocatalytic reactors using a CFD-Taguchi combined method. *Ind. Eng. Chem. Res.*, Vol. 50 No. 2 pp. 824-831.

Duran, J.E.; Taghipour, F. & Mohseni, M. (2010) Irradiance modeling in annular photoreactors using the finite-volume method. *J. Photochem. Photobiol. A: Chemistry*, Vol. 215 No. 1 pp. 81-89.

Dworkin, D. & Dranoff, J.S. (1978) Free radical transport in a photochemical reactor. *AIChE J.*, Vol. 24 pp. 1134-1137.

Elyasi, S. & Taghipour, F. (2006) Simulation of UV photoreactor for water disinfection in Eulerian Framework. *Chem. Eng. Sci.*, Vol. 61 No. 14 pp. 4741-4749.

Elyasi, S. & Taghipour, F. (2010) Simulation of UV photoreactor for degradation of chemical contaminants: Model development and evaluation. *Environ. Sci. Technol.*, Vol. 44 No. 6 pp. 2056-2063.

Fernandez, A.; Lassaletta, G.; Jimenez, V.Z.; Justo, A.; Gonzalez-Elipe, A.R.; Herrmann, J.M.; Tahiri, H. & Ait-Ichou, Y. (1995) Preparation and characterization of TiO_2 photocatalysts supported on various rigid supports (glass, quartz and stainless steel). Comparative studies of photocatalytic activity in water purification. *Appl. Catal. B: Environ.*, Vol. 7 No. 1-2 pp. 49-63.

Fogler, H.S. Elements of Chemical Reaction Engineering, third ed., Prentic Hall PTR, New Jersey, 1998.

Gaertner, R.F. & Kent, J.A. (1958) Conversion in a continuous photochemical reactor. *Ind. Eng. Chem. Res.*, Vol. 50 No. 9, pp. 1223-1226.

Gouvêa, C.A.K.; Wypych, F.; Moraes, S.G.; Durán, N.; Nagata, N. & Peralta-Zamora, P. (2000) Semiconductor-assisted photocatalytic degradation of reactive dyes in aqueous solution. *Chemosphere*, Vol. 40 No. 4 pp. 433-440.

Harada, J.; Akehata, T. & Shirai, T. (1971) Light intensity distribution in an elliptical photoreactor. *Kagaku Kogaku*, Vol. 35 pp. 233-239.

Harris, P.R. & Dranoff, J.S. (1965) A study of perfectly mixed photochemical reactors. *AIChE J.*, Vol. 11 pp. 497-502.

Herrmann, J.-M. (1999) Heterogeneous photocatalysis: Fundamentals and applications to the removal of various types of aqueous pollutants. *Catal. Today*, Vol. 53 No. 1 pp. 115-129.

Jacob, S.M. & Dranoff, J.S. (1966) Radial scale-up of perfectly mixed photochemical reactors. *Chem. Eng. Prog. Symp. Ser.*, Vol. 62 pp.47-55.

Jacob, S.M. & Dranoff, J.S. (1968) Design and analysis of perfectly mixed photochemical reactors. *Chem. Eng. Prog. Sym. Ser.*, Vol. 64 pp. 54-63.

Jacob, S.M. & Dranoff, J.S. (1969) Light intensity profiles in an elliptical photoreactor. *AIChE J.*, Vol. 15 pp. 141-144.

Jacob, S.M. & Dranoff, J.S. (1970) Light intensity profiles in a perfectly mixed photoreactor. *AIChE J.*, Vol. 16 pp. 359-363.

Jain, R.L.; Graessley, W.W. & Dranoff, J.S. (1971) Design and analysis of a photoreactor for styrene polymerization. *Ind. Eng. Chem. Prod. Res. Develop.*, Vol. 10 pp. 293-298.

Ji, P.; Zhang, J.; Chen, F. & Anpo, M. (2009) Study of adsorption and degradation of acid orange 7 on the surface of CeO_2 under visible light irradiation. *Appl. Catal. B: Environment.*, Vol. 85 No. 3-4 pp. 148-154.

Jing, D. & Guo, L. (2007) WS_2 sensitized mesoporous TiO_2 for efficient photocatalytic hydrogen production from water under visible light irradiation. *Catal. Commun.*, Vol. 8 No. 5 pp. 795-799.

Kansal, S.K.; Singh, M. & Sud, D. (2007) Studies on photodegradation of two commercial dyes in aqueous phase using different photocatalysts. *J. Hazard. Mater.*, Vol. 141 No. 3 pp. 581-590.

Kasanen, J.; Suvanto, M. & Pakkanen, T.T. (2009) Self-cleaning, titanium dioxide based, multilayer coating fabricated on polymer and glass surfaces. *J. Appl. Polym. Sci.*, Vol. 111 No. 5 pp. 2597-2606.

Kim, D.H.; Anderson, M.A. & Zeitner, W.A. (1995) Effects of firing temperature on photocatalytic and photoelectrocatalytic properties of TiO_2. *J. Environ. Eng.*, Vol. 121 No. 8 pp. 590-594.

Khodja, A.A.; Sehili, T.; Pilichowski, J.-F. & Boule, P. (2001) Photocatalytic degradation of 2-phenylphenol on TiO_2 and ZnO in aqueous suspensions. *J. Photochem. Photobiol. A: Chemistry*, Vol. 141 No. 2-3 pp. 231-239.

Kormann, C.; Bahnemann, D.W. & Hoffmann, M.R. (1988) Photocatalytic production of H_2O_2 and organic peroxides in aqueous suspensions of TiO_2, ZnO, and desert sand. *Environ. Sci. Technol.*, Vol. 22 No. 7 pp. 798-806.

Lim, T.H. & Kim (2004), S.D. Trichloroethylene degradation by photocatalysis in annular flow and annulus fluidized bed photoreactors. *Chemosphere*, Vol. 54 No. 3 pp. 305-312.

Lin, J. & Yu, J.C. (1998) An investigation on photocatalytic activities of mixed TiO_2-rare earth oxides for the oxidation of acetone in air. *J. Photochem. Photobiol. A: Chemistry*, Vol. 116 No. 1 pp. 63-67.

Lizama, C.; Freer, J.; Baeza, J. & Mansilla, H.D. (2002) Optimized photodegradation of reactive blue 19 on TiO_2 and ZnO suspensions. *Catalysis Today*, Vol. 76 No. 2-4 pp. 235-246.

López-Muñoz, M.-J.; Grieken, R.V.; Aguado, J. & Marugán, J. (2005) Role of the support on the activity of silica-supported TiO_2 photocatalysts: structures of the TiO_2/SBA-15 photocatalysis. *Catal. Today*, Vol. 101, pp. 307-314.

Magelli, F. & Santarelli, F. (1978) The modelling of batch photoreactors. *Chem. Eng. Sci.*, Vol. 33 pp. 611-614.

Marcus, R.J.; Kent, J.A. & Schenck, G.O. (1962) Industrial photochemistry. *Ind. Eng. Chem.*, Vol. 54, pp. 20-28

Marugán, J.; Hufschmidt, D.; López-Muñoz, M.-J.; Selzer, V. & Bahnemass, D. (2006) Photonic efficiency for methanol photooxidation and hydroxyl radical generation on silica-supported TiO_2 photocatalysts. *Appl. Catal. B: Environ*, Vol: 62, No.: 3-4, pp. 201-207.

Matsuura, T.; Cassano, A.E. & Smith, J.M. (1969) Acetone photolysis: kinetic studies in a flow reactor. *AIChE J.*, Vol. 15 pp. 495-501.

Matsuura, T. & Smith, J.M. (1970a) Light distribution in cylindrical photoreactors. *AIChE J.*, Vol. 16 pp. 321-324.

Matsuura, R. & Smith, J.M. (1970b) Kinetics of photocomposition of dodecyl benzene sulfonate. *Ind. Eng. Chem. Fund.*, Vol. 9 pp. 252-260.

Matsuura, R. & Smith, J.M. (1970c) Photodecomposition kinetics of formic acid in aqueous solutions. *AIChE J.*, Vol. 16 pp.1064-1071.

Matsuura, R. & Smith, J.M. (1971) Photodecomposition kinetics of formic acid in aqueous solutions. *Ind. Eng. Chem. Fundam.*, Vol. 10, pp. 316-318.

Mehrvar, M.; Anderson, W.A. & Moo-Young, M. (2002) Preliminary analysis of a tellerette packed-bed photocatalytic reactor, *Adv. Environ. Res.* Vol. 6 No. 4, pp. 411–418.

Mohajerani, M.; Mehrvar, M. & Ein-Mozaffari, F. (2011) Photoreactor design and CFD modelling of a UV/H_2O_2 process for distillery wastewater treatment. *Can. J. Chem. Eng.* (in press).

Mohajerani, M.; Mehrvar, M. & Ein-Mozaffari, F. (2010) CFD modeling of metronidazole degradation in water by the UV/H_2O_2 process in single and multilamp photoreactors. *Ind. Eng. Chem. Res.*, Vol. 49 No. 11 pp. 5367-5382.

Mohseni, M. & Taghipour, F. (2004) Experimental and CFD analysis of photocatalytic gas phase vinyl chloride (VC) oxidation. *Chem. Eng. Sci.*, Vol. 59 No. 7 pp. 1601-1609.

Neti, N.R.; Parmar, G.R.; Bakardjieva, S. & Subrt, J. (2010) Thick film titania on glass supports for vapour phase photocatalytic degradation of toluene, acetone, and ethanol. *Chem. Eng. J.*, Vol. 163 No. 3 pp. 219-229.

Nezamzadeh-Ejhieh, A. & Hushmandrad, S. (2010) Solar photodecolorization of methylene blue by CuO/X zeolite as a heterogeneous catalyst. *Appl. Catal. A: General*, Vol. 388 No. 1-2 pp. 149-159.

Ozisik, M.N. (1973) Radiative transfer and interactions with conduction and convection, Wiley, New York

Pal, B. & Sharon, M. (1998) Photocatalytic degradation of salicylic acid by colloidal Fe_2O_3 particles. *J. Chem. Technol. Biotechnol.*, Vol. 73 No. 3 pp. 269-273.

Pasquali, G. & Santarelli, F. (1978) Radiant energy transfer in batch photoreacting media. *Chem. Eng. Commun.*, Vol.2 pp. 271-274.

Piscopo, A.; Robert, D.; Marzolin, C. & Weber, J.V. (2000) TiO_2 supported on glass fiber for the photocatalytic degradation of benzamide. *J. Mater. Sci. Lett.*, Vol. 19 No. 8 pp. 683-684.

Qi, N.; Zhang, H.; Jin, B. & Zhang, K. (2011) CFD modelling of hydrodynamics and degradation kinetics in an annular slurry photocatalytic reactor for wastewater treatment. *Chem. Eng. J.*, (in press).

Reutergardh, L.B. & Langphasuk, M. (1997) Photocatalytic decolourization of reactive azo dye: A comparison between TiO_2 and CdS photocatalysis. *Chemosphere*, Vol. 35 No. 3 pp. 585-596.

Roger, M. & Villermaux, J. (1983) Modelling of light absorption in photoreactors. Part II. Density profile and efficiency of light absorption in a cylindrical reactor. Experimental comparison with complex kinetics. *Chem. Eng. Sci.*, Vol. 38 pp. 1593-1605.

Roger, M. & Villermaux, J. (1979) Modelling of light absorption in photoreactors. Part I. General formulation based on the laws of photometry. *Chem. Eng. J.*, Vol. 17, pp. 219-226.

Sabate, J.; Anderson, M.A.; Aguado, M.A.; Giménez, J.; Cervera-March, S. & Hill Jr., C.G. (1992) Comparison of TiO_2 powder suspensions and TiO_2 powder suspensions and TiO_2 ceramic membrans supported on glass as photocatalytic systems in the reduction of chromium. *J. Mol. Catal.*, Vol. 71 No. 1 pp. 57-68.

Saepurahman, Abdullah, M.A. & Chong, F.K. (2010) Preparation and characterization of tungsten-loaded titanium dioxide photocatalyst for enhanced dye degradation. *J. Hazard. Mater.*, Vol. 176 No. 1-3 pp. 451-458.

Sakthivel, S.; Shankar, M.V.; Palanichamy, M.; Arabindoo, B. & Murugesan, V. (2002) Photocatalytic decomposition of leather dye. Comparative study of TiO_2 supported on alumina and glass beads. *J. Photochem. Photobiol. A: Chemistry*, Vol. 148 No. 1-3 pp. 153-159.

Sakthivel, S.; Neppolian, B.; Shankar, M.V.; Arabindoo, B.; Palanichamy, M. & Murugesan, V. (2003) Solar photocatalytic degradation of azo dye: Comparison of photocatalytic efficiency of ZnO and TiO_2. *Sol. Energy Mater. Sol. Cells*, Vol. 77 No. 1, pp.65-82.

Santarelli, F. & Smith, J.M. (1974) Rate of photochlorination of liquid n-heptane. *Chem. Eng. Commun.*, Vol. 1, pp. 297-302.

Sathishkumar, P.; Sweena, R. & Wu, J.J. (2011) Anandan, S. Synthesis of CuO-ZnO nanophotocatalyst for visible light assisted degradation of a textile dye in aqueous solution. *Chem. Eng. J.*, Vol. 171 No. 1 pp. 136-14.

Sayama, K.; Hayashi, H.; Arai, T.; Yanagida, M.; Gunji, T. & Sugihara, H. (2010) Highly active WO_3 semiconductor photocatalyst prepared from amorphous peroxotungstic acid for the degradation of various organic compounds. *Appl. Catal. B: Environ.*, Vol. 94 No. 1-2 pp.150-157.

Schechter, R.S. & Wissler, E.H. (1960) Photochemical reactions in an isothermal laminar-flow chemical reactor. *Appl. Sci. Res.*, Vol. 9 No. 1, pp. 334-344

Shirotsuka, T. & Nishiumi, H. (1971) Theoretical basis for making an apparatus for photochemical reactions. *Kagaku Kogaku*, Vol. 35, pp. 1329-1338.

Song, S.; Xu, L.; He, Z.; Chen, J.; Xiao, X. & Yan, B. (2007) Mechanism of the photocatalytic degradation of C.I. reactive black 5 at pH 12.0 using $SrTiO_3/CeO_2$ as the catalyst. *Environ. Sci. Technol.*, Vol. 41 No. 16 pp. 5846-5853.

Sopyan, I.; Watanabe, M.; Murasawa, S.; Hashimoto, K. & Fujishima, A. (1996) An efficient TiO_2 thin-film photocatalyst: photocatalytic properties in gas-phase acetaldehyde degradation. *J. Photochem. Photobiol.*, Vol. 98 No. 1-2. pp. 79-86.

Taghiour, F. & Mohseni, M. (2005) CFD simulation of UV photocatalytic reactors for air treatment. *AIChE J.*, Vol. 51 No. 11 pp. 3039-3047.

Tang, W.Z. & Huang, C.P. (1995) Photocatalyzed oxidation pathways of 2,4-dichlorophenol by CdS in basic and acidic aqueous solutions. *Water Res.*, Vol. 29 No. 2 pp. 745-756.

Teramura, K.; Tanaka, T.; Kani, M.; Hosokawa, T. & Funabiki, T. (2004a) Selective photo-oxidation of neat cyclohexane in the liquid phase over V_2O_5/Al2O3. *J. Mol. Catal. A: Chem.*, Vol. 208 No. 1-2 pp. 299-305.

Teramura, K.; Tanaka, T.; Hosokawa, T.; Ohuchi, T.; Kani, M. & Funabiki, T. (2004b) Selective photo-oxidation of various hydrocarbons in the liquid phase over V_2O_5/Al_2O_3. *Catal. Today*, Vol. 96 No. 4 pp. 205-209.

Torres-Martínez, C.L.; Kho, R.; Mian, O.I. & Mehra, R.K. (2001) Efficient photocatalytic degradation of environmental pollutants with mass-produced ZnS nanocrystals. *J. Colloid Interface Sci.*, Vol. 240 No. 2 pp. 525-532.

Van Grieken, R.; Aguado, J.; López-Muoz, M.J. & Marugán, J. (2002) Synthesis of size-controlled silica-supported TiO_2 photocatalysts. *J. Photochem. Photobiol. A: Chemistry*, Vol. 148 No. 1-3, pp. 315-322.

Vincent, G.; Schaer, E.; Marquaire, P.-M. & Zahraa, O. (2011) CFD modelling of an annular reactor, application to the photocatalytic degradation of acetone. *Process Saf. Environ. Prot.*, Vol. 89 No. 1 pp. 35-40.

Waldner, G.; Brüger, A.; Gaikwad, N.S. & Neumann-Spallart, M. (2007) WO_3 thin films for photoelectrochemical purification of water. *Chemosphere*, Vol. 67 No. 4 pp. 779-784.

Wang, K.-H.; Tsai, H.-H. & Hsieg, Y.-H. (1998) The kinetics of photocatalytic degradation of trichloroethylene in gas phase over TiO_2 supported on glass beads. *Appl. Catal. B: Environ.*, Vol. 17 No. 1-2 pp. 25-36.

Wilkes, J.O. Fluid mechanics for chemical engineers with microfluidics and CFD, 2nd ed. Prentice Hall: NJ, 2005.

Williams, J.A. & Ragonese, F.P. (1970) Asymptotic solutions and limits of transport equations for tubular flow photoreactors. *Chem. Eng. Sci.*, Vol. 25 pp. 1751-1759.

Williams, J.A. (1978) The radial light intensity profile in cylindrical photoreactors. *AIChE J.*, Vol. 24 pp. 335-338.

Xu, Y.; Zheng, W. & Liu, W. (1999) Enhanced photocatalytic activity of supportedTiO_2: Dispersing effect of SiO_2. *J. Photochem. Photobiol. A: Chemistry*, Vol. 122, No.: 1, pp. 57-60.

10

Fluid-Structure Interaction of a Radial Turbine

Zied Driss, Sarhan Karray,
Hedi Kchaou and Mohamed Salah Abid
University of Sfax (US)
National Engineering School of Sfax (ENIS)
Laboratory of Electro-Mechanic Systems (LASEM)
Tunisia

1. Introduction

Fluid-structure interaction (FSI) plays a very important role in many industrial applications as diverse as aerospace, automotive, biomedical, civil and nuclear engineering. In the coupling between fluid and structure, the deforming structure modifies the boundary conditions for the fluid due to the motion of the fluid boundary. At the same time, the fluid flow causes varying loading conditions on the structure. Though efficient solvers for both the fluid and the structural dynamics exist, the development of tools for modelling various fluid-structure interaction problems remains a challenge. In order to describe FSI problems, an exchange of data has to take place between the fluid and structural fields. Based on this data exchanged, the methods for solving fluid-structure interaction problems can be divided into monolithic and partitioned couplings. For example, we can mention the researcher of Bobovnika et al. (2005). In this paper, the authors are interested to couple a finite-volume model to a finite-element model of the straight-tube Coriolis flowmeter. Michler et al. (2004) have developed a monolithic approach to fluid-structure interaction. Piperno and Farhat (2001) are interested to partitioned procedures for the transient solution of coupled aeroelastic problems. Sternel et al. (2008) are developed an efficiency and accuracy of fluid-structure interaction simulations using an implicit partitioned approach. Van Brummelen et al. (2003) are studied the energy conservation under incompatibility for fluid-structure interaction problems. The monolithic coupling consists of solving both parts of fluid and structure in the same system of equations. The choice of time step is only limited by the required precision. However, this type of algorithm requires a development of a computer code. Moreover, the numerical methods employed for the fluid and the structure domains are different and are difficult to place within the same code, hence the utility of the partitioned algorithm methodology. In this method, the fluid and structural parts are resolved separately. This method has been introduced by Park and Felippa (1983), and further investigated by Wood (1990). Problems dealing with FSI involve the coupling of Computational Fluid Dynamics (CFD) and Computational Structure Dynamics (CSD) codes. Various numerical simulations study the fluid behaviour, when the movement of

the structure is prescribed analytically (Ralph and Pedley, 1985; Natarajan and Mokhtarzadeh-Dehghan, 2000). Other researchers concentrated on the fluid part, while a simple structural model for a rigid body was used (Cossu and Morino, 2000). Further simplifications have been done by neglecting the dynamic effects and simulating static FSI. For example, Beckert et al. (2000) used a multivariate interpolation scheme for coupling fluid and structural models in three dimensional spaces. They applied it to static aeroelastic problems, in order to predict the equilibrium of elastic wing models in transonic fluid flow. The main advantage of the partitioned approach is that it allows already developing efficient and well validated solvers for both the fluid and structure subtasks to be combined. Some theoretical and numerical studies of partitioned coupling algorithms for one and two dimensional problems can be found (Cebral et al., 1995). Commercial codes for combining existing solvers have been developed. Seiber (2002) developed an efficient coupling algorithm combining FASTEST-3D code and FEAP code for solving various fluid-structure interaction problems in three-dimensional domains for arbitrary elastic structures. Glück et al. (2003) applied a partitioned coupling between the CFD code (FASTEST-3D) and the CSD code (ASE) to thin shells and membranous structures with large displacements. The latter method have been modified and coupled by MpCCI. Bucchignani et al. (2004) presented a numerical code to study the problem of an incompressible flow in a stirred vessel. It was based on a method of a partition treatment type, with the fluid and structural fields resolved by coupling two distinct models. Wang (2008) provided an effective new idea to solve aeroelasticity problems, in which the tools Fluent and ABAQUS/ANSYS are employed.
In this chapter, a computational analysis of the fluid-structure interaction in a stirred vessel equipped with a radial turbine is presented. The hydrodynamic behavior of the turbine was studied previously (Nagata, 1975; Suzukwa et al., 2006). However, they did not take into account the effect of the structural deformations. For this reason, we are interested to study the turbulent flow in a stirred vessel as well as the mechanical deformation of the structure. In a stirred vessel, the fluid can have a significant effect on the deformation of the blades during mixing depending on the turbine speed and the fluid flow. So, a coupling algorithm has been developed to predict the equilibrium of the elastic blade of the radial turbine in a turbulent flow.

2. FSI method

The FSI problems involve the coupling between the CFD and the CSD codes. This coupling is used in the partitioned method (Bobovnika et al., 2005 ; Driss et al., 2007 ; Sternel et al., 2008) when the fluid and structural parts are resolved separately. The coupling interface allows the exchange of the pressures at interface, when the mesh in each fluid and structure field is different and is difficult to place within the same code, hence the utility of the partitioned algorithm methodology. The coupling interface organizes the transfer of information and makes an effective coupling between the fluid and the structure fields. At the fluid-structure interface, the results from the CFD code are applied as a load to the CSD code. The displacements are calculated using the CSD code to characterize the turbine blade deformation. To predict numerically the FSI characteristics of the radial turbine, we have developed a numerical method to simulate the flow depending of the structure displacements. This method involves three modules: the pre-processor, the processor and the post-processor. Data is interchanged between these three modules. The implementation of the computational simulation requires many hypotheses such as the geometrical concept, the choice of the meshing type, the physical models, the boundary conditions and the discretization method (Driss et al., 2010).

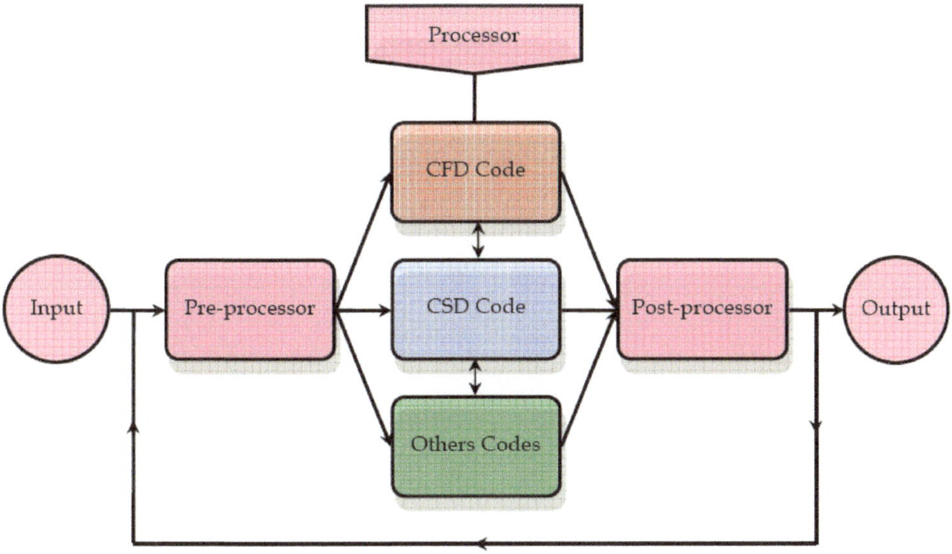

Fig. 1. Schematic presentation of the FSI method

2.1 Pre-processor

The pre-processor serves as a source of input for the CFD or CSD codes by means of an operator interface. This is transformed into a suitable form for the processor. The user activities at the pre-processing stage involve definition of the geometry, mesh generation and specification of the appropriate boundary conditions. The computational model requires that the volume occupied by the fluid inside the vessel is described by a computational grid or cells. In these cells, variables are computed and stored. The computational grid must fit the contours of the vessel and its internals, even if the components are geometrically complex. In our CFD applications, a new meshing method is developed to study the impellers turbines with complex geometries. This method permits the numerical analyses of turbines with simple and complex geometries (Driss et al., 2005, 2007, 2010). In this method, the computer aided design (CAD) is used at first to construct the turbine shape. After that, a list of nodes is defined to belong to the interface separating the solid domain from the fluid domain passing through the computational structure dynamics (CSD) code. Using this list, the mesh in the flow domain is automatically generated for the three-dimensional simulations. After that, a staggered mesh is used in such a way that four different control volumes are defined for a given node point, one for each of the three vector components and one for the scalar variable. Therefore, the region to be modeled is subdivided into a number of control volumes defined on a cylindrical coordinates system. As application to this method, we are interested to study a stirred tank equipped by a radial turbine (Fig. 2). The tank is a vertical cylindrical vessel with a height-to-diameter ratio H/D of 1. The shaft is placed concentrically with a diameter ratio s/D of 0.04. The tip to tip impeller diameter ratio d/D is a 0.5. The turbine has four blades characterised by the length equal to d/2 and the height equal to h. The turbine is placed in a mid-height tank (z=0.5 H). The geometry system resembles that already studied by Suzukawa et al. (2006) and Nagata (1975). For the numerical computations, the Young's modulus and the Poisson's ratio are equal to $E=2.1\ 10^{11}$ MPa and v=0.3 respectively. In the following investigations, the fluid is assumed to be incompressible and Newtonian. The structure is assumed to be isotropic linear and elastic material law is applied.

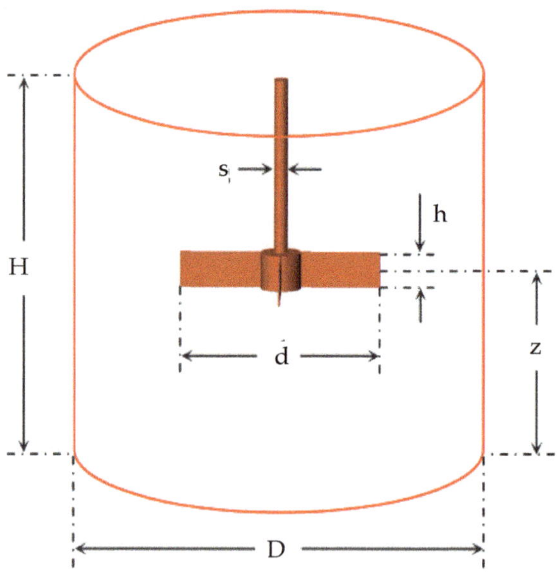

Fig. 2. Stirred tank equipped with a radial turbine

2.2 Processor

In the processor, the coupling interface is used to allow the transfer of the pressure and the displacement between the CFD and the CSD codes. The pressure provided by the CFD code is used to calculate the forces applied to the structure. The displacements of the structure are used to update the control volume and the boundary conditions.

2.2.1 CFD code

The CFD code resolves the incompressible Navier-Stokes equations in conjunction with the standard k-ε turbulence model. The continuity equation is written as follows:

$$div\,\vec{V}_f = 0 \tag{1}$$

The momentum equation is given in the following form:

$$\frac{\partial(\rho_f\,\vec{V}_f)}{\partial t} + \mathrm{div}\left(\rho_f\,\vec{V}_f \otimes \vec{V}_f + \mathrm{p}\bar{\bar{\mathrm{I}}}\right) = \mathrm{div}\left(\bar{\bar{\tau}} + \bar{\bar{\tau}}_R\right) + \rho_f\left(\mathrm{N}^2\,\vec{r} - 2\vec{N} \wedge \vec{V}_f\right) \tag{2}$$

The standard k-ε turbulence equations are given in the following form:

$$\frac{\partial k}{\partial t} + div\left[\vec{V}_f\,k - \frac{2}{\pi}\left(\frac{d}{D}\right)^2 \frac{1}{\mathrm{Re}}\left(1 + \frac{\upsilon_t}{\sigma_k}\right)\overrightarrow{grad}\,k\right] = \frac{2}{\pi}\left(\frac{d}{D}\right)^2 \frac{1}{\mathrm{Re}}\,\mathrm{G} - \varepsilon \tag{3}$$

$$\frac{\partial \varepsilon}{\partial t} + div\left[\vec{V}_f\,\varepsilon - \frac{2}{\pi}\left(\frac{d}{D}\right)^2 \frac{1}{\mathrm{Re}}\left(1 + \frac{\upsilon_t}{\sigma_\varepsilon}\right)\overrightarrow{grad}\,\varepsilon\right] = \frac{\varepsilon}{k}\left[\mathrm{C}_{1\varepsilon}\,\frac{2}{\pi}\left(\frac{d}{D}\right)^2 \frac{1}{\mathrm{Re}}\,G - \mathrm{C}_{2\varepsilon}\,\varepsilon\right] \tag{4}$$

The turbulent kinetic energy production takes the following form:

$$G = \upsilon_t\left[2\left[\left(\frac{\partial U}{\partial r}\right)^2 + \left(\frac{\partial V}{r\partial\theta} + \frac{U}{r}\right)^2 + \left(\frac{\partial W}{\partial z}\right)^2\right] + \left[\frac{\partial V}{\partial r} - \frac{V}{r} + \frac{\partial U}{r\partial\theta}\right]^2 + \left[\frac{\partial W}{r\partial\theta} + \frac{\partial V}{\partial z}\right]^2 + \left[\frac{\partial U}{\partial z} + \frac{\partial W}{\partial r}\right]^2\right] \tag{5}$$

The k-ε model is based on the concept of turbulent viscosity where k and ε are calculated by solving the equation above. The value of the turbulent viscosity is deduced from the following equation:

$$\upsilon_t = \mathrm{C}_\mu\,\frac{\pi}{2}\left(\frac{D}{d}\right)^2 \mathrm{Re}\,\frac{k^2}{\varepsilon} \tag{6}$$

Numerical methods consist on an approximation of the unknown flow variables by means of simple functions and a discretisation by substitution of the approximation into the governing flow equations and subsequent mathematical manipulations (Figure 3). The finite volume method consists on the formal integration of the governing equations of fluid flow over all the control volumes of the solution domain. The transport equations are integrated over its own control volume using the hybrid scheme discretization method. The pressure-velocity coupling is handled by the SIMPLE algorithm of Patankar (1980). The algebraic equations solutions are obtained in reference to the fundamental paper published by Douglas and Gunn (1964). The control volume integration, distinguishes the finite volume method from all other computers techniques. The resulting statements express the conservation of relevant properties for each finite size cell.

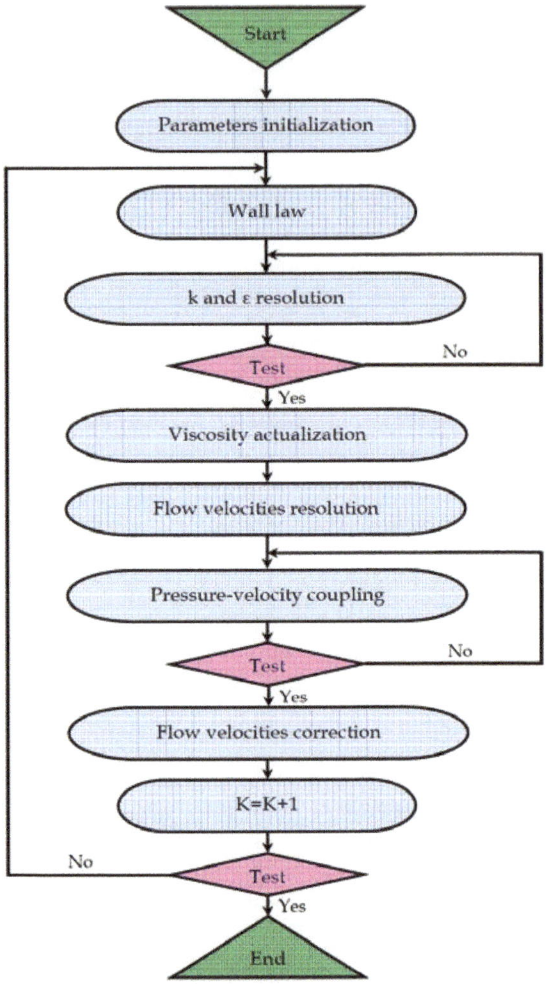

Fig. 3. Schematic presentation of the CFD code

2.2.2 CSD code

The CSD code allows the static calculation of the structure and solves the partial differential equations of the structure model. In this study, elastic isotropic structures were considered. The continuity equation and the balance of the momentum equation are given in the following form:

$$\frac{D\rho_s}{Dt} + div\left(\rho_s \vec{V}_s\right) = 0 \tag{7}$$

$$\rho_s \frac{D\vec{V}_s}{Dt} - \overrightarrow{div}\,\overline{\overline{\sigma}} = \vec{f}_{ex} \tag{8}$$

To achieve this goal, the CSD code must read the input file already created. After that, the implementation edition can be preceded. At this stage, the new parameters of our application are introduced. Particularly, the analysis type, the mesh shape, the boundary conditions and the result field of the problem are redefined. After the program execution, different results can be visualized like the nodes displacement, the strain and the Von Mises stress (Fig. 4).

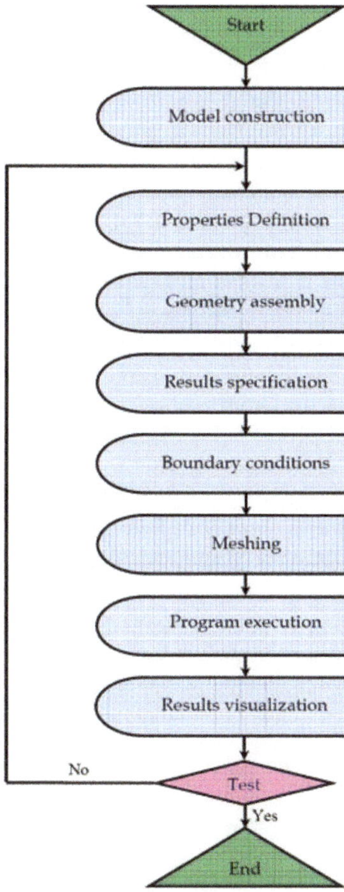

Fig. 4. Schematic presentation of the CSD code

2.2.3 Coupling algorithm

In this chapter, we have used a partitioned coupling algorithm between the fluid and structure solver. The fluid-structure interface represents the contact mechanics problem between an elastic structure and the fluid flow. This requires the characterization of boundary conditions exchanged and describe the interaction between the fluid and the structure. Two conditions on the level of the interface are used. These conditions are given in the following form:

$$\vec{V}_s \vec{n} = \vec{V}_f \vec{n} \tag{9}$$

$$\overline{\overline{\sigma}}\,\vec{n} = \overline{\overline{\tau}}\,\vec{n} \tag{10}$$

With:

$$\vec{n} = \vec{n}_f = -\vec{n}_s \tag{11}$$

In the case of no-slip walls, the compatibility condition for partitioned coupling states that the velocity at the fluid-structure interface must be shared by both the fluid and the structure at the interface. The coupling interface ensures the transfer of the pressure and displacement mentioned above, when the mesh in each fluid and structure codes is different. It organizes the transfer of information in time and makes an effective coupling between the fluid and structure codes. Forces equilibrium must also exist at the interface. The force applied by the fluid to the structure and vice versa should be of the same magnitude but in opposing directions. The structural and fluid equations are solved independently. Loads and boundary conditions are exchanged after each converged increment. The quantities exchanged from the CFD and the CSD codes are respectively the pressure and displacement modes. Then, the coupling algorithm is employed in which the CFD and CSD solvers run concurrently and exchange solution quantities at the end of each converged time step. The advantage here is that in order to solve the basic equations, different numerical solution schemes and analysis software can be applied. The coupling algorithm is based on the sequential process (Fig. 5). In each iteration step, the pressure, calculated in CFD code, is used to determine the force exerted by fluid flow. The geometry is then updated after the calculation of the structural displacement by the CSD code. Therefore, the role of the coupling algorithm is to exchange nearly any kind of data between the coupled codes. The iteration steps are stopped when the convergence is reached.

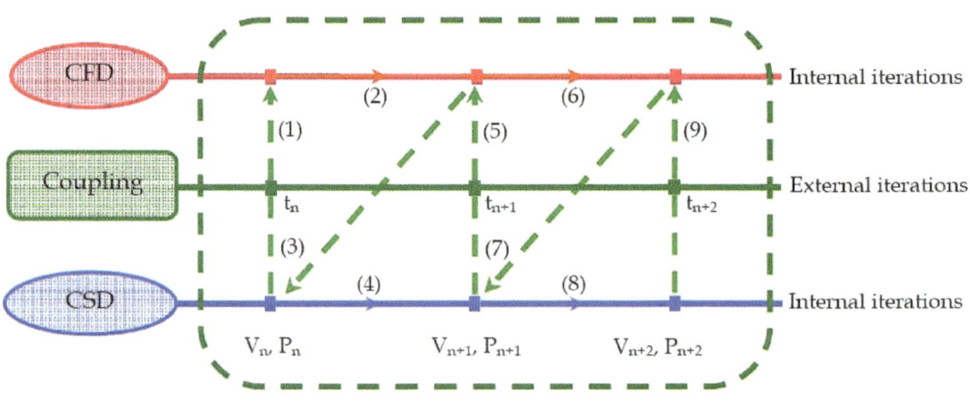

Fig. 5. Coupling algorithm

2.3 Post-processor

The post-processor defines the third module presenting the results obtained after the execution of the numerical codes. A huge resource has been devoted to the development post-processing techniques and display. These include geometry domain, grid display, vector plots, surface plots and view manipulation. Several results can be gotten such as the velocity field, the viscous dissipation rate and the turbulence characteristics. Also, it's possible to exchange data through the post-processor with other codes or other interfaces [4].

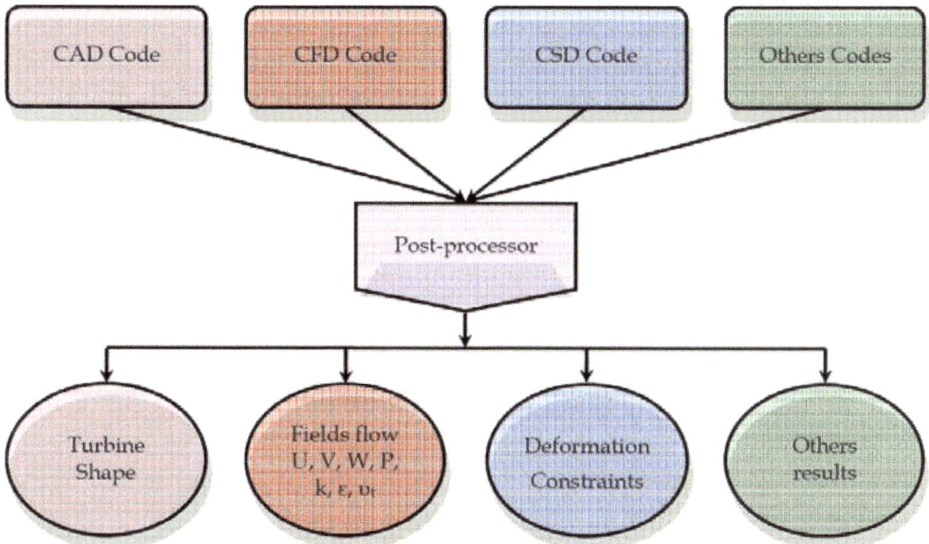

Fig. 6. Schematic presentation of the post-processor

3. Fluid-structure results

3.1 CFD code results

The hydrodynamic results are obtained after execution of our CFD code. Particulary, we are interested to the velocity fields, the turbulent kinetic energy, the dissipation rate of the turbulent kinetic energy and the turbulent viscosity. In these conditions, the flow simulation are defined by the Reynolds number Re=100000 and the Froude number Fr=0.19.

3.1.1 Flow patterns

Figures 7 and 8 show the velocity vector plot (U,W) in two r-z planes defined respectively by the angular coordinate equal to $\theta=35°$ and $\theta=57°$. These presentation planes are situated respectively in the upstream and in the downstream of the blade plane. The distribution of the velocity field shows the presence of a radial jet on the level of the turbine which changes against the walls of the tank with two axial flows thus forming two recirculation zones on the two sides of the turbine. Far from the region swept by the turbine, we note a deceleration of the flow. Even if the turbine attains a shape deformation due to the flexion of the blade, the velocity field slightly differs from iteration to another. In fact, during the first three iterations the hydrodynamic structure of the flow varies from iteration to another. From the fourth iteration, the field velocity starts to be preserved and the turbine preserves his deformed shape.

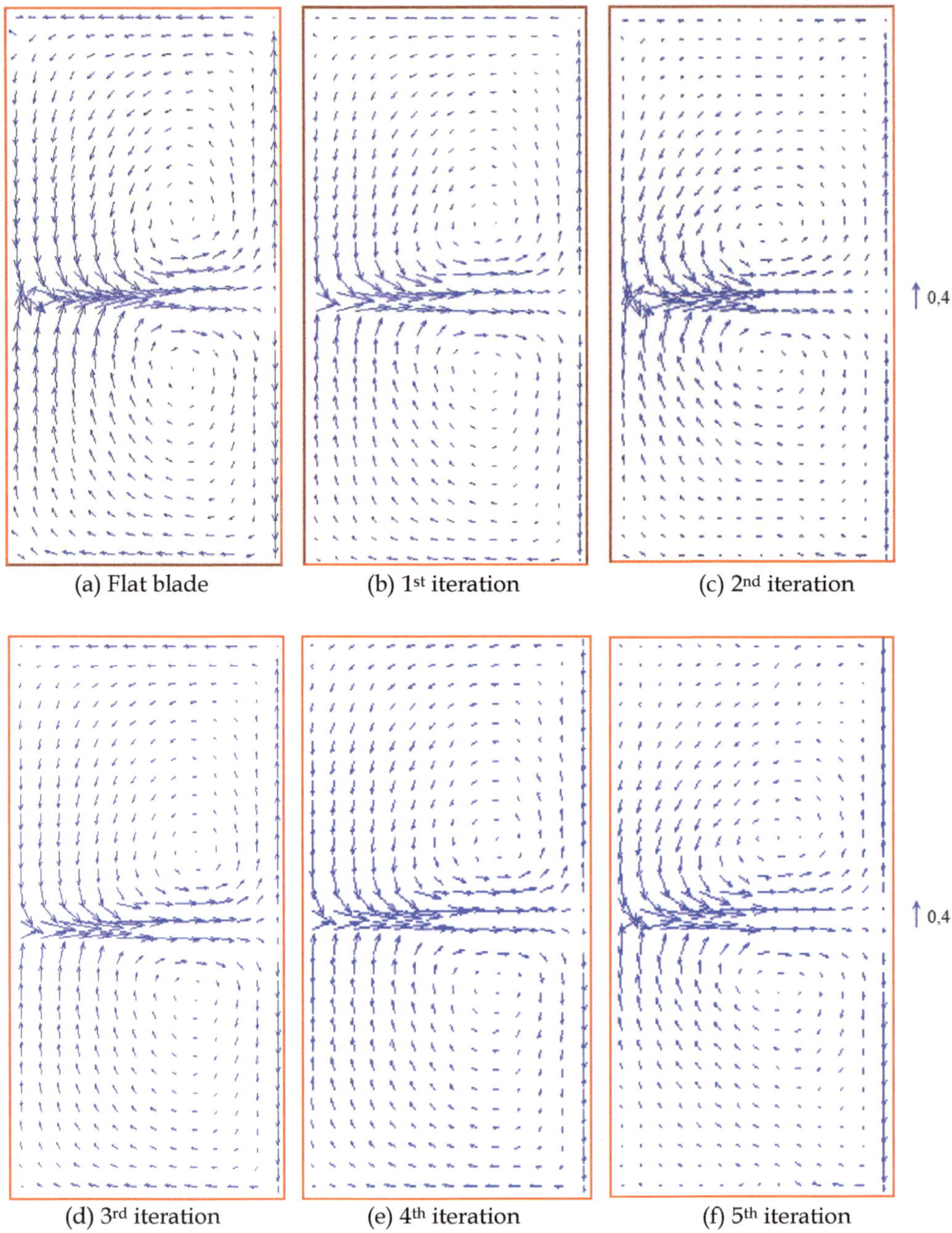

Fig. 7. Flow patterns induced in r-z plane (θ=33°)

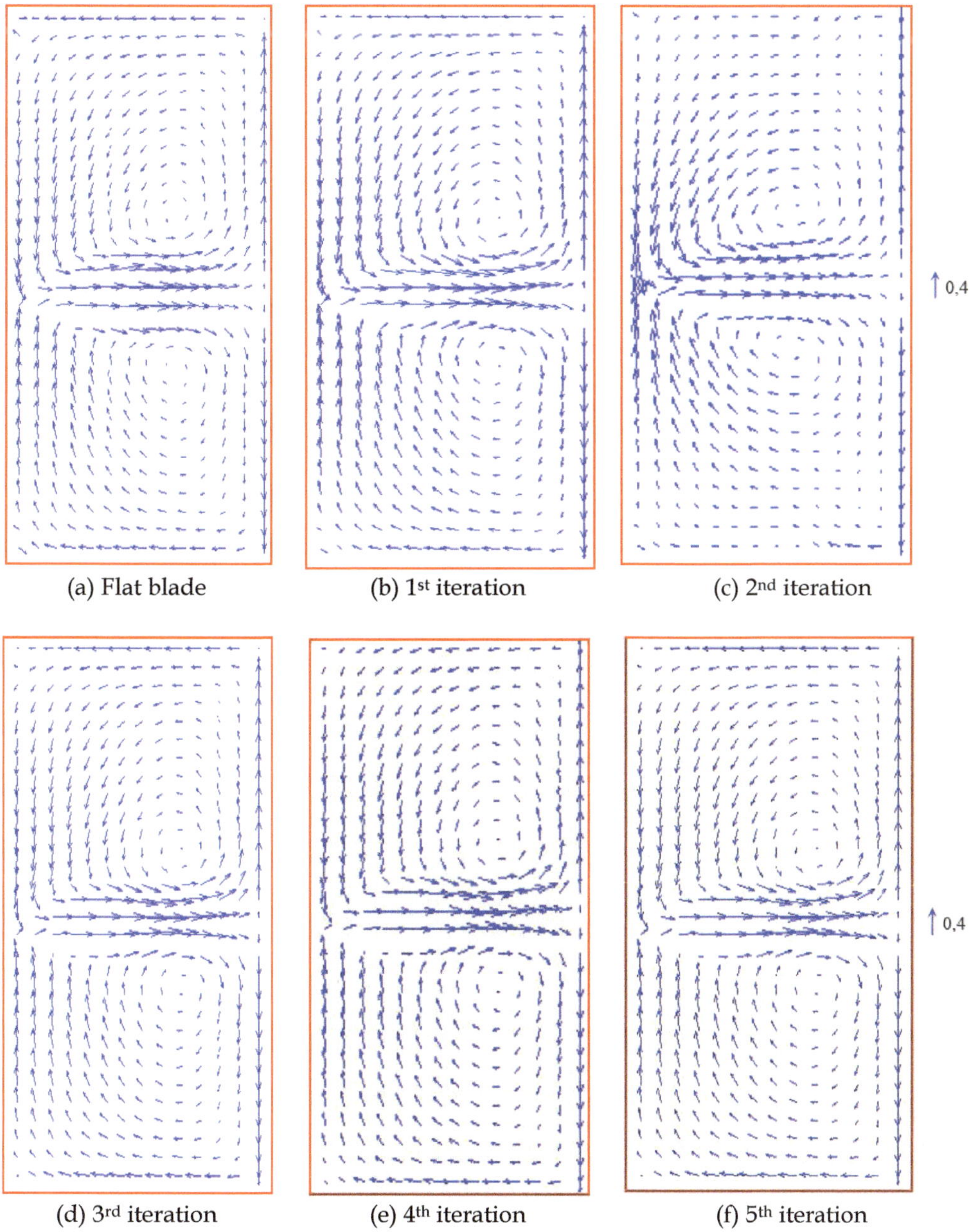

Fig. 8. Flow patterns induced in r-z plane (θ=57°)

3.1.2 Axial profile of the radial velocity

Figure 9 presents the axial profiles of the dimensionless radial velocity component U(z). These profiles are defined in the flat-blade downstream. Under the case, the dimensionless radial coordinates are equal to r=0.33 (Figure 9.a), r=0.5 (Figure 9.b) and r=0.75 (Figure 9.c). These figures adequately portray the swirling radial jet character. Within a flat blade, the axial profile shows a parabolic pace. The extremer is defined at the mid-height of the blade. The same fact is observed within a retreated blade in the 5th iteration. But, the radial velocity component value decreases. These results are also confirmed in the research paper results published by Driss et al. (2005). In this case, we superimposed the experimental results founded within a flat blade by Nagata (1975). The good agreement between the experimental results and the numerical results confirms the validity of the analysis method.

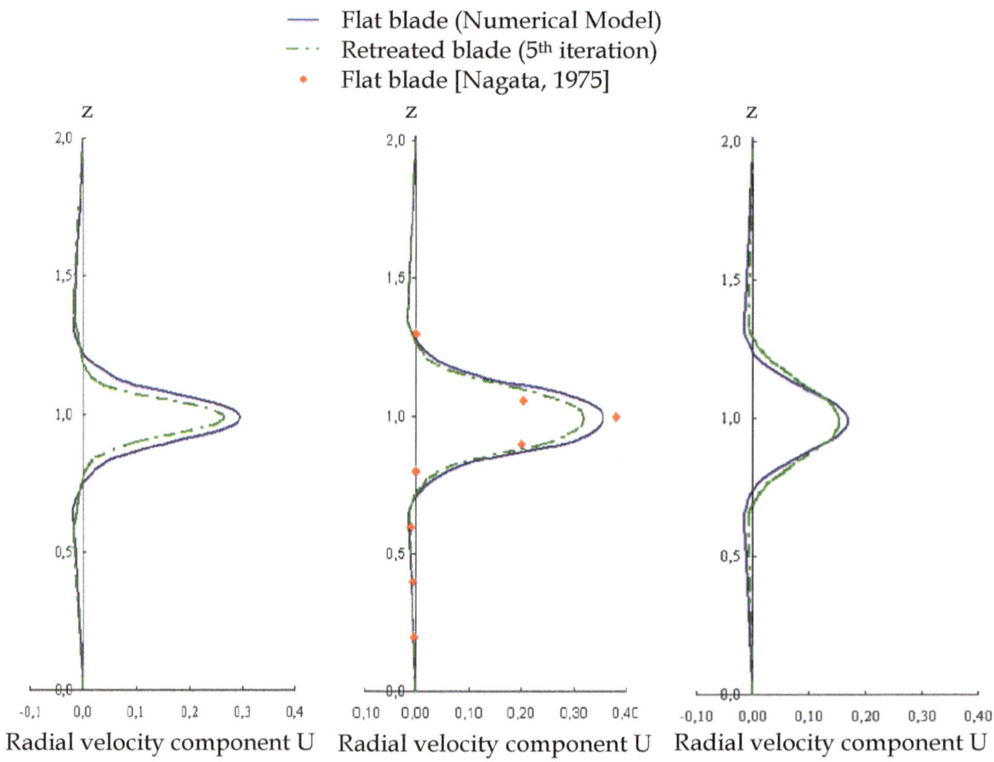

(a) Radial position r=0.33 (b) Radial position r=0.5 (c) Radial position r=0.66

Fig. 9. Axial profiles of the radial velocity component

3.1.3 Turbulent kinetic energy

Figure 10 presents the distribution of the turbulent kinetic energy in r-z plane defined by the angular coordinate equal to θ=35°. The presentation plane is situated in the upstream of the blade plane. During five iterations of the coupling algorithm, it's noted that the area of the maximum values are located in the wake which develops an upstream of blades. In all fields swept by the turbine blades, the turbulent kinetic energy remains rather high. Out of this field, the turbulent kinetic energy becomes very weak. The comparison of these results confirms that the maximum values are reached during the second iteration.

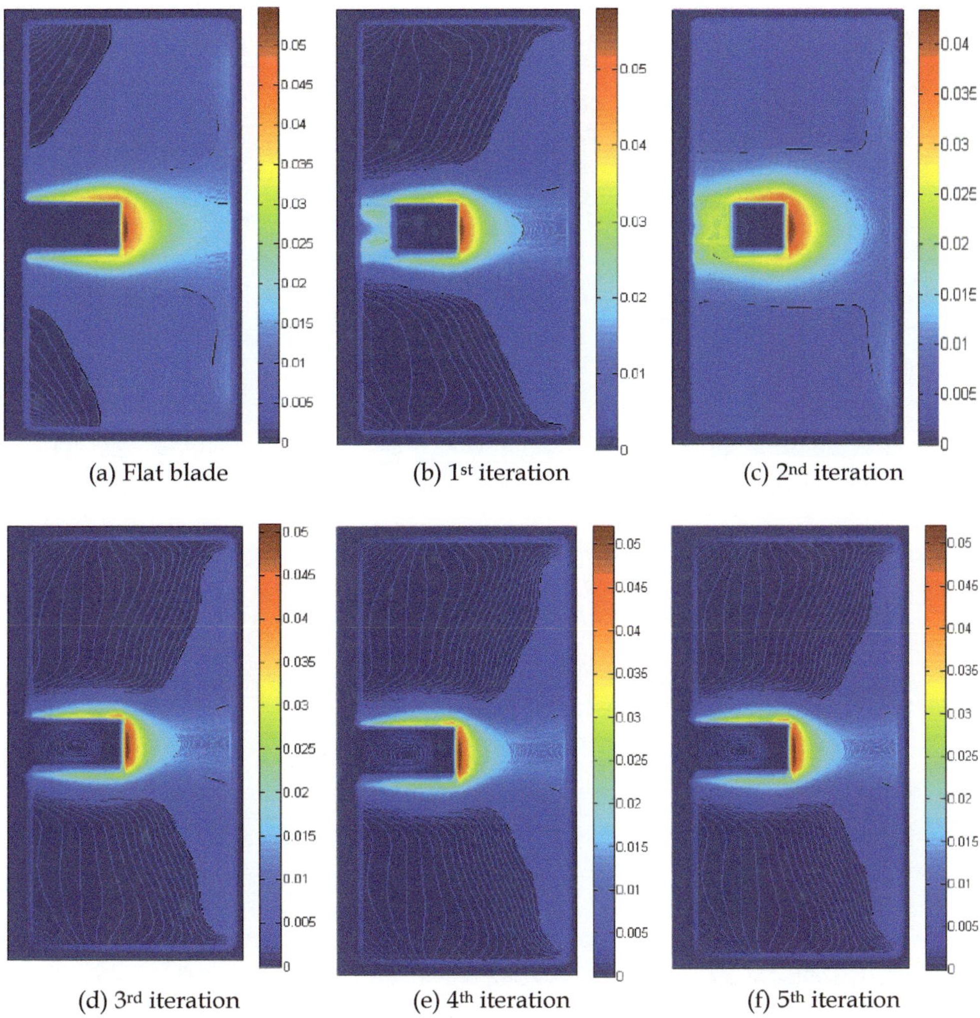

Fig. 10. Distribution of the turbulent kinetic energy k in r-z plane (θ=35°)

3.1.4 Dissipation rate of the turbulent kinetic energy

Figure 11 presents the evolution of the distribution of dissipation rate in r-z plane defined by the angular coordinate equal to θ=35° during iterations of the coupling algorithm. Globally, we observe a distribution similar to that already obtained with the turbulent kinetic energy. Indeed, it is noted that the area of the maximum values are located in the wake which develops nearly to the blades end. However, the dissipation rate becomes very weak outside the field swept by the blades. The greatest rate is reached in the fourth and the fifth iterations (Fig. 11-d and 11-e).

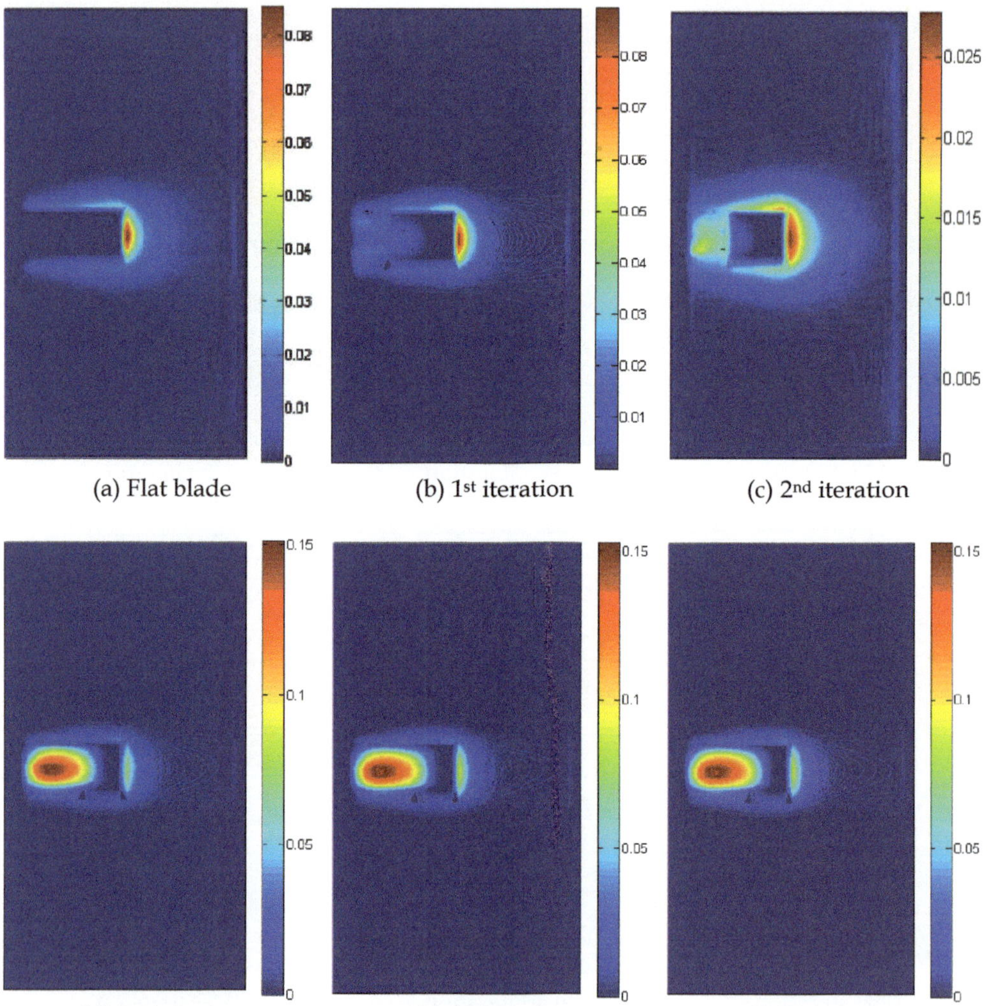

(a) Flat blade (b) 1st iteration (c) 2nd iteration

(d) 3rd iteration (e) 4th iteration (f) 5th iteration

Fig. 11. Distribution of dissipation rate ε in r-z plane (θ=35°)

3.1.5 Turbulent viscosity

Figure 12 presents the evolution of the turbulent viscosity distribution in r-z plane defined by the angular coordinate equal to θ=35° during iterations of the coupling algorithm. In the field swept by the turbine blades, the turbulent viscosity remains rather high. Near the walls and around the axis of the turbine, the turbulent viscosity undergoes a very fast fall. This is due to the deceleration of the flow. The comparison of these results confirms that the maximum value is reached in the second iteration (Fig. 12-c).

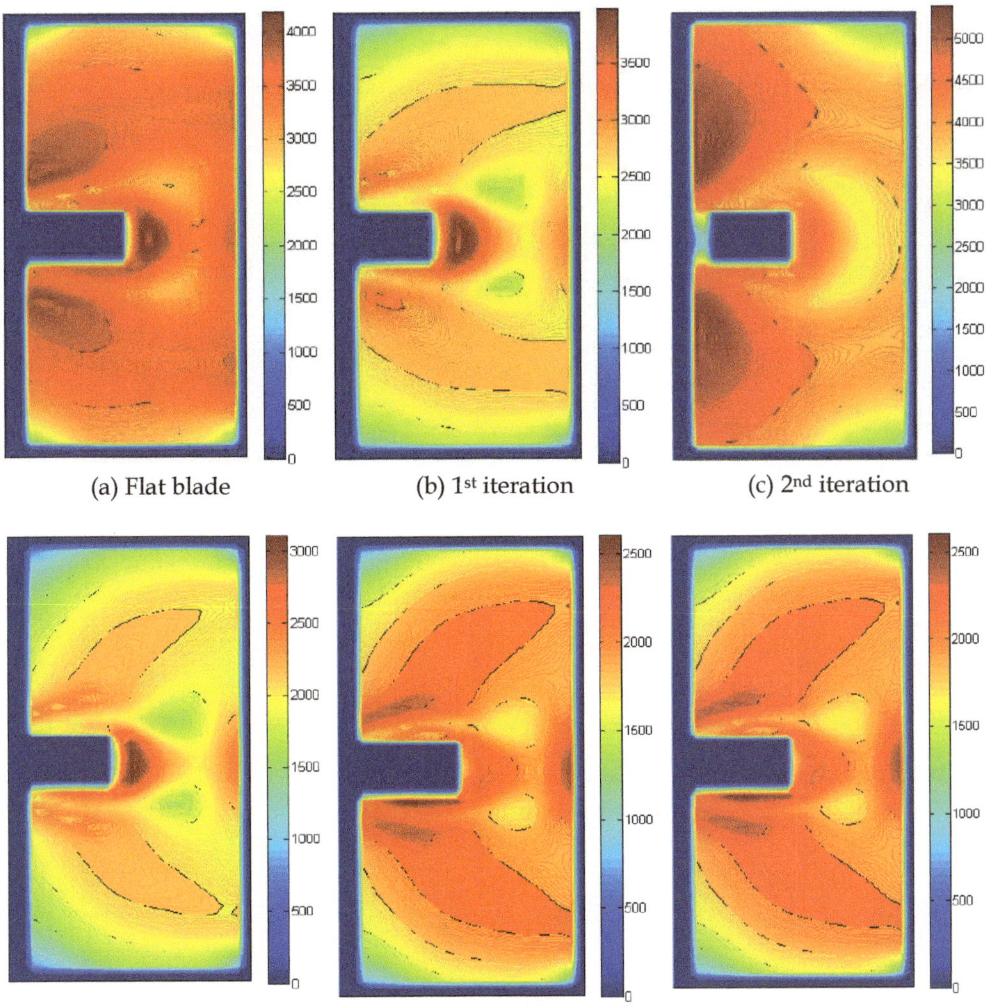

(a) Flat blade (b) 1st iteration (c) 2nd iteration

(d) 3rd iteration (e) 4th iteration (f) 5th iteration

Fig. 12. Distribution of the turbulent viscosity in r-z plane (θ=35°)

3.2 CSD code results

In this part, we are interested to the radial turbine structure in order to determine the displacement field, the deformed turbine shape and the Von Mises Stress. These results are conducted from the CSD code. The discretization of the domain is ensured by a finite elements method.

3.2.1 Displacement field

Figure 13 presents the evolution of the displacement field of the blades during various iterations of our coupling algorithm. Particularly, we are interested to the displacement of the nodes belonging to the blade. This step is fundamental in the coupling algorithm process to can deduce the new meshing domain. This operation is repeated several times until obtaining the equilibrium. Under these conditions, it is supposed that the deformation of the turbine axis due to torsion is negligible.

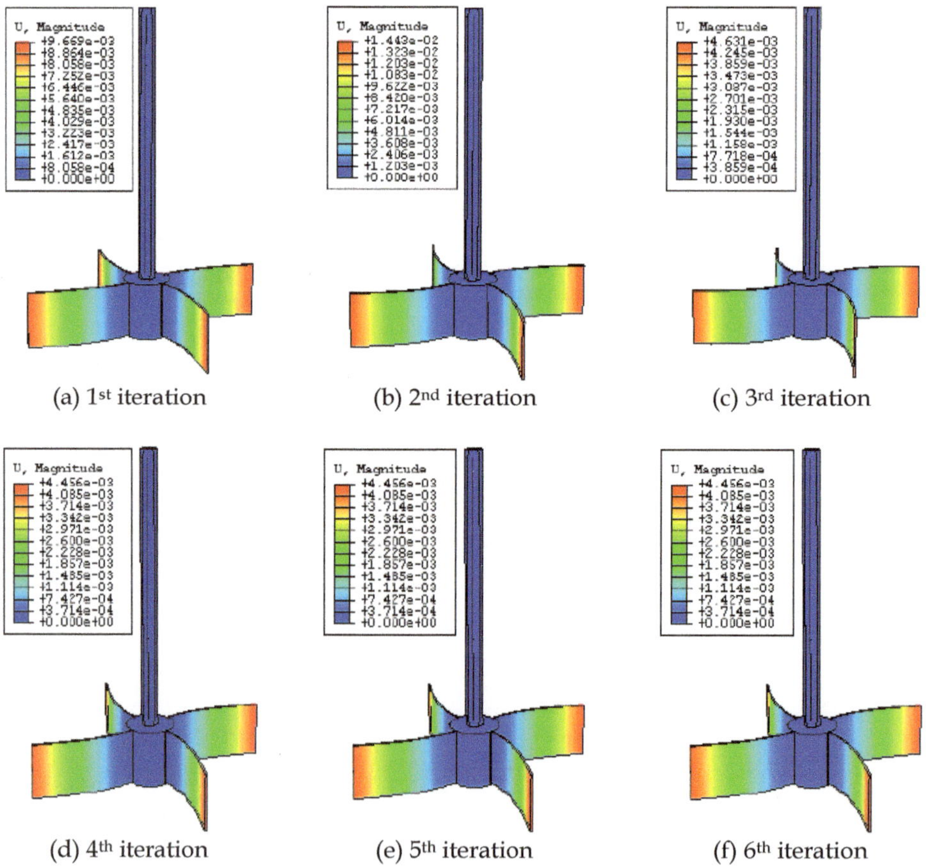

(a) 1st iteration (b) 2nd iteration (c) 3rd iteration

(d) 4th iteration (e) 5th iteration (f) 6th iteration

Fig. 13. Displacement field of the radial turbine

3.2.2 Deformed turbine shape

Figure 14 presents the evolution of the deformed turbine shape during iterations of the coupling algorithm. These results are obtained after the execution of the CSD code. The comparison between the results from iteration to another, it's noted that the maximum value is reached in the third iteration (Fig. 14-c). In this case, the retreated angle, defined between the tangent plane at the end of the blade and the radial plane passing by the same end and the axis of rotation to the turbine (Driss et al., 2007), is equal to ξ=16° 40''.

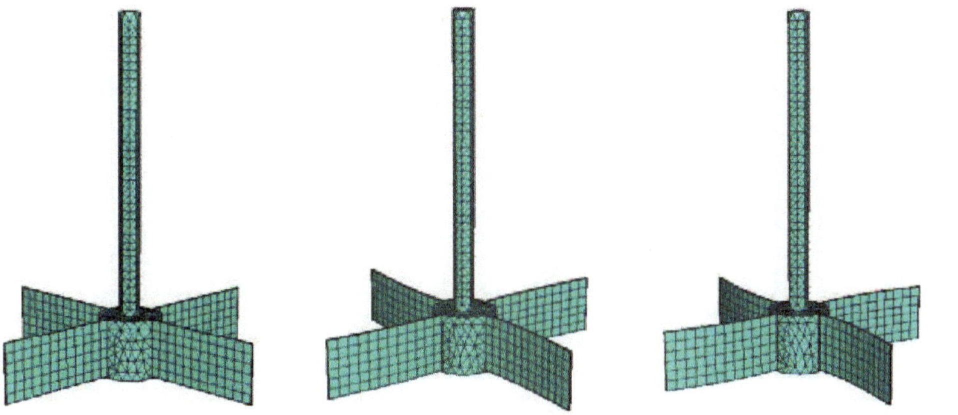

(a) 1st iteration: ξ=0° (b) 2nd iteration: ξ=11° 12'' (c) 3rd iteration: ξ=16° 40''

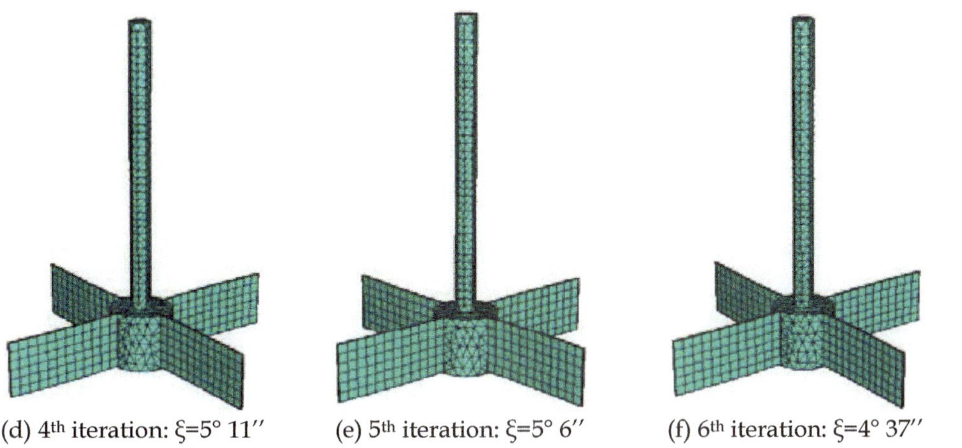

(d) 4th iteration: ξ=5° 11'' (e) 5th iteration: ξ=5° 6'' (f) 6th iteration: ξ=4° 37''

Fig. 14. Deformed shape of the radial turbine

3.2.3 Von Mises stress

Figure 15 presents the distribution of the Von Mises stress during iterations of the coupling algorithm. These results are obtained after the execution of the CSD code. The comparison of the results from iteration to another, it's noted that the maximum value is reached in the second iteration (Fig. 15-b).

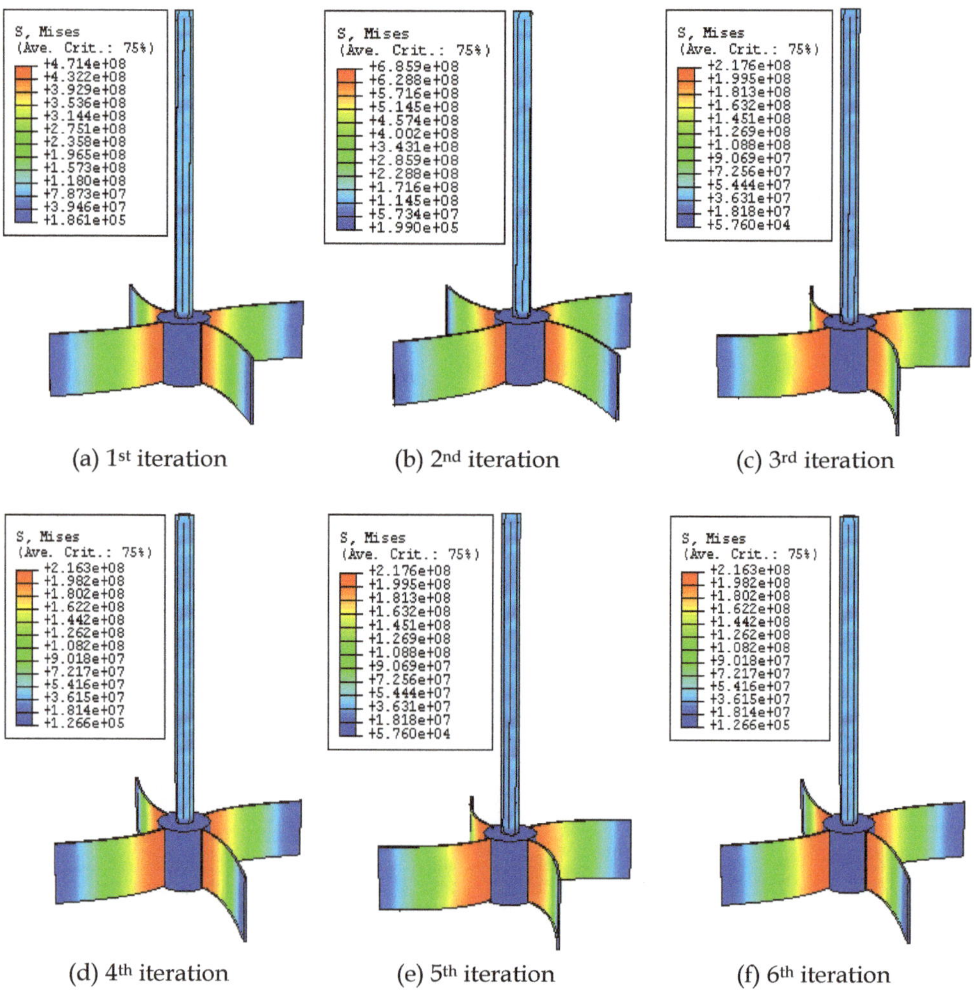

Fig. 15. Distribution of the Von Mises stress

4. Conclusion

In this chapter, we have coupled a computational fluid dynamics (CFD) code with a computational structural dynamics (CSD) code by using an efficient coupling interface for solving fluid-structure interaction (FSI) problems in the stirred tank equipped by a radial turbine. Specific techniques of rearrangement of grid and treatment of the boundary conditions are used to follow the behavior of the system. This method takes advantage of the parallel process involved within each analysis code. This allows both parts of the fluid-structure interaction problem to be solved in the best possible way: a Finite Volume Method for the fluid dynamics and a finite element method for the structure. The CFD results obtained allow a visualizing of the velocity field, the turbulent kinetic energy, the dissipation rate of the turbulent kinetic energy, the turbulent viscosity. The CSD results permit to obtain the displacement field, the deformed shape and the Von Mises stress. These results proof that the fluid flow can have a significant effect on the deformation of the turbine blades. So, we can conclude that it's fundamental to take consideration of this phenomenon in the design of the industrial process. In the future, we propose to develop the numerical method to can study instationary FSI problem.

5. Nomenclature

$C_{1\varepsilon}$	constant in the standard k-ε model
$C_{2\varepsilon}$	constant in the standard k-ε model
C_{μ}	constant in the standard k-ε model
d	turbine diameter, m
D	tank diameter, m
E	Young's modulus, MPa
Fr	Froude number
G	turbulent kinetic energy production
h	blade height, m
k	turbulent kinetic energy
N	velocity of the turbine, s^{-1}
p	pressure
r	radial position
Re	Reynods number
s	shaft diameter, m
t	time
U	radial velocity components, dimensionless
V	tangential velocity components, dimensionless
W	axial velocity components, dimensionless
z	axial position
ε	dissipation rate of the turbulent kinetic energy
ρ	density
υ_t	turbulent viscosity, dimensionless
v	Poisson's ratio

σ_k	constant in the standard k-ε model
σ_ε	constant in the standard k-ε model
θ	angular coordinate, rad
ξ	retreated angle
$\vec{n}$	normal vector
$\vec{r}$	vector position
$\vec{V}$	velocity vector, $m.s^{-1}$
$\overline{\overline{\tau}}$	stress tensor
$\overline{\overline{\tau}}_R$	Reynolds tensor
$\overline{\overline{I}}$	identity tensor
$\overline{\overline{\sigma}}$	stress tensor of the structure
$\vec{f}_{ex}$	exterior force

Indices

f	fluid
s	structure

6. References

Bobovnika, G.; Mole, N.; Kutina, J.; Stokb, B. & Bajsic, I. (2005). Coupled finite-volume/finite-element modeling of the straight-tube Coriolis flowmeter, Journal of Fluids and Structures, 20, 785-800.

Bucchignani, E.; Stella F. & Paglia F. (2004). A partition method for the solution of a coupled liquid-structure interaction problem, Applied Numerical Mathematics, 51, 463-475.

Cebral, C.; Piperno, S. & Larrouturou B. (1995). Partitioned procedures for the transient solution of coupled aeroelastic problems. Part 1: Model problem, theory and two-dimensional application, Journal of Computer Methods in Applied Mechanics and Engineering, 124, 79-112.

Cossu C. & Morino L., On the instability of a spring-mounted circular cylinder in a viscous flow at low Reynolds numbers, Journal of Fluids and Structures, 14, 183-196.

Driss, Z.; Kchaou H., Baccar, M. & Abid M.S. (2005). Numerical investigation of internal laminar flow generated by a retreated-blade paddle and a flat-blade paddle in a vessel tank, International Journal of Engineering Simulation, 6, 10-16

Driss, Z. (2008). Contribution in studies of the turbines in an agitated vessel, PhD thesis, National School of Engineers of Sfax, University of Sfax, Tunisia

Driss, Z.; Bouzgarrou, G.; Kchaou, H.; Abid, M.S. (2011). Computer simulation of the laminar flow in stirred tanks generated by the proximity impellers of a mono and double screws type with simple and modified profiles, Mechanics & Industries, 12, 109-121

Driss, Z.; Karray, S.; Kchaou, H.; Abid, M.S. (2007). Computer Simulations of Fluid-Structure Interaction Generated by a Flat-Blade Paddle in a Vessel Tank, International Review of Mechanical Engineering, 1, 608-617

Driss, Z.; Bouzgarrou, G.; Chtourou, W.; Kchaou, H.; Abid, M.S. (2010). Computational studies of the pitched blade turbines design effect on the stirred tank flow characteristics, European Journal of Mechanics B/Fluids, 29, 236-245

Douglas, J. & Gunn, J. E. (1964). A general formulation of alternating direction implicit methods, Num. Math. 6, 428-453

Glück, M.; Breuer, M.; Durst, F.; Hlfmann A. & Rank, E. (2003). Computation of wind-induced vibrations of flexible shells and membranous structures. Journal of Fluids and Structures 17 739-765

Karray, S.; Driss, Z.; Kchaou, H. & Abid M.S. (2011). Hydromechanics characterization of the turbulent flow generated by anchors impellers, Engineering Applications of Computational Fluid Mechanics, (in press), 5, 1-14

Karray, S.; Driss, Z.; Kchaou, H. & Abid M.S. (2011). Numerical simulation of fluid-structure interaction in a stirred vessel equipped with an anchor impeller, Journal of Mechanical Science and Technology, (in press), 25, 1-12

Michler, C.; Hulshoff, S. J., Van Brummelen E. H. & De Borst R. (2004). A monolithic approach to fluid-structure interaction, Computers & Fluids, 33, 839-848

Nagata, S. (1975). Mixing: principles and applications, Halstead press, Japan.

Natarajan S. & Mokhtarzadeh-Dehghan M. R. (2000). Numerical prediction of a (potential) soft acting peristaltic blood pump. International Journal for Numerical Methods in Fluids, 32, 711-724

Patankar, S. V. (1980). Numerical heat transfer and fluid flow. Series in Computational Methods in Mechanics and Thermal Sciences, McGraw Hill, New York.

Piperno, S. & Farhat C. (2001). Partitioned procedures for the transient solution of coupled aeroelastic problems, Part II: energy transfer analysis and three-dimensional applications, Computer Methods in Applied Mechanics and Engineering. 190, 3147-3170

Ralph, M.E. & Pedley T.J. (1989). Viscous and inviscid flows in a channel with a moving indentation, Journal of Fluid Mechanics. 209, 543-566

Sternel, D. C.; Schäfer, M.; Heck M. & Yigit, S. (2008). Efficiency and accuracy of fluid-structure interaction simulations using an implicit partitioned approach, Comput Mech., 43, 103-113

Sieber, G. (2002). Numerical Simulation of Fluid-Structure Interaction Using Loose Coupling Methods, PhD thesis, at the Department of Numerical Methods in Mechanical Engineering, Darmstadt University of Technology

Suzukawa, K.; Mochizukib, S. & Osaka, H. (2006). Effect of the attack angle on the roll and trailing vortex structures in an agitated vessel with a paddle impeller, Chemical Engineering Science, 61, 2791- 2798

Van Brummelen, E.H.; Hulshoff S.J. & De Borst, R. (2003). Energy conservation under incompatibility for fluid-structure interaction problems, Computer Methods in Applied Mechanics and Engineering, 192, 2727-2748

Wang, Y. (2008). Combination of CFD and CSD packages for fluid-structure interaction, Journal of Hydrodynamics, 20, 756-761

Wood, W.M. (1990). Practical Time-stepping Schemes, Oxford University Press, New York

11

Industrial Sprays: Experimental Characterization and Numerical Modeling

Avinash Khopkar[1], Michael D. Cloeter[2] and Quan Yuan[2]
[1]*Dow Chemical International Pvt. Ltd., Pune*
[2]*The Dow Chemical Company, Freeport, TX*
[1]*India*
[2]*USA*

1. Introduction

Sprays in industrial processes are becoming more common as well as more challenging for design. Spray formation processes are critical to the performance of a number of technologies and applications. The spray applications can be quite varied and include combustion systems (gas turbine engines, internal combustion engines, incinerators, furnaces, rocket engines), agriculture (pesticide and herbicide treatments), paints and coatings (furniture, automobiles), consumer products (cleaners, personal care products), fire suppression systems, spray cooling (materials processing, computer chip cooling), gas treatment (thermal quench, acid neutralization), medicinal/pharmaceutical, reactor feed injections, liquid distribution into chemical units (defoaming, uniform flux over column packing), and spray drying (foods, drugs, materials processing).

The process requirement varies depending upon the application. For example, making small drops are important for internal combustion engines and other applications where rapid mass transfer into the gas phase is critical. Large drops may be desired where entrainment in a counter-current gas flow must be minimized. In another case, creating a narrow drop size distribution, neither too small or too large, is desired for agricultural sprays when trying to achieve low driftable fines while at the same time maintaining good target coverage.

Controlling the spread of drop distribution plays very important role in paints and coatings, consumer products, medicinal delivery, and spray drying applications. On the other hand, distributing liquid mass more uniformly throughout the spray and increasing the fraction of liquid that impacts a target are key requirements in paints and coatings, spray cooling, and fire suppression applications.

Injector designers face challenges in designing the atomization process for achieving the desired process output. Despite immense progress in laser diagnostics for spray flows, the reliable experimental data on atomization characteristics are still scarce due to measurement difficulties associated with the dense spray zones formed around the liquid jet core. As a result, application of empirical correlations for atomization flow characteristics has a very narrow range of applicability. It is therefore, essential to develop and apply new tools to enhance our understanding of the atomization process. Such an understanding will be useful in devising cost effective and reliable scale-up of spraying operations.

Research support for industrial processes and products must advance with these challenges. This involves both experimental and numerical capabilities. It is important to critically evaluate the current understanding of experimental and numerical capabilities and translate this understanding into improving spraying operation. Such an attempt is made here.

2. Experimental methods for spray analysis

Imaging methods such as laser-induced fluorescence are used to track velocity profiles before, during, and after liquid film breakup, as well as to track distribution of mass flux and drop size distribution (DSD). Governing liquid break-up regimes depend on level of turbulence and interactions with gas phase, and can include unstable jets, waves, perforations, and secondary atomization. The interaction of micro-structured liquids with turbulence can have a profound impact on breakup dynamics. Even slight formulation changes can have unexpectedly large impact on the spray properties.

Multiple examples have been chosen for illustrating the variation in spray problems encountered in practice. Certain commonalities in the analysis are evident regardless of the spray problem.

In industrial problems, not all sprays can be studied in a lab setting using state-of-the-art laser diagnostics. Frequently, the spray may consist of a relatively large flowrate (100+ gpm) that is not amenable to laboratory research. The researcher needs facilities to characterize such large sprays, including gas-assisted sprays. The authors' facilities have a nozzle test stand for evaluating such sprays. See Figure 1.

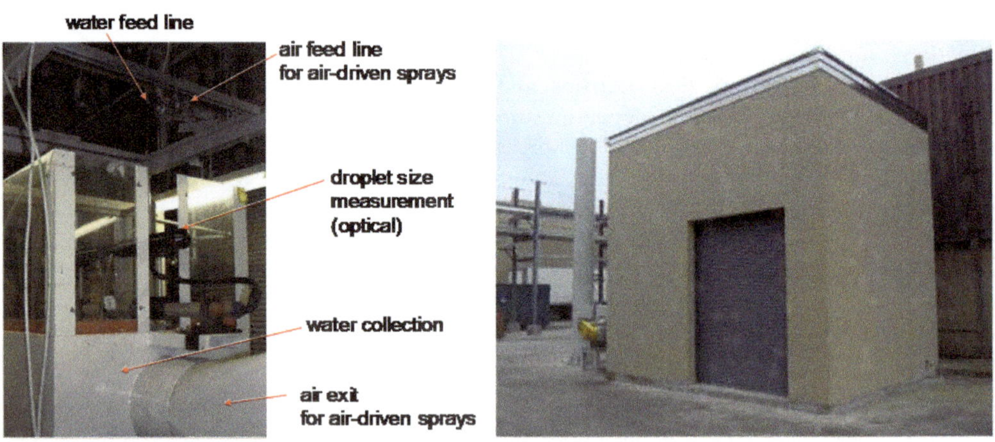

Fig. 1. Interior and exterior view of the nozzle test stand at The Dow Chemical Company, Freeport, Texas, USA.

The use of an enclosed structure is necessary due to the dangerously loud noise levels generated by certain sprays, particularly industrial gas-assisted sprays. A blower and stack is included to assist air flow and prevent mist accumulation inside the test stand building. Rented pumps and compressors are brought to the stand when the needs of the project exceed the capability of permanent pumps and the plant air utility.

At other times, the researcher must go to the spray, which can be in a plant or pilot facility. Thus the tools of researcher must include an array of capabilities. The tools described herein vary from the very simple and portable to the more complex and fixed methods that have been used by the authors. The following descriptions start with the simplest options and continue to describe those that are more complex.

2.1 Flash photography

Flash photography is a basic method that allows one to capture a view of a spray over the small duration (< 0.1 ms) of the flash. This is often a short enough exposure that the spray appears frozen in time. Important basic features of the spray can be revealed such as spray angle, symmetry of the bulk flow, distance of the breakup zone, breakup mechanism, and basic shape of the mass distribution (when viewed from several angles). The effect of flow rate on each of these variables can also be studied.

This method, along with video, has been extremely useful to the authors for initial characterization of many sprays. It can be used in "out-of-lab" environments such as the test stand shown in Figure 1 and production plants. Figure 2, for instance shows comparisons from the test stand and a plant thermal quench arrangement. The flow rate exceeds 100 gpm.

Fig. 2. Left, image of an axially-vaned thermal quench nozzle in operation in the authors' test stand. The aim of this test was to quantify the spray angle and asymmetries of the plant nozzle. Several flow rates and nozzle designs were tested. Right, field test in plant.

This method has been used for viscous and non-Newtonian sprays as well. Figure 3 shows images from a tangential whirl nozzle with flow of a 200 cP corn syrup-water solution. While this solution can be considered Newtonian, the effect of shear-thinning or other non-

Newtonian behavior on nozzle performance can also be quantified by testing solutions with a low-shear viscosity near 200 cP.

The Reynolds number for the conditions of Figure 3 is 500. It can be seen that the observed spray angle of 60 degrees is substantially less than the published (water) spray angle of 90 degrees. It is also seen that an "S"-shaped pattern is observed as opposed to the hollow cone observed for water. The sheet breakup starts at 6 inches from the nozzle tip. Perforated sheet breakup is observed, as well as Rayleigh-Plateau instabilities near the edge of the spray.

All of these characteristics are observed through a standard digital SLR camera. Many nozzles can be tested in this way in a single afternoon in order to find the right nozzle for a given application.

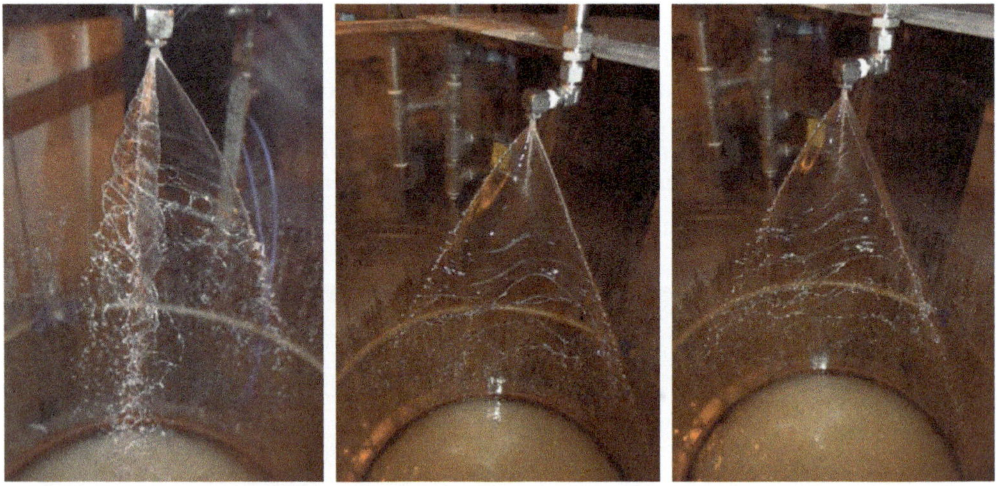

Fig. 3. Tangential whirl nozzle with 200 cP corn syrup/water solution. Flow conditions 27 psi, 1.45 gpm, Reynolds number based on the orifice diameter is 500.

2.2 High-speed imaging

High-speed imaging is a method that can give an estimate of the velocity of the spray by comparing images that are a known time increment apart. For dense sprays, the analysis may be limited to the exterior portions. The individual droplets may not be accessible, but we have found that movement of *structures*, consisting of groups of drops, is sufficient to get a reasonable estimate. High-speed imaging has also been extremely useful for discovery of non-idealities such as pulsations in the flow. These can arise, for instance, from cavitation inside the nozzle or from sub-optimal interior contacting in gas-assisted flows.

An example occurred in a spray nozzle used for a liquid injection of an additive into a gas-phase reactor. By the naked eye, the spray (as reproduced in the lab) looked perfectly acceptable. High-speed imaging, however, showed pulsations (or "surges") in the spray flow that were beyond the ability of the naked eye to perceive. Example images from 1000 Hz imaging are shown in Figure 4. The images of the figure are sequential, showing that the duration of pulses was less than 2 ms. Examination of several seconds of data showed that the pulses occur at a frequency of 50 Hz, above the naked eye perception of ca. 20 Hz. More will be covered on the problem in the *PLIF Imaging* section.

Fig. 4. Consecutive high speed images showing the pulsation effect from an injection nozzle. A frequency of 1000 Hz was used in the imaging. The pulsation frequency was measured as 50 Hz.

Although high-speed cameras are sensitive equipment, they have sufficient portability that we have used them in outside-the-lab environments at reasonable ambient conditions. Figure 5, for instance, shows consecutive images of a dense air-assisted spray that was running in the nozzle test stand. The velocity profile was needed for boundary conditions of a CFD model. Velocities of 60 - 100 ft/s were measured, with the magnitude dependent on the location. In dense sprays, reasonable estimates for velocity can be provided by watching the motion of flow features in the spray, then applying velocimetric principles [Raffel *et al.*, 1998]. Near the edge, tracking the motion of groups of individual drops is also useful.

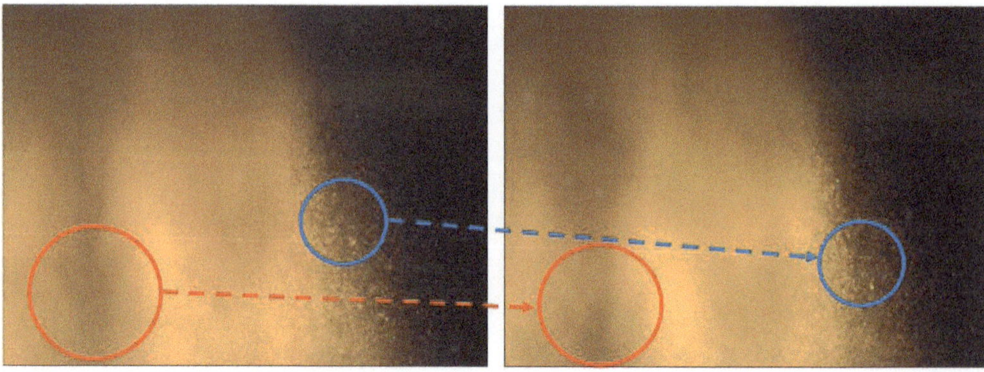

Fig. 5. High speed images (1000 Hz) showing how flow features and collections of drops are used to estimate velocity locally using velocimetry principles. This example is from a dense air-assisted spray located outside the authors' laboratory, hence outside the realm of our laser diagnostics.

2.3 PLIF imaging

When a small amount of a fluorescent dye is added to the feed liquid, Planar Laser-Induced Fluorescence (PLIF) can be used to capture detailed images of the spray as it issues from the nozzle. A plane of pulsed laser light (duration 10 ns) is aligned to pass through the axis of the spray. A camera is aligned perpendicular to the laser plane. The system is synchronized so that the laser pulsates while the camera shutter is open. Figure 6 shows a photographic

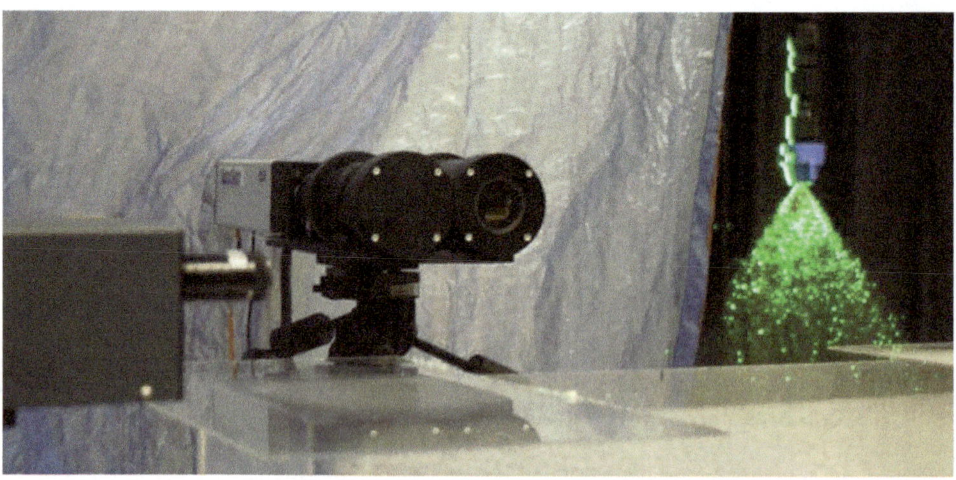

Fig. 6. Photograph taken during a PLIF imaging experiment utilizing a Dantec Dynamics FlowMap System Hub.

image taken during a PLIF experiment to best illustrate the method. In velocimetry mode, dual images using dual laser pulses are taken on the order of 10 to 100 μs apart. The movement of the spray structure or droplets themselves over the known time increment is used to calculate the local velocity. The method has been shown to give consistent velocity results before, during, and after the droplet formation. The velocity distribution can give some information on expected mass flux pattern. However, the method doesn't distinguish between large and small droplets or account for the variable spatial concentration of droplets, so strictly speaking an actual flux cannot be computed from the velocity vector results. Averaging of many PLIF images can give useful information on mass flux within the slice of the spray that is illuminated.

Figure 7 shows a PLIF image of an agricultural spray and the corresponding averaged velocimetric data taken from 500 image pairs. The vectors form a fairly continuous contour that is consistent with the characteristic shouldering of the spray flux.

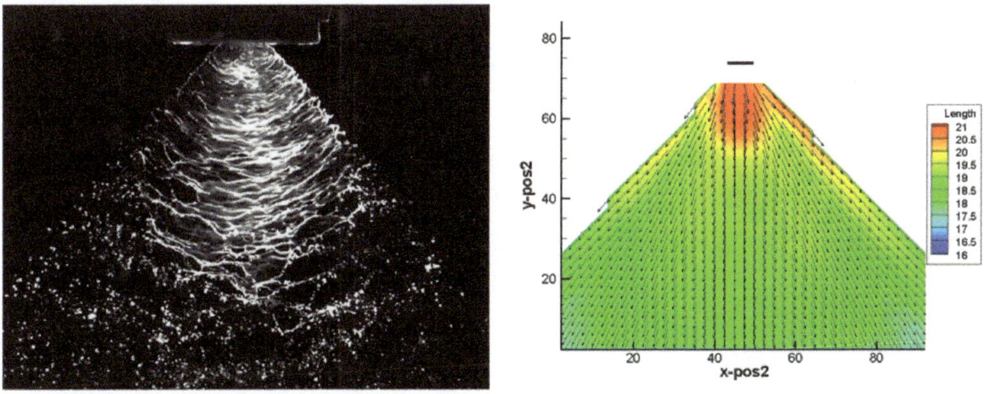

Fig. 7. Left, example PLIF image of a flat fan agricultural spray, Spraying Systems Co. TeeJet 8002. Right, velocity contour plot with overlaid vectors, average of 500 image pairs.

Figure 8 shows PLIF imaging of the same spray as is shown in the high-speed images of Figure 4. The short 10 ns duration of the laser is sufficient to provide rich detail on the spray structure. The pulsating nature of the spray is again evident by comparing images. The PLIF image allows quantitative calculations to estimate the amount of mass issued in the pulses (or "surges") compared to the baseline flow between pulses.

Figure 9 shows a comparison of Mie Scattering and PLIF imaging. The Mie Scattering is sensitive to surface area whereas PLIF imaging is proportional to volume. It is easily noted that a large fraction of the spray volume is present on the outside edges of the spray, and at the same time there are many small droplets present in the interior portion.

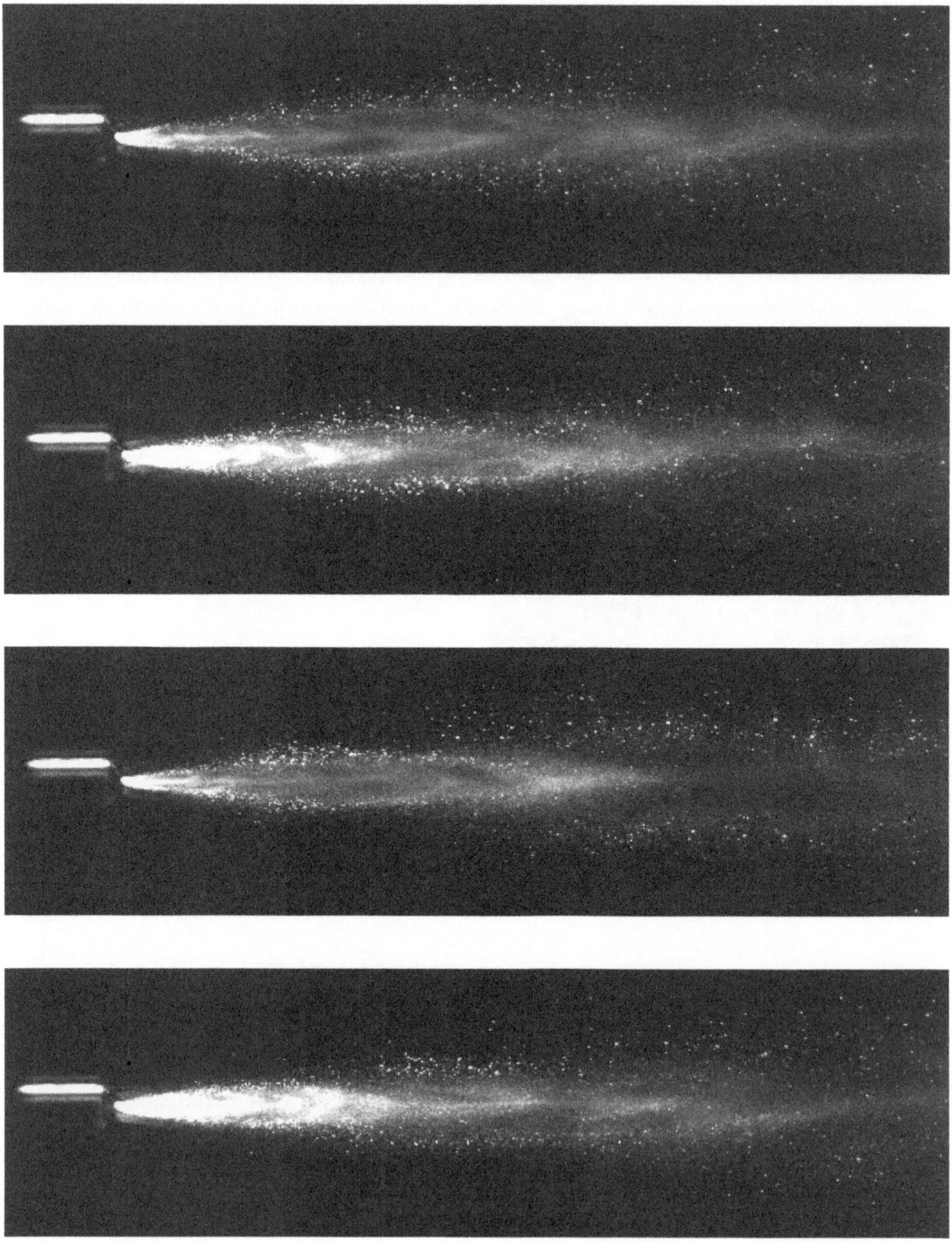

Fig. 8. For the identical nozzle as shown in Figure 4, PLIF imaging at 5 Hz was used to better understand the flow characteristics during and between pulsations.

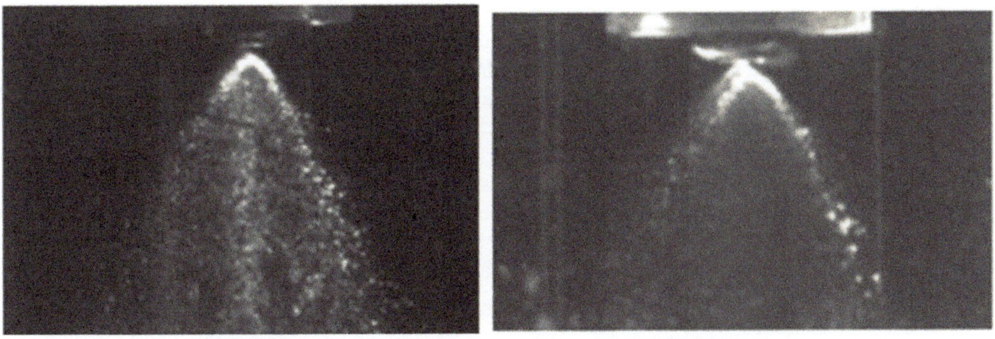

Fig. 9. Mie Scattering (left) and PLIF (right) images, showing surface area versus volume effect.

2.4 PLIF patternation

Similar to the PLIF velocimetry approach, the laser is turned 90 degrees to be orthogonal to the axial direction of the spray. Approximately 10,000 images are taken and averaged, as is shown in Figure 10. Depending on the density of the spray, it may not be possible to permeate the entire spray cross-section, making it necessary to assume symmetry.

Optical patternation techniques are particularly useful when the fluid to be sprayed is hazardous. This can occur, for instance, when spraying active herbicide formulations. The authors' lab uses a transparent ventilated spray enclosure, constructed of acrylic and glass, to protect personnel from contact with the hazardous fluid. The sprayed liquid is fully collected and optionally recycled in order that the amount of waste generated in an experimental series is reduced. The enclosure is also useful for high mist-generating sprays to prevent mist accumulation onto sensitive electronic equipment.

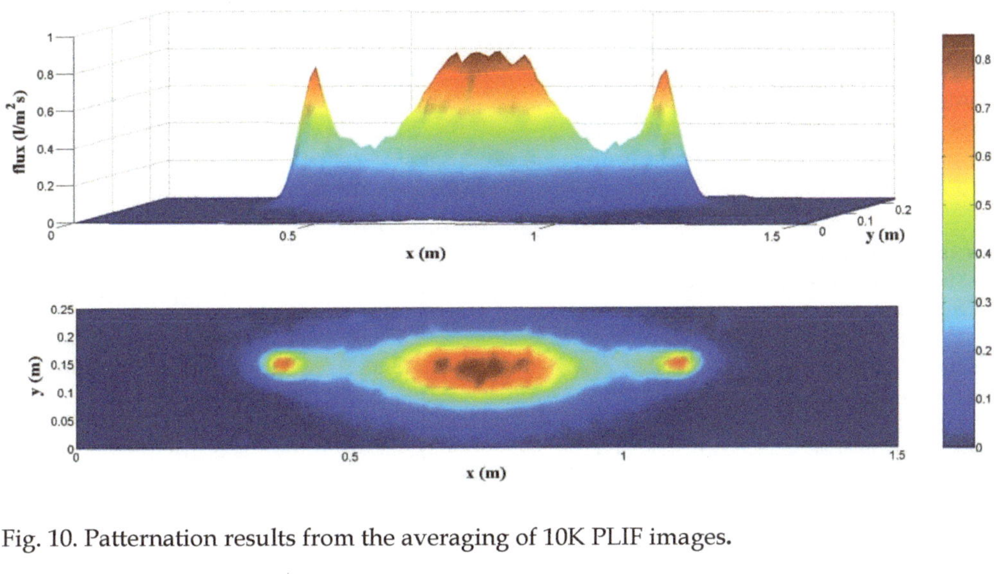

Fig. 10. Patternation results from the averaging of 10K PLIF images.

2.5 Mie scattering for scalar mixing

In cases where the measurement of gas motion and scalar mixing are important to characterize the spray, a fog has been employed to mark the gas flow. In one such method, air is fed to the system which has first been passed sequentially through aqueous solutions of hydrochloric acid (HCl) and ammonium hydroxide (NH_3OH), which efficiently generates a fog of ammonia droplets. This was employed in the apparatus shown in Figure 11. In this case, the fog was introduced in a shroud around the spray nozzle. Commercial oil-atomizing or glycerin-atomizing devices are also available to generate fogs. Smoke-generating testing devices, used to test for negative pressure, can also be employed to qualitatively view the air flow induced by a spray.

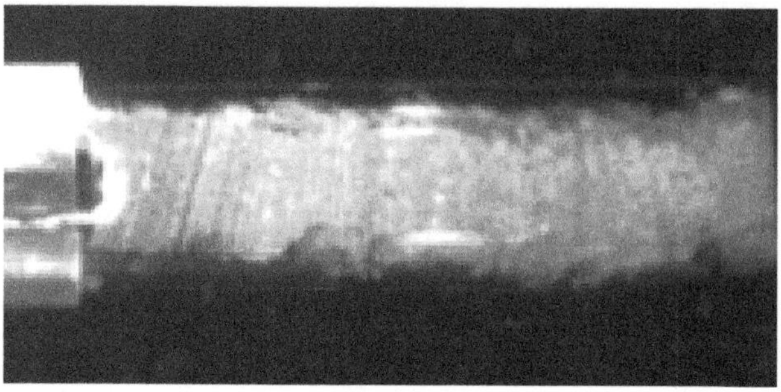

Fig. 11. Mie scattering of a shroud fog introduced around a spray nozzle to examine scalar mixing. The spray nozzle was off when this image was taken.

2.6 Mechanical patternation

The authors have utilized both one-dimensional and two-dimensional mechanical patternation techniques. For sprays with relatively small flowrates (< 0.5 gpm), an arrangement of ½″ x ½″ cuvettes has been utilized to look at two-dimensional patterns. A hinged door covers the cuvettes until the desired flow conditions through the spray nozzle have been established, then the door is rapidly opened. As the fastest-filling cuvette(s) approach full level, the door in returned to its original position and flow is stopped. The contents of each cuvette is then carefully weighed. Figure 12 shows the method and results. Sometimes, 2-D patternation is desired to look at subtleties in the flow pattern at the outside edges and to look at, for instance, turbulence history indicators in the sprayed solution. One such example is spraying of emulsions to study drop size distribution of an oil phase before and after spraying as a function of position in the patternator.

2.7 Dynamic image analysis

We have used this method to capture images of the spray droplets which are then processed in real time. A strobe is used to backlight the imaged area, which is typically of dimensions near 1 cm x 1 cm. Out-of-focus drops are removed from the analysis by defining a sharpness criterion for the droplet. Partial drops, extending beyond the edge of the images, are also removed. Non-spherical droplets are converted to an equivalent diameter by matching the projected area. A sufficient number of images are taken at 30 Hz to achieve converged drop size distribution, commonly 50,000 to 100,000 validated drops. The frequency can be adjusted based on expected drop velocity to assure that the same drops are not counted more than once. Figure 13 shows a sample image taken from dynamic image analysis of a flat fan agricultural spray.

This type of size analysis is commonly used in the authors' lab for nozzles with relatively small-scale flows, although the nozzle test stand of Figure 1 also uses this type of drop size analysis.

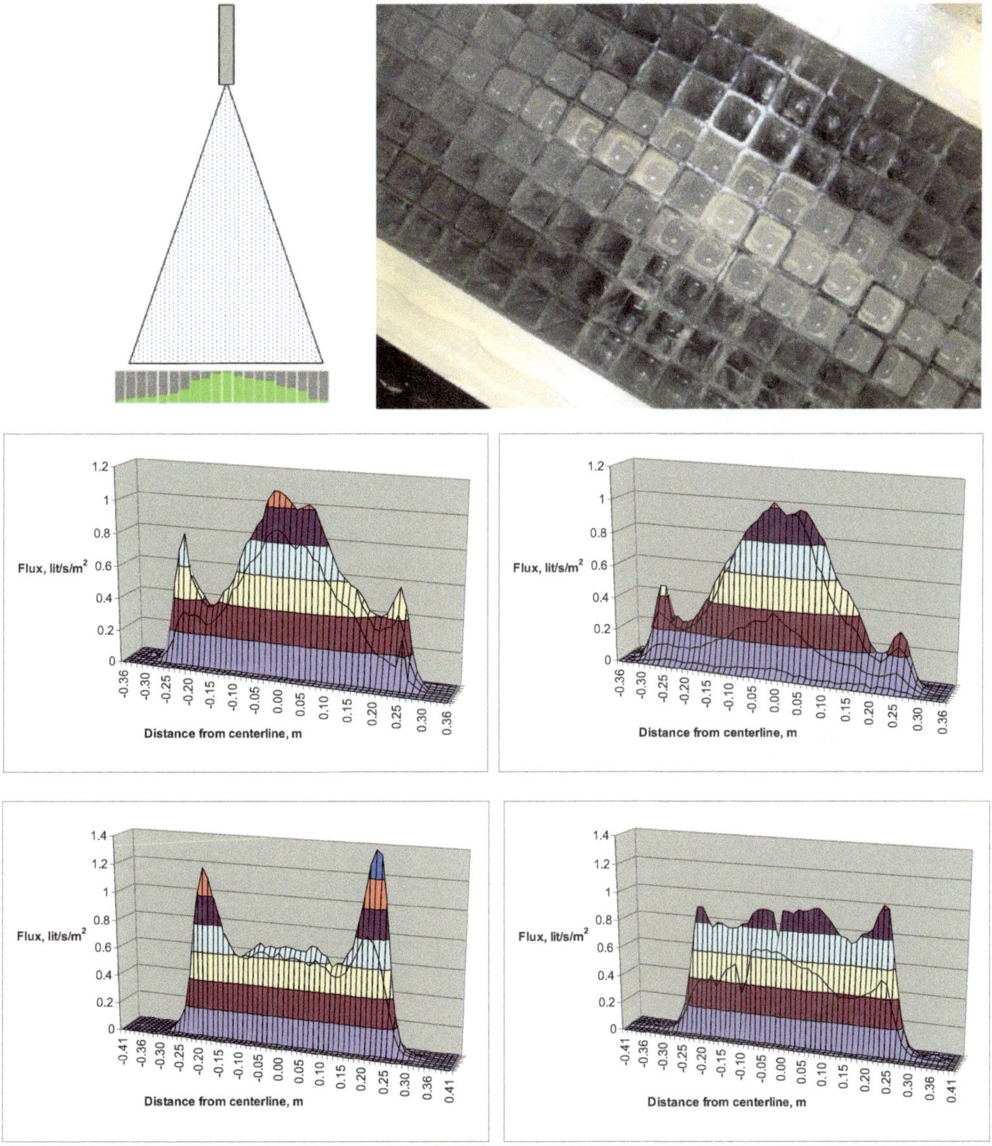

Fig. 12. Mechanical patternation. Upper left, general concept. Upper right, filled cells from a 2-D patternation measurement of a flat fan spray. Middle images, 2-D patternation results from a Spraying Systems Co. Teejet 8002 nozzle under identical conditions except for a formulation change to the sprayed liquid. Bottom images, Teejet 8002E nozzle at same conditions as the Teejet 8002 in the middle images.

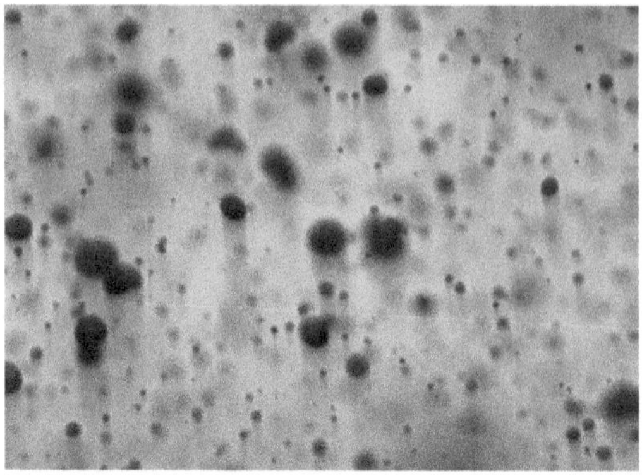

Fig. 13. Sample dynamic image from an agricultural spray.

2.8 Summary of experimental methods

The analytical needs of the researcher varies from spray to spray and project to project. In some cases, only verification of spray angle is needed for a custom industrial installation or for a series of similar but slightly different nozzles. At the other end of the extreme, there are situations where information on velocity profile, drop size, and mass flux distribution are needed to parameterize the inputs to a CFD model. In other cases, the impact of changes in formulation need to be elucidated based on changes to film breakup dynamics and drop size. The experimental tools described here have been able to provide the required information for a wide variety of industrial problems encountered.

3. Computational methods for spray analysis

In last two decades, several attempts have been made using Eulerian-Lagrangian (E-L) computational fluid dynamics (CFD) based models for simulating spraying. However, well established E-L CFD models for fluid flow and evaporative/reactive spray dynamics are sensitive to spray origin parameters, such as initial droplet size distribution, initial velocity and droplet number density. These spray origin parameters are controlled by primary atomization. Advanced mathematical models for primary atomization are needed to compute time-averaged and transient characteristics of droplet formation. Recent advances have been made using computational model for developing better understanding on the primary atomization process (Gorokhovski and Herrmann, 2008; Bini et al., 2009; Herrmann et al. 2011). However, these studies demand significant computational time and resources even for idealized flow conditions (e.g. nonreactive system).

3.1 Primary atomization

3.1.1 Eulerian-Eulerian approach for modeling primary atomization

In the Eulerian-Lagrangian approach for modelling spray behavior, the atomization process is typically modelled using a surface wave instability model that predicts spray parameters such as the spray angle and droplets' diameters. The surface wave growth models widely used include the surface wave instability model [Reitz and Diwakar], the Kelvin-Helmholtz/Rayleigh-Taylor (KHRT) instability model [Patterson and Reitz] and the Taylor Analogy Breakup (TAB) model [O'Rourke and Amsden]. The effect of nozzle geometry on the primary atomization is approximated using an empirical nozzle-dependent constant, which require experimental data or experience to give reliable predictions. With the high performance computing (HPC) system becoming more powerful and less costly in recently years, it is becoming more feasible to directly model the primary atomization by resolving the gas-liquid interface evolution and breakup use an Eulerian-Eulerian approach [Gorokhovski and Herrmann].

Immiscible fluids interface tracking can be done using several approaches in the Eulerian framework: the front tracking method [Unverdi and Tryggvason, 1992], the volume of fluid (VOF) method [Hirt and Nichols, 1981] and the level set method [Osher and Sethian 1988, Sussman 1994]. The front tracking method, which tracks the motion of the surface marker particles attached to the gas-liquid interface, has numerical limitations for three-dimensional irregular interfaces. It is therefore not popularly used for predicting primary atomization. In the VOF method, the volume fraction of a fluid (i.e. the liquid phase) α is solved for each of the computational cells:

$$\frac{\partial \alpha}{\partial t} + \mathbf{V} \cdot \nabla \alpha = 0 \tag{1}$$

$$\alpha = \frac{1}{V_{cell}} \int_{V_{cell}} H(x, y, z, t) dxdydz \tag{2}$$

The nature of the VOF model ensures mass conservation. Because the method takes a volumetric approach to track the interface motion, numerical diffusion is inevitably inherent in the model. Special numerical algorithms are available for reconstructing the interface shape and minimizing the numerical diffusion [Youngs 1982, Ubbink 1997, Muzaferija et. al. 1998]. Surface tension is typically implemented as a volumetric force using the continuum surface force (CSF) method [Brackbill et. al. 1992].

Figure 14 shows the predicted jet breakup to form droplets for a laminar jet using the VOF method, showing good agreement with the measured droplet sizes.

In the level set method, a scalar equation representing the signed distance from the interface is solved:

$$\frac{\partial \phi}{\partial t} + \mathbf{V} \cdot \nabla \phi = 0 \tag{3}$$

The interface is described with the zero level curve of the scalar. For ensuring the scalar function to remain the signed distance to the gas-liquid interface, a redistancing algorithm is typically applied. This can generate mass loss in under-resolved regions. This can be reduce by refining the computational mesh for avoiding under-resolved regions or taking the coupled level set VOF (CLS) approach [Mikhael Gorokhovski and Herrmann, 2008]. Figure 15 gives a snapshot of the predicted drops using the coupled level set VOF method [Liu,Garrick and Cloeter, 2011].

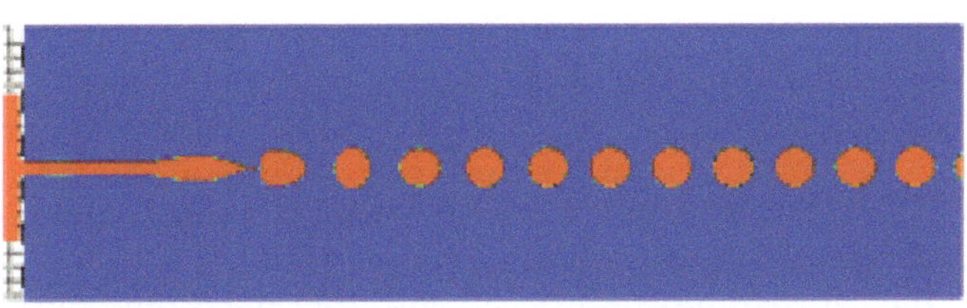

Fig. 14. Predicted droplets formed due to the Rayleigh-Taylor instability using the VOF method, showing good agreement with measured droplet sizes.

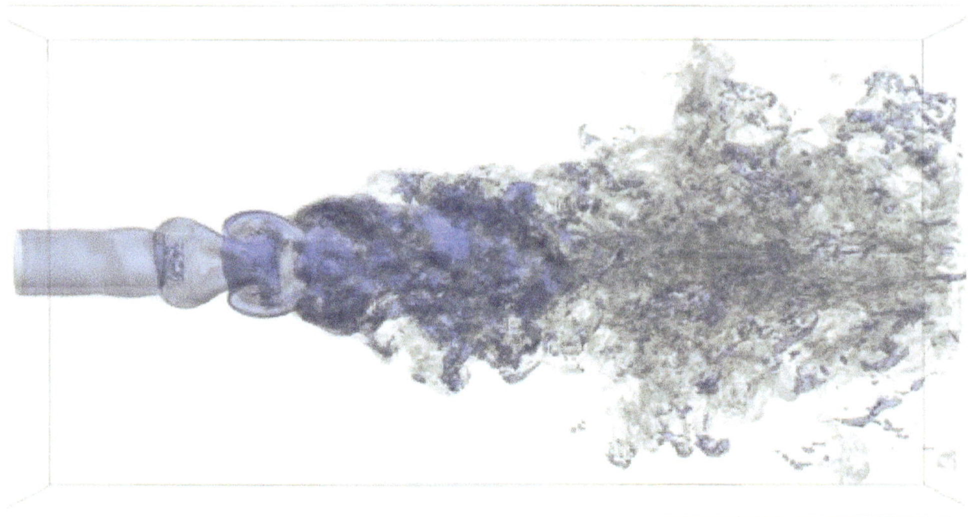

Fig. 15. Predicted primary atomization using the coupled Level Set - VOF method [Liu, Garrick and Cloeter, 2011].

3.2 Coupling reaction, flow and spray dynamics

Computational implementation of coupled primary atomization, two-phase spray model, simultaneous heat and mass transfer and subsequent reaction is a challenging and difficult task. The major difficulty in primary atomization simulation is disparity between the time and length scales in liquid core and gas phase domain. Efficient computational approaches need to be developed to resolve the numerical stiffness due to the coupling between the jet transport equation and gas phase dynamics near nozzle. Furthermore, adequate topological representation of the three dimensional jet core shape and the surrounding gas computational grid needs to be established. The volume of fluid (VOF) method is used to capture gas-liquid interface (Liang and Ungewitter). However, VOF requires very high grid resolution for resolving jet surface. Therefore, including primary atomization in simulating reacting sprays become a huge challenge with respect to computational demand. In the majority of CFD simulation of spray combustion, the atomization process is presented in terms of a point or surface source of droplets with a prescribed size distribution introduced at the nozzle exit. The subsequent breakup and coalescence of the droplets are modelled based on the droplet Weber number criteria. This approach uses the assumption that the dynamics and breakup of liquid jet indistinguishable from those of a train of drops. In this work, a CFD model is presented using the above described approach to simulate reacting spray dynamics.

The particle formation in the fluid bed reactor is a complicated process involving multiphysics. It is affected by many factors such as the dispersion of catalyst drops, the solvent evaporation rate (if it is used for dissolving solid catalyst), the mass transfer of reactant gas into the catalyst drops and the reaction kinetics. The nozzle operating conditions and design of the injection nozzle control the dispersion of catalyst droplets in the dilute zone. Whereas, the mass transfer to and from the droplets and the interaction between the droplets are controlled by the quality of droplet dispersion in the dilute zone. This indicates complex interactions between the operating conditions (shroud velocity), reactor hardware (nozzle design) and performance (dispersion of droplets and rates of transport processes). Due to the complex physics involved, it is advisable to break the complex process into multiple simpler sub-processes for better understanding of their effects on the overall process. The above discussed problem is broke into two parts. In the first part, CFD model was developed to understand the influence of the operating conditions and prevailing fluid dynamics on the dispersion of catalyst droplets. Whereas, in the second part CFD model was developed to understand the influence of prevailing fluid dynamics on the mass transfer and on the growth of polymer particles. The methodology is summarized in Figure 16. The findings from both the parts generate more understandings on how to control the final particle size distribution in the reactor.

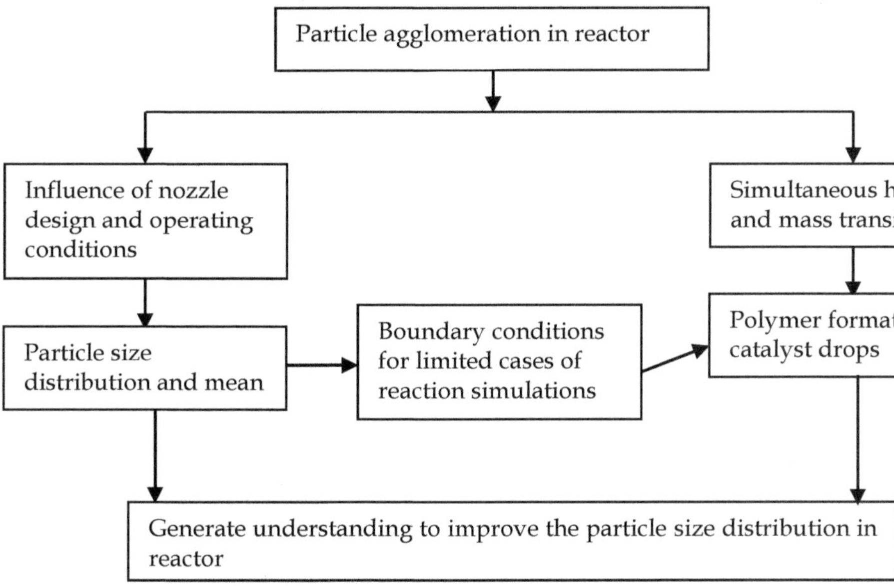

Fig. 16. Problem Solving Methodology

3.2.1 CFD model and boundary conditions

Flow simulations were conducted using the commercial CFD package FLUENT 6.4. A few approximations were made while formulating the CFD model. Flow of different gas phase components was modeled using species transport equations. The mass flow rates of each feed stream are listed in Table 1. Ideal gas densities of each gas streams were calculated before the simulation for the given reactor pressure and temperature and were then specified in the CFD model. This approach allows us to use the incompressible flow formulation. Accurate prediction of turbulent quantities is important for reasonable prediction of the spreading rate of the jet. In the present work, the realizable k-ε turbulence model was used to simulate the turbulent flow for its better prediction of turbulent jets.
Simulations were carried out in a section of the fluidized bed with the height equal to the diameter (H/D= 1). The schematic of the computational domain is shown in Figure 17. The

bottom cross section of the domain was modelled as a velocity inlet with a uniform distribution of all gas phase components. The top surface of the computational domain was modelled as a zero gradient boundary condition. The gradients normal to the outlet boundary were set to zero for all the variables except pressure.

Feed stream	Velocity, m/s
Bottom feed	0.5
Support tube	16 32
Catalyst injection	150
Catalyst drop	150

Table 1. Mass Flow Rates for Different Inlet Streams

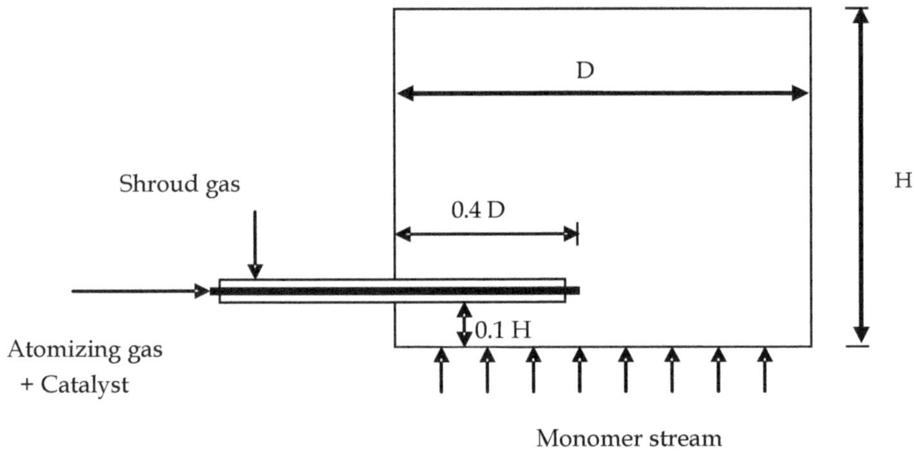

Fig. 17. Schematic of the Computational Domain

The dispersion of catalyst droplets in the domain was simulated using the Lagrangian approach. In addition to solving transport equations for the continuous phase, FLUENT allows you to simulate a discrete second phase in a Lagrangian frame of reference. This second phase consists of spherical particles (which may be taken to represent droplets or bubbles) dispersed in the continuous phase. This modelling approach is commonly called as discrete phase modelling (DPM). DPM computes the trajectories of these discrete phase entities, as well as heat and mass transfer to/from them. Trajectories of the secondary phase were calculated using discrete phase inertia, hydrodynamic drag, gravity force and other interphase forces.

$$\frac{du_p}{dt} = F_D\left(u - u_p\right) + \frac{g_x\left(\rho_p - \rho\right)}{\rho_p} + F_x \tag{4}$$

Where F_x is the additional acceleration (force/unit particle mass) and $F_D(u\text{-}u_p)$ is the drag force per unit particle mass.

The dispersed phase particles are not passive contaminants and the presence of these particles may influence the flow of the continuous phase. The level of interaction becomes especially complex for a turbulent flow field. Secondary phase loading and the number density of particles determine the level of interaction or degree of coupling between the continuous phase and dispersed phase. Depending on the degree coupling (one way, two way or four way), solutions of the both phases interact with each other (see Ranade, 2002 for more details). Due to the strong interaction of between the droplets and the surrounding fluid, in the present work, four way coupling: continuous phase-dispersed phase particles-dispersed phase particles-continuous phase, was used.

When the continuous phase flow field is turbulent, its influence on particle trajectories needs to be represented in the model. The situation becomes quite complex in the case of two-way and four-way coupling between continuous phase and dispersed phase, since the presence of dispersed phase can affect turbulence in the continuous phase and vice versa. For such cases, it is necessary to calculate the trajectories of a sufficiently large number of particles using the instantaneous local velocity to represent the random effects of turbulence on particle dispersion. The instantaneous velocity is equal to the mean velocity (which is known) plus a fluctuation velocity (which is unknown). Predictions using a turbulence model may give values of variance of the fluctuating velocity. The assumption of a Gaussian distribution for the fluctuations and a random value for fluctuation to be added to the mean velocity is not sufficient to obtain the instantaneous value. Instantaneous values should satisfy the Lagrangian correlation coefficient along the trajectory. Several models have been proposed to estimate the instantaneous velocity. See, for example, reviews by Sommerfeld (1993) and Gouesbet and Berlemont (1999). Two commonly used models to estimate instantaneous fluid velocity from the time-averaged flow field of the continuous phase are discrete random walk (DRW) model (Sommerfeld, 1990) and continuous random walk (CRW) model (Thomson, 1987). In the present study the DRW model was used. In this model, the fluctuating component of the velocity is assumed to have a Gaussian distribution and is calculated by multiplication of a normally distributed random number and the local root mean square (rms) value of the velocity fluctuations. The same value of random number is used for the eddy lifetime (integral time scale). For each eddy lifetime, new value of random number is used. This stochastic process generates a correlation coefficient which linearly decreases from 1 at a delay equal to zero, to 0 at a delay equal to twice the eddy lifetime. Despite such a crude approximation for the correlation coefficient, this approach leads to reasonable results (Sommerfeld, 1990).

It is necessary to model the droplet collisions, breakup and coalescence to simulate the spray realistically. In the present work, the droplet collision model, available with FLUENT™, was used to simulate the droplet collisions. The O'Rourke collision algorithm was used for computationally efficient spray calculation. The O'Rourke method is a stochastic estimate of collisions. It makes the assumption that two parcels may collide only if they are located in the same continuous-phase cell (O'Rourke, 1981). This assumption is valid only when the continuous-phase cell size is small compared to the size of the spray. Once it is decided that two parcels of droplets collide, the algorithm further determines the type of collision. Only coalescence and bouncing outcomes are considered. The probability of each outcome is calculated from the collisional Weber number. This collision model assumes that the frequency of collisions is much less than the particle time step. Also, the model is most applicable for low-Weber-number collisions where collisions result in bouncing and coalescence. FLUENT offers two spray breakup models: the Taylor Analogy Breakup (TAB) model and wave-model. In the present study, the TAB model was used to simulate the droplet breakup. This model is applicable for Weber number less than 100. Accurate determination of droplet drag coefficient is crucial for accurate spray modelling. In this project work, a dynamic drag model was used to account for variations of the droplet shape. More details of droplet collision model, TAB model and dynamic drag model can be found in the FLUENT™ User Guide.

3.2.2 Heat and mass transfer models

Three models corresponding to following three different regimes of heat and mass transfer between the dispersed and continuous phases were considered.

- Inert heating
- Vaporization
- Boiling

Inert heating

For the liquid droplet when its temperature is less than the vaporization temperature T_{vap} of the droplet and/or when all the volatile mass of the drop is evaporated, the inert heating was considered. These conditions may be written as (Sommerfeld, 1993):

$$\begin{aligned} &T_D \le T_{vap} \\ &m_D \le (1 - f_{v0}) m_{D0} \end{aligned} \tag{5}$$

Where T_D is droplet temperature, T_{vap} is the vaporization temperature, m_{D0} is the initial mass of the droplet, m_D is the current mass of drop and f_{v0} is mass fraction of the volatile components. When these conditions are satisfied, the following heat balance equation was used to relate the drop temperature $T_D(t)$ to the convective heat transfer:

$$m_D C p_D \frac{dT_D}{dt} = h A_D \left(T_\infty - T_D \right) \tag{6}$$

Where h is the heat transfer coefficient calculated using Ranz and Marshall (Gouesbet and Berlemont, 1999; Sommerfeld, 1990) and T_∞ is the continuous phase temperature.

Vaporization from drop

When the evaporation of liquid phase starts, it is essential to consider the mass transfer as well as heat transfer from the drop surface. This regime of droplet heating was initiated when the temperature of the droplet reaches the vaporization temperature, T_{vap} and continues until droplet reaches the boiling point, T_{bp}, or until the droplet's volatile fraction is completely consumed (Sommerfeld, 1990):

$$\begin{aligned} &T_D < T_{bp} \\ &m_D > (1 - f_{v0}) m_{D0} \end{aligned} \tag{7}$$

Mass transfer: The rate of vaporization was modelled by relating the flux of droplet vapor into the gas phase to the gradient of the vapor concentration between the droplet surface and the bulk gas as (Gouesbet and Berlemont, 1999):

$$N_i = k_c a\left(C_{i.s} - C_{i,\infty}\right) \tag{8}$$

The concentration of vapor at the droplet surface is evaluated by assuming that the partial pressure of vapor at the interface is equal to the saturated vapor pressure, P_{sat}, at the particle droplet temperature, T_D:

$$C_{i,s} = \frac{P_{sat,T_D}}{RT_D} \tag{9}$$

where R is the universal gas constant.

$$C_{i,\infty} = X_i \frac{P_{op}}{RT_\infty} \tag{10}$$

where Xi is local bulk mole fraction of species i, P_{op} is the operating pressure and T_∞ is the local bulk temperature in the gas. The mass transfer coefficient k_c was calculated using the correlation of Sherwood and Pigford (Sommerfeld, 1993):

$$Sh = 2 + 0.5252 * \mathrm{Re}^{0.5}\, Sc^{0.33} \tag{11}$$

The mass of droplet reduced to

$$m_D(t + \Delta t) = m_D(t) - N_i A_D M_i \Delta t \tag{12}$$

Where M_i is the molecular weight of species i and m_D is the mass of the droplet.
Heat transfer: The droplet temperature was updated according to a heat balance that relates the sensible heat change in the droplet to the convective and latent heat transfer between the droplet and the gas:

$$m_D Cp_D \frac{dT_D}{dt} = hA\left(T_D - T_\infty\right) + \frac{dm_D}{dt}\lambda \tag{13}$$

Boiling from the drop

This phase was initiated when the temperature of the droplet has reached the boiling temperature, T_{bp}, and while the mass of the droplet exceeds the non-volatile fraction, $(1 - f_{v0})$:

$$\begin{aligned} T_D &\geq T_{bp} \\ m_D &\geq \left(1 - f_{v0}\right) m_{D0} \end{aligned} \tag{14}$$

When the droplet temperature reaches the boiling point, a boiling rate equation is applied (Sommerfeld, 1993)

$$\frac{d(d_D)}{dt} = \frac{4k_\infty}{\rho_D Cp_\infty d_D}\left(1 + 0.23\sqrt{\mathrm{Re}_d}\right) \ln\left[1 + \frac{Cp_\infty\left(T_\infty - T_D\right)}{hf_g}\right] \tag{15}$$

Where, Cp_∞ is the specific heat of gas (J/kg-K), ρ_D is the density of drop (kg/m^3) and k_∞ is the thermal conductivity of the gas (W/m-K). Note that the model requires $T_\infty > T_{bp}$ in order for boiling to occur and that the droplet remains at fixed temperature (T_{bp}) throughout the boiling law.

Mass transfer of gas phase component to drop

The rate of C_2 mass transfer was modelled by relating the flux of C_2 into the droplet to the differential of the C_2 concentration between the bulk gas and droplet surface as

$$N_i = k_g a\left(C_{i.\infty} - C_{i,s}\right) \tag{16}$$

The concentration of vapor at the droplet surface is evaluated by assuming that the partial pressure of vapor at the interface is equal to the saturated vapor pressure, P_{sat} , at the particle droplet temperature, T_D:

$$C_{i,s} = \frac{P_{sat,T_D}}{RT_D} \tag{17}$$

where R is the universal gas constant.

$$C_{i,\infty} = y_i \frac{P_{op}}{RT_\infty} \tag{18}$$

where y_i is local bulk mole fraction of species i, P_{op} is the operating pressure and T_∞ is the local bulk temperature in the gas. The mass transfer coefficient k_g was calculated using Equation 8.

3.2.3 Polymerization reaction

Some polymerizations can be characterized by chain initiation, chain propagation, chain transfer, chain termination, and catalyst deactivation. Obtaining their *absolute* rates as a function of process conditions is the ultimate goal of the kinetic studies. However, in some solution polymerization processes, the majority of the kinetic studies are on the *relative* polymerization kinetic rates, which enables modelling of polymer properties as well as quantification of the overall material balance for the reactors. Different and more accurate experimental methods are required to estimate the absolute kinetics than are used to develop the relative ones. In many instances, the absolute chain propagation and the catalyst deactivation rate constants are used for the catalyst efficiency. The reaction rate is obtained by multiplying the catalyst flow rate by catalyst efficiency. The catalyst efficiency expression used in the present work is:

$$E = K^* \frac{k_p}{k_d}[C_{monomer}]\theta \tag{19}$$

where, E is the catalyst efficiency [lb polymer/lb catalyst], K^* is an empirical constant, θ is the reactor residence time [s], k_p is the propagation rate constant [1/s/(mol/L)], and k_d is the catalyst deactivation rate constant [1/s].

3.2.4 Influence of shroud velocity

Shape and extent of the dilute zone formed by a jet in a reactor is controlled by different process parameter such as support tube exit velocities and nozzle design [11]. The shape of dilute zone and local fluid dynamics in the dilute zone determines the dispersion of the liquid catalyst in the dilute zone. Simulations were therefore carried out for two different support tube exit velocities, viz. 16 m/s and 32 m/s. The predicted velocity contours at the vertical plane cutting through the nozzle are shown in Figure 18. It can be seen from Figure 18 that the jet coming out of the nozzle first travels horizontally and then turns to the upward direction. Momentum balance between the nozzle jet and bottom feed determines the shape and extent of jet and therefore the dilute zone in the reactor. The model predictions show better shroud flow at a higher support tube exit velocity.

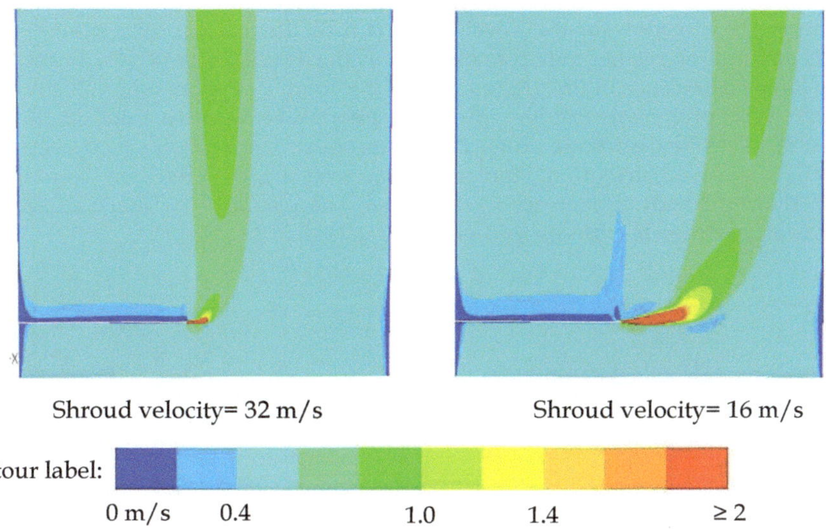

Fig. 18. Predicted Velocity Magnitude Distribution at *r-z* Plane Cutting through Nozzle for Different Shroud Flow Rates

The CFD model was then extended to simulate the dispersion of catalyst droplets. DPM simulations were carried out to simulate the trajectories of the catalyst droplets without evaporation of solvent. Catalyst droplets were injected into reactor from the catalyst inlet of the nozzle. Inlet boundary conditions such as the drop size distribution, mass flow rate and velocity of the drops are very important. In the present work, experimentally measured drop size distribution fitted using the Rosin Rammler distribution was specified at inlet. The distribution parameters are thus: minimum diameter equal to 15 microns, maximum diameter equal to65 microns, mean diameter equal to 31.25 microns, and width of distribution equals to 1.6. Drops were injected into the reactor with a velocity equal to the atomizing gas velocity and a velocity of 150 m/s. The predicted snapshots of droplet dispersion for both tube velocities are shown in Figure 19.

A few conclusions can be drawn from the predicted snapshots shown in Figure 19. The shape of dilute zone determines the trajectories of the droplets. For the 16 m/s tube velocity, the injected drops first move in the horizontal direction and are accumulated at the end of the dilute zone before moving up. Strong deceleration of the droplets along the jet and sudden upward movement of jet might be responsible for this kind of dispersion. Such dispersion may cause strong interactions between droplets and lead to coalescence of the droplets or agglomeration of particles. On the other hand, the quality of dispersion is significantly improved when the support tube velocity is increased to 32 m/s. The increase in the tube velocity not only improves the shroud flow, but also significantly minimizes the extent of accumulation of droplets in the dilute zone. The predicted drop size distributions were converted into volumetric distributions. Such information is helpful for comparisons with the inlet distribution and gives insights into the chemical reaction in the dilute zone. The comparison of the predicted droplet size distributions at two different flow times (20 milliseconds and 40 milliseconds respectively) for both support tube exit velocities is shown in Figure 20 together with the inlet distribution. It can be seen from Figure 20 that compared to the higher support tube flow, the predicted diameter distribution significantly moves toward the right direction for the lower support tube flow. It was also observed that the predicted distribution further moves to the right as the droplets proceeds with the flow. This suggests more coalescence in the dilute zone for the 16 m/s support tube velocity. Figure 20 also shows very promising results with the support tube velocity equal to 32 m/s. The predicted droplet size distribution after 20 milliseconds for the 32 m/s feed velocity is very close to the inlet size distribution. Similar to 16 m/s results, the predicted drop distribution moves to the right with an increase in flow time. The results show much smaller volume fraction of larger droplets with the feed rate of 32 m/s than 16 m/s.

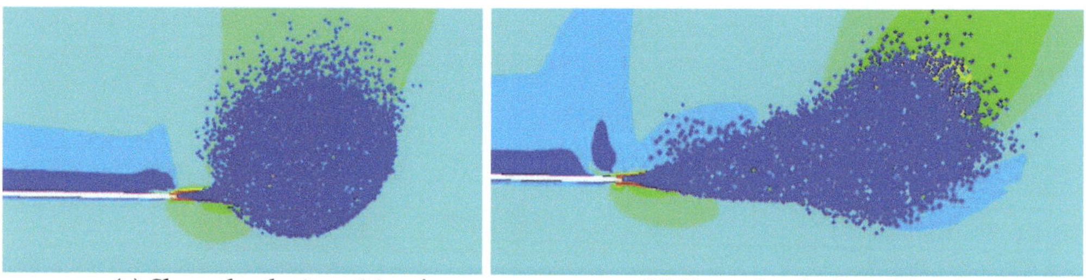

(a) Shroud velocity= 16 m/s (b) Shroud velocity= 32 m/s

Fig. 19. Predicted Snapshots of Droplet Dispersion in Dilute Zone at Three Different Time Intervals

A polymerization reaction in a drop is a bulk or volumetric reaction. Therefore, the rate of polymer formation is directly dependant on the volume of the drop. It is possible to have different polymer formation rate for different droplet sizes. A separate model was developed to simulate the mass transfer of monomer from gas phase to droplet and simultaneous polymerization reaction. This model was then incorporated into the CFD model using user defined functions (UDF's). The predicted results were used to shed some light on the particle growth process. The particle growth rate was estimated using predicted particle size. The predicted values of particle growth rate at different time intervals are shown in Figure 21. Few conclusions can be drawn from Figure 21. First, for both the particles, the rate of particle growth is very fast at the beginning. The particle growth rate then slows down. Second, the particle growth rate is comparatively very fast for bigger droplets than smaller droplets. This indicates that the larger droplets grow much faster than the smaller droplets and can form bigger particles. This phenomenon will influence the individual particle growth when different size droplets are fed into the reactor. Different particle growth rates also highlight the possibility of having different rate controlling steps for polymerization reaction. The predicted growth rate suggests the possibility of having kinetically controlled reaction in the smaller drops due to smaller drop volume. Whereas, it is possible that reaction kinetics may not control the growth of bigger droplets.

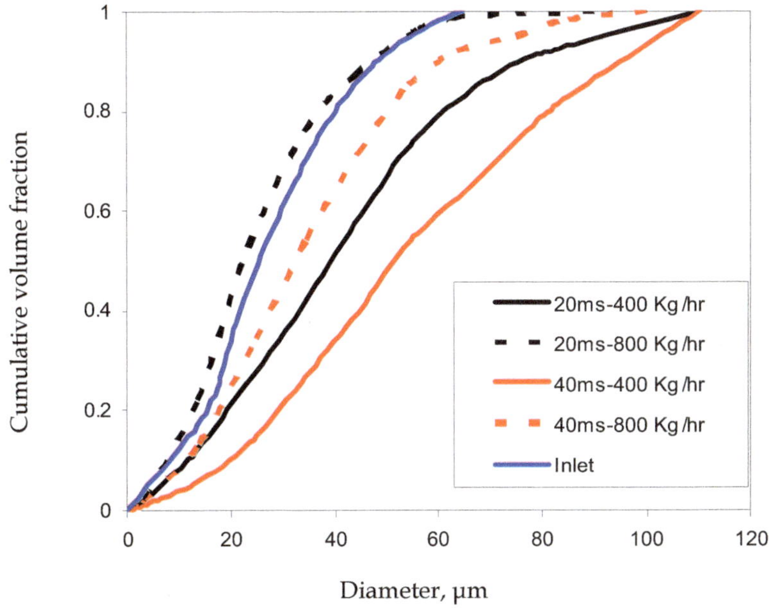

Fig. 20. Influence of Support Tube Exit Velocity on Drop Diameter Distribution at 20 Milliseconds and 40 Milliseconds

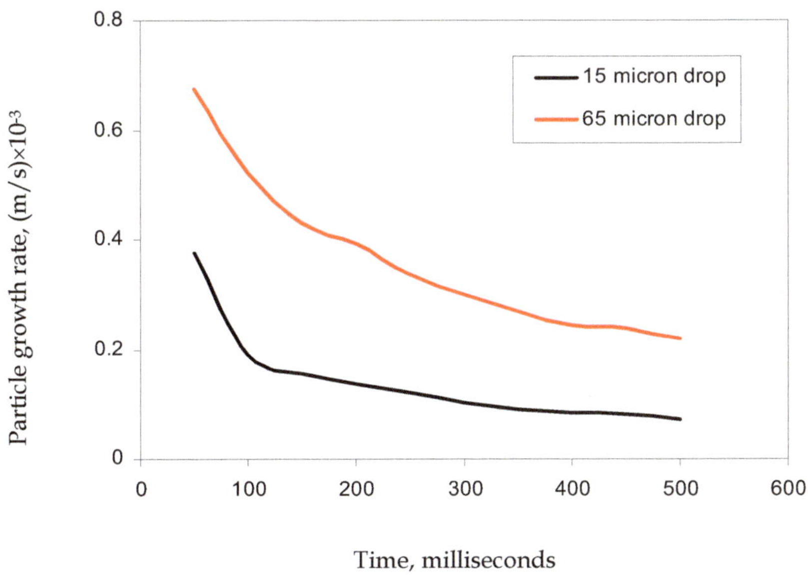

Fig. 21. Predicted Influence of Drop Size on Particle Growth Rate

It is of the interest to understand the influence of support tube velocity on the particle growth process due to polymerization reaction. The predicted particle size distributions at different time intervals for both the shroud velocity were then compared. The comparison of predicted particle size distribution for both the support tube flow rates is shown in Figure 22. It can be seen from Figure 22 that the higher support tube flow rate significantly influenced the particle

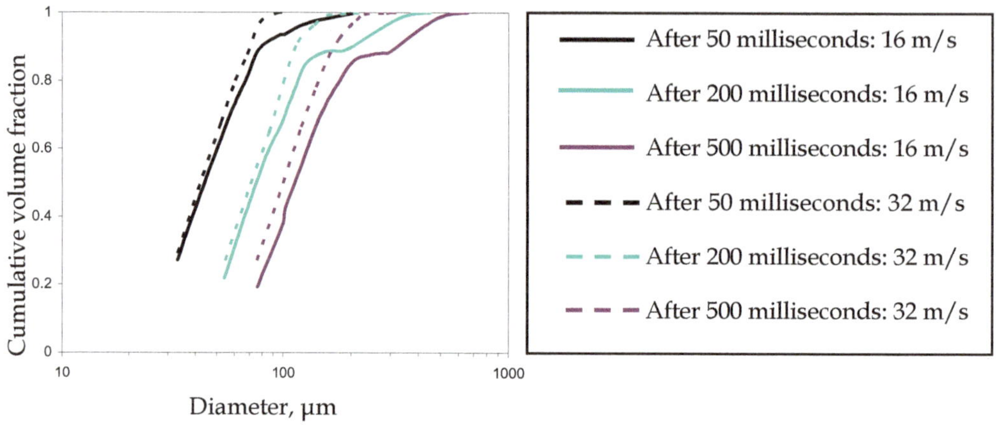

Fig. 22. Influence of Support Tube Flow Rate on Predicted Particle Diameter Distribution

size distribution in reactor. It was observed that the better shroud flow not only controlled the rapid growth of the bigger particles but also significantly minimized the total volume fraction of the bigger drops present in the distribution. The obtained narrow particle size distribution at higher support tube flow rate suggests the possibility of having kinetic controlled polymerization reaction in almost all the fed droplets. Lower coalescence rate and reaction into kinetically controlled regime with higher shroud velocity will help to control the particle size distribution in the reactor.

3.3 Summary of numerical modeling

In this chapter, we have demonstrated the application of computational fluid dynamics (CFD) models for simulating reactive sprays. The present work highlighted the applicability of CFD models based on the Eulerian-Eulerian and Eulerian-Lagrangian approaches for simulating reactive sprays. Despite some of the limitations, computational models were shown to provide useful information on the transport processes occurring in the dilute zone. The computational models were able to predict the implications of operating conditions and dispersion of droplets in the dilute zone on the transport process and subsequent particle growth. Careful numerical experiments using these CFD models can be used for better understanding the characteristics of existing spraying operation, to enhance its performance, assess different configurations and greatly assist the engineering decision making process. The approaches, sub-models and the results discussed here will provide a useful basis for practical applications as well as for further developments.

Although the models discussed here are capable of providing valuable and new insights which hitherto were unavailable, there is still significant scope to improve the fidelity of these models.

- Important and intellectually challenging fluid dynamic and transport phenomena occur at different length and time scales with sprays. On the micron scale, boundary layers and wake develop because of relative motion between drops and the gas. In addition to this, there are abundant complicated fluid dynamics factor at this length scale: shear driven internal circulation of the liquid in the droplet, Stefan flow due to vaporization and condensation, hydrodynamic interfacial instabilities leading to droplet shape distortion and subsequent droplet shattering and droplet interaction with the free stream turbulence of gas phase. Whereas, on the coarser scale, we have complexities associated with the exchanges of mass, momentum and energy of many droplets in the some sub-volume of interest and process by which a coherent liquid flow disintegrates into droplets. The problem becomes very complicated because of the strong coupling possesses by these different scales. One cannot explain the processes on the coarser scale without detail knowledge of the fine-scale phenomena. Note that in some practical problems, the disparity of length scale differ by several orders of magnitude so that a challenging subgrid modeling problem results. Actually, direct numerical simulation (DNS), where all the flow scales are resolved, could provide valuable insight. However, carrying out DNS of a complete spraying operation is a huge challenge due to computational cost. Large eddy simulation (LES), where only large scale turbulence is resolved can improve the accuracy of the spray simulation compared to the Reynolds averaged approaches (RANs). However, the required models for simulating multi-physics of spraying operations are often non-existent or lack the desired accuracy.
- Characterization of primary atomization represents a challenge for experiments. Numerical modeling should provide a much needed alternative. However, numerical studies of primary atomization are also sparse. No clear gold standard has emerged on how to conduct a numerical simulation of complex two-phase flows. Therefore, several key issues such as the discontinuous nature of the flow properties across the phase-interface, the singularity of the surface tension forces, and the very large range of scales involved in atomization remain unanswered.

Adequate attention to key issues mentioned above and the creative use of computational flow modeling will hopefully make useful contributions to industrial spraying operation.

4. Acknowledgements

The following individuals are gratefully acknowledged for their help in acquiring the various images presented throughout this chapter: Kuide Qin, Charles Lipp, Pramod Patil, Billy Smith, Santhosh Ramalingam, Mike Trippeer, and Rugved Pathare.

5. References

Bini, M., Jones, W.P. and Lettieri, C. 2009, Large Eddy Simulation of spray atomization with stochastic modelling of breakup, *Proceedings of the European Combustion Meeting*.

Brackbill, U., Kothe, D.B., Zemch, C.: "A Continuum Method for Modeling Surface Tension", Journal of Computational Physics, 1992(100), pp.335-354.

Gorokhovski, M. and Herrmann, M. 2008, Modeling primary atomization, *Annual Rev. Fluid Mech.*, 40, 343-366.

Gouesbet, G. and Berlemont, A., Eulerian and Lagrangian approaches for predicting the behavior of discrete particles in turbulent flows, *Prog. Energy Combustion Sci.* (1999), 25, pp. 133–159.

Hirt, C.W., Nichols, B.D.: "Volume of Fluid (VOF) Method for The Dynamics of Free Boundaries", Journal of Computational Physics, 1981(39), pp.201-225.

Liu, W., Garrick, S., and Cloeter, M.D., *Large-scale Simulation of Sprays with Adjuvants and Non-Newtonian Fluids,* Poster ILASS2011-P06, ILASS-Americas 23rd Annual Conference on Liquid Atomization and Spray Systems, Ventura, CA, May 2011.

Muzaferija, S., Peric, M., Sames, P. and Schellin, T.: "A Two-Fluid Navier-Stokes Solver to Simulate Water Entry", in Proc. 22nd Symposium on Naval Hydrodynamics, Washington D.C., 1998, pp.277-289.

Osher, S., Sethian, J.A.: "Fronts Propagating with Curvature-Dependent Speed: Algorithms Based on Hamilton-Jacobi Formulations", Journal of Computational Physics 1988(79), pp.12-49.

O'Rourke, P.J., *Collective Drop Effects on Vaporizing Liquid Sprays,* PhD thesis, Princeton University, Princeton, New Jersey, 1981.

Patterson, M.A., Reitz, R.D.: "Modeling the Effects of Fuel Characteristics on Diesel Engine Combustion and Emission", SAE 980131, 1998.

Raffel, M., Willert, C.E., and Kompenhans, J., *Particle Image Velocimetry. A Practical Guide,* Springer-Verlag, Berlin, 1998.

Ranade, V.V., *Computational Flow Modelling for Chemical Reactor Engineering,* Academic Press, New York, 2002.

Reitz, R.D., Diwakar, R.: "Structure of High Pressure Fuel Spray", SAE Technical Paper Series 870598, 1987.

Sommerfeld, M., Reviews in numerical modeling of dispersed two phase flows, *Proceedings of 5th Int. Symp. on Refined Flow Modeling and Turbulence Measurements,* Paris, 1993.

Sommerfeld, M., Numerical simulation of the particle dispersion in turbulent flow: the importance of particle lift forces and particle/wall collision models, in *Numerical Methods for Multiphase Flows,* Vol. 91, ASME, New York, 1990.

Sussman, M., Smereka, P. and Osher, S.: "A Level Set Method for Computing Solutions to Incompressible Two-Phase Flow", Journal of Computational Physics, 1994(114), pp.146-159.

Thomson, D.J., Criteria for the selection of stochastic models of particle trajectories in turbulent flows, *J. Fluid Mech.* (1987), 180, pp. 529–556.

Ubbink, O.: "Numerical Prediction of Two Fluid Systems with Sharp Interface", Ph.D. thesis, Imperial College of Science, Technology and Medicine, London, England, 1997.

Unverdi, S.O., Tryggvason, G.: "A Front Tracking Method for Viscous, Incompressible, multi-fluid flows", Journal of Computational Physics, 1992 (100), pp.25-37.

Youngs, D.L.: "Time-Dependent Multi-Material Flow with Large Fluid Distortion", Numerical Methods for Fluid Dynamics, Academic Press, 1982.

12

Multidimensional Design of Hydraulic Components and Systems

Massimo Milani, Luca Montorsi and Fabrizio Paltrinieri
DISMI – University of Modena and Reggio Emilia
Italy

1. Introduction

The importance of numerical simulation in hydraulic components design is rapidly increasing. In fact, the computational simulations of compressible and incompressible flows can relevantly support experiments, and the CFD is becoming a valuable tool also for the design and the pre-prototyping processes (Yang, 2002, 2005; Barman, 2005). Moreover, the human and computational resources to be involved in the numerical analysis are now not only acceptable, but also advantageous, thanks to the continuous development of computational platforms, as well as of the CFD tools.

Nevertheless the accuracy and reliability of the numerical results must be addressed when approaching a new problem, since they demonstrated to be very sensitive to the fluid-dynamics characteristics of the case studied (such as the Reynolds number in the critical sections, the geometry complexity and the boundary conditions). For example, the widely used cylindrical orifices (metering valves, directional valves, injectors, instruments, etc.) show that the fluid dynamic performance is significantly affected by the geometrical details; consequently, a sharp-edge inlet presents an efflux coefficient lower than a rounded one (Ohrn et al., 1991), while small modifications in the curvature can determine remarkable differences both in the flow field, and in pressure losses.

Furthermore, a key feature in the design process of fluid power components and systems is the capability of controlling or avoiding cavitation and aeration. The problems caused by the cavitating phenomena are widely known (Oshima & Ichikawa, 1985, 1986) and they can lead to efficiency loss, vibrations and noise, unexpected change in the characteristics of flow rate and flow forces and even erosion of the components when the phenomenon becomes particularly aggressive (Oshima e al., 2001).

Therefore, the possibility of predicting cavitation is fundamental in hydraulic components design and both experiments and numerical simulation have to be employed and integrated in order to help the understanding of the basics of cavitation occurrence. Particularly, numerical analysis can provide a great amount of data that can extend the experimental area and deepen the insight of the physical phenomenon (Yang, 2002, 2005). Nevertheless, the accuracy and reliability of the numerical models have to be addressed when approaching a new problem or when they are modified to account for a more detailed description of the physical process. The model sensitivity with respect to the fluid-dynamics characteristics of the case studied (such as the Reynolds number in the critical sections, the geometry complexity and the boundary conditions) must be properly addressed before adopting the numerical tool as a predictive design tool.

When addressing numerically the flow through the metering section of hydraulic components, particular care should also be devoted to the operating conditions accounted for in the simulations. In fact, the behaviour of the flow is usually highly time dependent, as well as the geometry of the component varies according to the working conditions. Thus, the choice of carrying out a steady state or a fully transient CFD simulation of the component is a critical issue in the design methodology. On one hand, the former approach is characterized by a significantly lower computational effort as well as a shorter case setup time, but it evaluates the fluid-dynamics performance of the hydraulic component only under the assumption of steady state operation, which is often a very limiting restriction compared to the real operations. Conversely, the fully transient CFD calculation predicts more accurately the real behaviour of the flow within the hydraulic component, but the required computational effort is remarkably large and both the setup time and the simulation execution duration can be considerably long. Therefore, it is important to highlight the quality of the results that can be obtained by performing these two types of CDF simulations, in order to select the correct approach for each analysis that has to be carried out.

In this chapter, the above mentioned critical aspects in the application of multidimensional numerical analysis for the design of mechanical devices and components for hydraulic systems are addressed. The objective of the chapter is to provide a roadmap for the multidimensional numerical analysis of the hydraulic components to be used effectively in the design process. In particular, two examples of hydraulic systems are accounted for in the application of the CFD analysis: a proportional control valve and a fuel accumulator for multi-fuel injection systems. These test cases have been selected due to their representativeness in the field of hydraulic applications and to the complexity and variety of the physical phenomena involved.

2. Prediction capabilities of the numerical models

The Navier–Stokes equations for isothermal, incompressible flows are solved by means both of the open source multidimensional CFD code OpenFOAM (OpenCFD, 2010, OpenFOAM®-Extend Project, 2010), and the commercial CFD code ANSYS CFX. Bounded central differencing is used for the discretization of the momentum, second-order upwind for subgrid kinetic energy. Pressure–velocity coupling was achieved via a SIMPLE similar procedure. The second-order implicit method is used for time integration scheme. No time integration scheme is involved, and all the calculations are performed under steady state conditions.

In the modeling of the turbulent cases the performance of three turbulence models are investigated:

1. The standard k-ε model (KE), including wall functions for the near-wall treatment;
2. the a low-Reynolds number model developed by Launder and Sharma (LS), (Launder & Sharma, 1974) in which the transport equations for the turbulent quantities are integrated to the walls;
3. the two zonal version of the k-ω model, known as the shear stress transport model (SST) (Menter, 1993).

Since the objective of this investigation is the analysis of confined flows, the attention is paid to near-wall behaviour, to the damping functions and to the boundary conditions. Hence the choice of addressing the predictive capabilities of turbulence models with a different near wall treatment is made. In the k-ε model, the transport equations are not integrated to the walls. Instead the production and dissipation of kinetic energy are specified in the near-wall cell, using the logarithmic law-of-the-wall. The validity of the near wall treatment requires that the values of the y^+ must be in the range between 30 and 100. In the k- ε model formulation proposed by Launder and Sharma the definition of the turbulent kinetic energy dissipation is modified to account for the low Reynolds regions close to the wall. In fact, this equation is solved for $\tilde{\varepsilon}$ instead of ε, and a new term E is added to compensate for additional production and to further balance diffusion and dissipation in the vicinity of the walls. In this way, the Launder and Sharma approach has the advantage of the natural boundary condition $\tilde{\varepsilon}=0$ at the walls. The model's constants are the same as the standard formulation, but the damping factor functions are evaluated as a function of the turbulent Reynolds number. The boundary conditions at the wall are $k_{\omega}=0$ and $\tilde{\varepsilon}=0$. For the low Reynolds number turbulence models the requirement on the y^+ values is more stringent, in fact y^+ should never be larger than one. The. shear stress transport model is derived from the original two zonal version of the k-ω model proposed by Wilcox (Wilcox, 1988), and based on the transport equation proposed by Kolmogorov. In this model the damping functions in near-wall regions are not necessary. The eddy viscosity is calculated as the ratio between the turbulent kinetic energy and the the specific dissipation, ω. The original version of the model demonstrated to be very sensitive to ω specified in the free-stream (Menter, 1994). Therefore a new formulation was proposed by Menter combining the the k- ω model by Wilcox in the inner region of the boundary layer, and the standard k - ε in the outer region and the free-stream. The ω equation differs from the one of the original k-ω model because of an additional cross-diffusion term. In the inner part of the boundary region the model's constants are the same as the k-ω model, while in the outer region they become similar the k-ε ones. The boundary conditions do not change from the original k-ω model formulation, and similarly require y^+ smaller than 3. In this analysis careful attention was paid in order to comply with the requirements on the y^+ values for the different turbulence models adopted.

2.1 Flow through a circular pipe

First phase of the analysis is the simulation of the fundamental test case represented by the flow through a circular pipe. The geometrical domain is assumed to be a straight circular pipe, characterized by an axial length equal to 1.0 m and a diameter of 10 mm, pure water is considered as the operating fluid and both laminar and turbulent conditions are accounted for. Due to the axial symmetry of both the geometry and the boundary conditions, the simulations assume a two dimensional domain. Therefore, the meshes are defined as circular sectors with an angular amplitude equal to 5° (see Fig.1).

A preliminary sensitivity analysis with respect to the grid resolution is carried out, in order to assess the best compromise between the accuracy of the results and the computational cost. Different grids are constructed varying the cell spacing both in the axial and radial direction. For the laminar case, the cell axial dimensions equal to 1.0, 2.5, and 10 mm are compared keeping constant the resolution along the cylinder radius (i.e. 0.3 mm); whereas the grids used in the turbulent case are characterized by a radial cell size equal to 0.3, 0.1

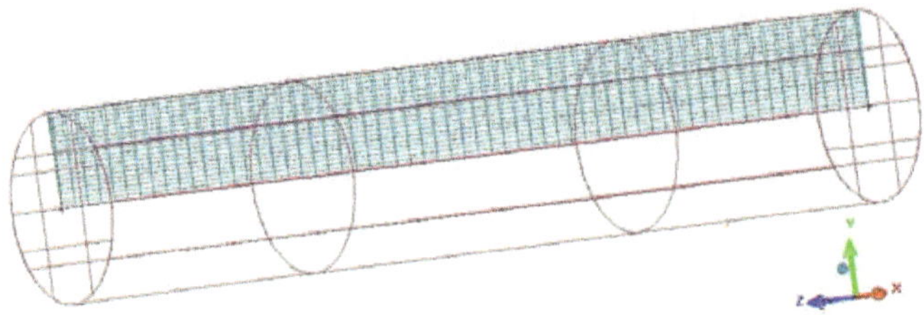

Fig. 1. 2D mesh for the straight circular pipe

Case #	CFD code	Grid (ax. & rad.)	Turbulence Model
A_1; A_2; A_3; A_4	FOAM	1.0x0.3; 2.5x0.3; 5.0x0.3; 10.0x0.3	Laminar
A_5; A_6; A_7; A_8	CFX	1.0x0.3; 2.5x0.3; 5.0x0.3; 10.0x0.3	Laminar
A_9; A_10; A_11	FOAM	5.0x0.05; 5.0x0.1; 5.0x0.3	LS
A_12; A_14; A_16	CFX	5.0x0.05; 5.0x0.1; 5.0x0.3	KE
A_13; A_15; A_17	CFX	5.0x0.05; 5.0x0.1; 5.0x0.3	SST

Table 1. List of simulations for the laminar and turbulent flow through the circular pipe

and 0.05 mm and a constant axial dimension (i.e. 5 mm). The meshes consist only in hexahedrons and prisms along the cylinder axis; the height of the cells near the walls is carefully chosen in order to comply with the requirements of the turbulence models used in the simulations on the distance to the wall of the first node. Table 1 shows the comprehensive list of the simulations performed. For the laminar cases a fluid axial velocity of 0.1 m/s is set at the inlet boundary, while a static pressure boundary condition was considered at the pipe outlet. The numerical results for the laminar regime were compared with the Hagen-Poiseuille theory (Pao, 1995), that is applicable when no relative motion between the fluid and the wall appears, and an incompressible Newtonian fluid undergoes to a isothermal laminar flow. Under these assumptions, the axial velocity profile versus the distance from the axis can be determined from the expression:

$$v = \frac{1}{4\mu} \cdot \left(-\frac{\partial p}{\partial z} \right) \cdot \left(r_0^2 - r^2 \right) \tag{1}$$

As well known, the laminar velocity assumes the fully developed profile at a distance from the pipe inlet on the basis of the Langhaar length, defined as:

$$L' = 0.058 \cdot \mathrm{Re} \cdot D \tag{2}$$

In Figs. 2 a) and c) the theoretical axial velocity profile as a function of the distance from the cylinder axis is compared with the calculated profiles for the different grid resolutions. Both codes used in the simulations demonstrated a good agreement with the axial velocity profile derived from the Hagen-Poiseuille, and also demonstrated a negligible influence on the cell dimension. Similar behaviour could be noticed in the velocity profile prediction, see Figs. 2 b) and d), and practically no differences could be seen among the simulated cases.
When investigating the flow field under fully turbulent conditions a fluid axial velocity of 5.0 m/s was set at the inlet boundary. Calculations were performed for three grids having

different cell dimensions, and the performance of the turbulence models described in the previous sections (i.e. the standard k - ε model, the k - ω shear stress transport model and the low Reynolds Launder-Sharma models) is investigated. The results were compared with the experimental measurements performed by Nikuradse available in literature (Pao, 1995). In Figs. from 3 to 5 the axial velocity versus the distance from the cylinder axis and the cylinder outlet are depicted. The KE model demonstrated a good agreement with the measurements and a scarce sensitivity to the grid size; in particular the logarithmic law-of-the-wall was able to capture the curvature of the axial velocity in proximity to the wall. Whereas, the SST model resulted to be remarkably more influenced by the mesh resolution. In fact Fig. 4 b) shows that the predicted axial velocity differs significantly for the three simulated grids. In particular the coarser grid calculation underestimates the velocity magnitude, while the discrepancy between the curves relating the fine and intermediate meshes are less evident and closer to the results found for the KE model. The more evident sensitivity of the SST model to the grid size is due to the stricter requirement on the distance of the first grid node to the wall. The sensitivity to the mesh resolution was found to be even more marked for the LS model (see Fig. 5). In this model, in fact, the transport equations are integrated up to the wall and the effect of the near wall cell is more directly transferred to the inner domain. This model demonstrated to overestimate the axial velocity in the core region as a consequence of the velocity magnitude tendency when approaching the wall boundary. The dissipation of the turbulent kinetic energy is therefore overestimated and the model constant should be tuned accordingly. For the present study the constants of the considered models were kept the same as the suggested values.

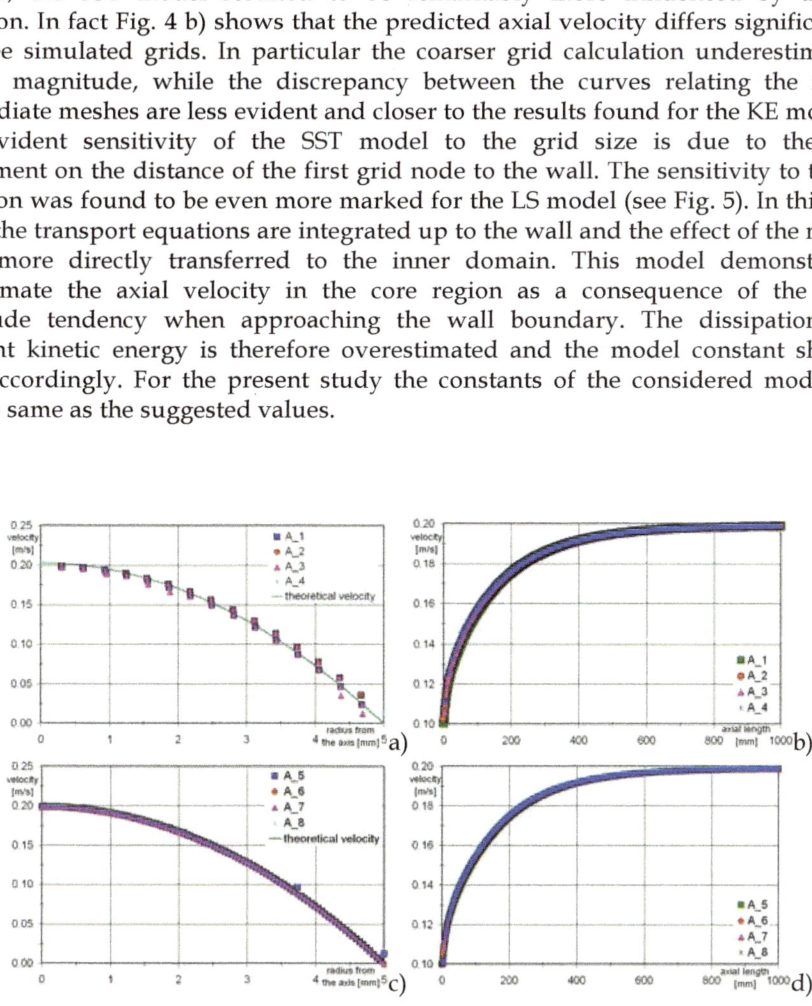

Fig. 2. a) and c) Axial velocity profile vs. the distance from the axis and b) and d) axial velocity profile along the cylinder axis vs. the distance from the pipe inlet: comparison among different grid resolutions (laminar regime)

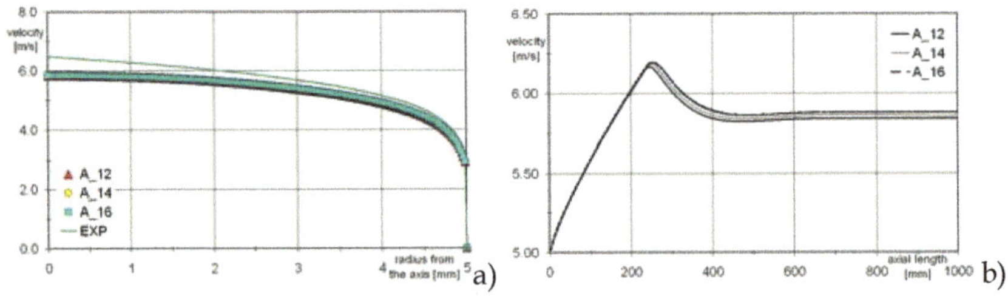

Fig. 3. a) Axial velocity profile vs. the distance from the axis and b) Axial velocity profile along the cylinder axis vs. the distance from the pipe inlet: comparison among different grid resolutions (KE model)

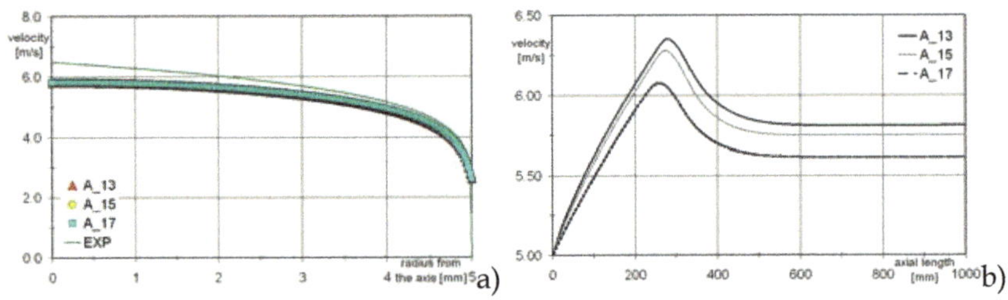

Fig. 4. a) Axial velocity profile vs. the distance from the axis and b) Axial velocity profile along the cylinder axis vs. the distance from the pipe inlet: comparison among different grid resolutions (SST model)

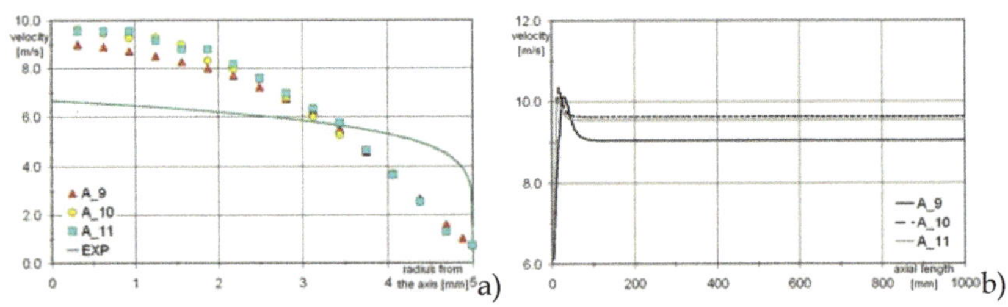

Fig. 5. a) Axial velocity profile vs. the distance from the axis and b) Axial velocity profile along the cylinder axis vs. the distance from the pipe inlet: comparison among different grid resolutions (LS model)

2.2 Flow through a small sharp-edged cylindrical orifice

Further step of the analysis is the investigation of the flow pattern through a sharp-edged cylindrical orifice. Although the geometry of this test case is very simple, it well represents the fluid-dynamics conditions which can be found in the critical sections of many hydraulic components. In addition, experimental measurements of the flow through abrupt section change in circular pipes are available in literature. Therefore, it is possible to address the predictive capabilities of the numerical calculations by comparing the main flow characteristics with experiments. In the present work the measurements carried out by Ramamurthi and Nandakumar in 1999 were taken as a reference. In the experiments, with reference to pure water, the effects of varying the diameters of the sharp-edge orifice and the ratio of the inlet diameter (*d*) and the orifice length (*l*) are investigated for different Reynolds numbers of operation. The experimental set-up consisted of a tank for the water, a nitrogen gas source for pressurizing the tank and a feed-line fitted with flow control valves for supplying water to the orifice. The modelling of the experimental device focuses only on the orifice geometry, and particular care is paid to assure that the flow at the orifice extremes is not influenced by the inlet and outlet of the CFD domain. To do this, two plenums are considered at the orifice inlet and outlet, each one having a diameter and an axial length ten times larger than the orifice diameter. Thanks to the axial symmetry of the geometry and of the boundary conditions, it is possible to account only for a 5° sector of the whole domain, and to simulate the experiments as two dimensional cases. Fig. 6 shows an example of the CFD domain used. The simulations accounted for three orifice diameters (i.e. 0.5, 1 and 2 mm), three aspect ratios (i.e. 1, 5 and 20) and different Reynolds numbers. The complete list of the simulations, together with their main features, is detailed in Table 2.

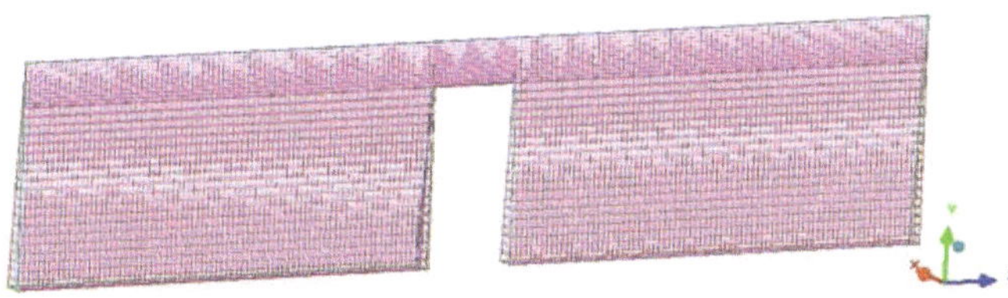

Fig. 6. CFD domain for the small sharp-edged cylindrical orifice test case

Case #	Section diameter	l/d	Reynolds number	Turbulence Model
B_1; B_4; B_7	0.5; 1.0; 2.0 mm	1	5x103; 1x104; 2x104	KE
B_2; B_5; B_8	0.5; 1.0; 2.0 mm	5	1.1x104; 2x104; 4x104	KE
B_3; B_6; B_9	0.5; 1.0; 2.0 mm	20	2x104; 4x104; 6x104	KE
B_10; B_13; B_16	0.5; 1.0; 2.0 mm	1	5x103; 1x104; 2x104	SST
B_11; B_14; B_17	0.5; 1.0; 2.0 mm	5	1.1x104; 2x104; 4x104	SST
B_12; B_15; B_18	0.5; 1.0; 2.0 mm	20	2x104; 4x104; 6x104	SST
B_19; B_22; B_25	0.5; 1.0; 2.0 mm	1	5x103; 1x104; 2x104	LS
B_20; B_23; B_26	0.5; 1.0; 2.0 mm	5	1.1x104; 2x104; 4x104	LS
B_21; B_24; B_27	0.5; 1.0; 2.0 mm	20	2x104; 4x104; 6x104	LS

Table 2. List of simulations for the turbulent flow through a small sharp-edged cylindrical orifice

In the modelling, the above mentioned standard k-ε model, the shear stress transport model and the low Reynolds Launder-Sharma model are compared. Moreover, similarly to the flow through a circular pipe case, different grid resolutions are tested to reduce the grid sensitivity. Finally, a ratio between the radial cell dimension and the orifice radius of 0.02 is found to be a good compromise between computational cost and results accuracy. The numerical results are compared with measurements mainly in terms of the orifice discharge coefficients. In the experiments the discharge coefficient is calculated using the relation:

$$Q = C_d A_o \sqrt{\frac{2\Delta p}{\rho}} \tag{3}$$

where Δp is the pressure drop across the orifice, ρ the fluid density, A_0 the orifice geometrical area and Q the volumetric flow rate.

Particular care was devoted in calculating the pressure drop and, in order to assess the influence of the abrupt section change only, it was defined as the difference between the inlet static pressure, p_{in} , and the static pressure at the vena contracta position, p_{vc} .

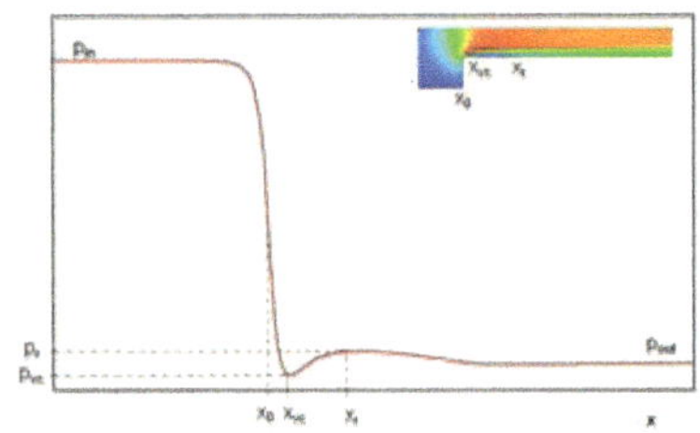

Fig. 7. Axial pressure for the small sharp-edged cylindrical orifice test case

With reference to Fig. 7, the fluid pressure in the vena contracta section, and the axial position of both the vena contracta (x_{vc}) and the reattachment section (x_r) are accurately estimated using the pressure distribution along the orifice length. In particular, the latter parameter is assumed as the axial coordinate of the relative maximum of the pressure profile downstream the vena contracta position. Fig. 8 a) collects, for all the cases reported in Table 3, the comparison between the experimental and the predicted discharge coefficients. As shown, for a given test case and for a wide variation of the Reynolds Number, each graph directly compares the experimental discharge coefficient with the numerical predictions obtained by using all the turbulence models previously depicted. The agreement of the calculated discharge coefficients and the experiments varied significantly among the test cases. The simulations demonstrate a better predictive capability for larger aspect ratios and lower Reynolds number. Under these conditions, the values calculated by using the KE and SST models are sufficiently close to the measured ones. On the other hand, a worse agreement is found for lower aspect ratios. Both turbulence models are able to capture the overall trend of the experiments, even though the SST model demonstrated a better accuracy.

While the behaviour of the KE and SST resulted to be similar, the values calculated by the LS model are found to be significantly different from measurements, and in particular for cases characterized by a small diameter and a low aspect ratio. The tendency of the LS turbulence model of underestimating the velocity close to the wall (highlighted in the flow through a circular pipe analysis also) is therefore confirmed. Consequently, the high turbulent kinetic energy dissipation rate causes a reduced velocity in proximity to the wall, thus forcing a smaller orifice effective area. This behaviour can be noticed also when comparing the axial coordinate of the vena contracta and of the reattaching point calculated by using the three turbulence models (see Fig. 8 b) and c)). A similarity can be noticed for the KE and SST models: while the reattaching points calculated by the LS model result to be

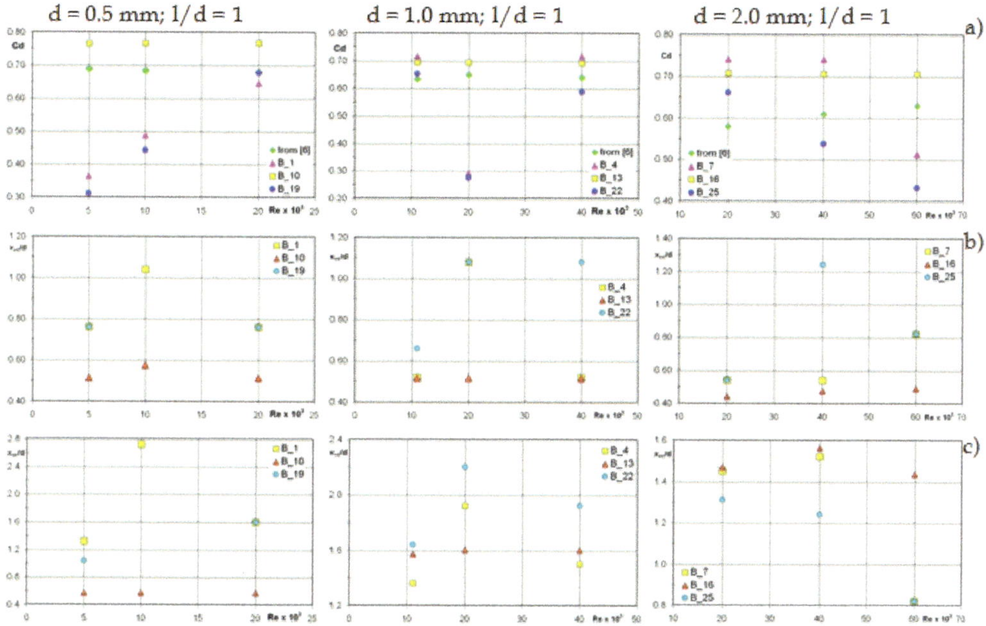

Fig. 8. a) Experimental and calculated discharge coefficients and axial coordinate of b)the vena contracta position and c) the reattaching point for different Reynolds number

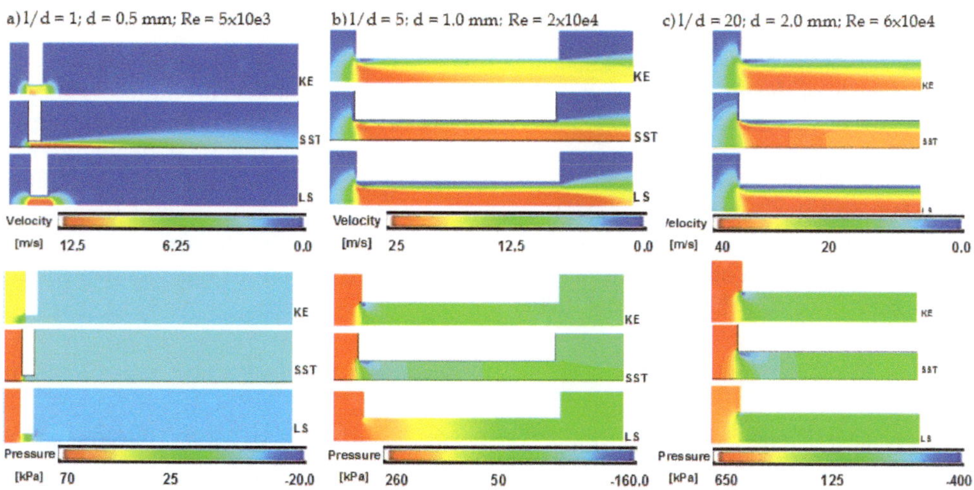

Fig. 9. Turbulence models influence on the velocity magnitude and pressure distribution

more distant from the orifice inlet. In many cases the relative maximum in the pressure profile along the geometry axis is found to be downstream the orifice outlet. This behaviour can be explained by investigating the velocity and pressure distribution along the channel (see Figs. 9). Finally, it is possible to conclude that the standard k-ε model and the k-ω shear stress transport models demonstrate a sufficient accuracy in calculating the discharge coefficients for most of the analyzed cases. Contrarily, the low Reynolds Launder-Sharma model is found to overestimate the turbulent kinetic energy dissipation, particularly in the wall proximity, thus causing an underestimation in the orifice effective area prediction. For the small diameter and low aspect ratio cases, the cavitating phenomena cannot be neglected in order to predict with a reasonable accuracy the discharge coefficients.

3. Modelling of cavitation and aeration

Once the predictive capabilities of the numerical models for the simulation of the incompressible flow are investigated, further step of the analysis is the modelling of the cavitation and aeration phenomena. The cavitation model used in this work is based on a barotropic equation of state approach and it assumes that the two phases (gas and liquid) have equal velocities and temperatures at any point of the flow (i.e. homogeneous equilibrium) (Yadigaroglu & Lahey, 1976). The assumption of isothermal flow is also made. The selected barotropic equation used in the modelling reads as follows:

$$\frac{d\rho}{dt} = \psi \frac{dp}{dt} \tag{4}$$

where ψ is the mixture compressibility. The parameter that accounts for the amount of fluid which is in the liquid or gaseous phase is the following:

$$\gamma = \frac{\rho - \rho_{l,sat}}{\rho_{ref} - \rho_{l,sat}} \tag{5}$$

where the reference gas density (ρ) is determined as the maximum value between the vapour density of the liquid fluid and the dissolved gas density at the current local pressure. The amount of dissolved gases which are locally released is calculated by introducing the Henry law for the equilibrium condition and accounting for the time dependence of solubility on a Bunsen coefficient basis. The released gas amount is firstly calculated from Henry law and the initial volume fraction of dissolved gas. The influence of time on gas absorption and dissolution is then accounted for by means of the Bunsen coefficient, c_B; thus the time variation of the released gas volume fraction, ε_c, can be calculated as follows (Weiss, 1970; Zarotti, 1998; Payri et al., 2002):

$$\frac{d\varepsilon_c}{dt} = \frac{\varepsilon - c_B p - \varepsilon_c}{\tau_c} \tag{6}$$

where ε is the gas - liquid saturated mass ratio, and τ_c is the time constant of the absorption or dissolution of the gas into the liquid medium. The Bunsen coefficient can range approximately from 0.02 for water to 0.09 for mineral oil (Zarotti, 1998), while it is very difficult to find in literature typical values of the time constant τ_c. Unfortunately, this lack of data is a stringent restriction for the gas release estimation. In fact, the amount of gas released during an expansion process is greatly influenced by the time constant, as well as the amount of gas which is dissolved during compression. Fig. 10 shows the volume fraction between the released gas and the liquid during an expansion from 1 bar to the liquid vapour pressure, a settling phase and a subsequent compression back to 1 bar. The time for the expansion and compression phases is set to 0.2 s while the settling phase is equal to 0.8 s.

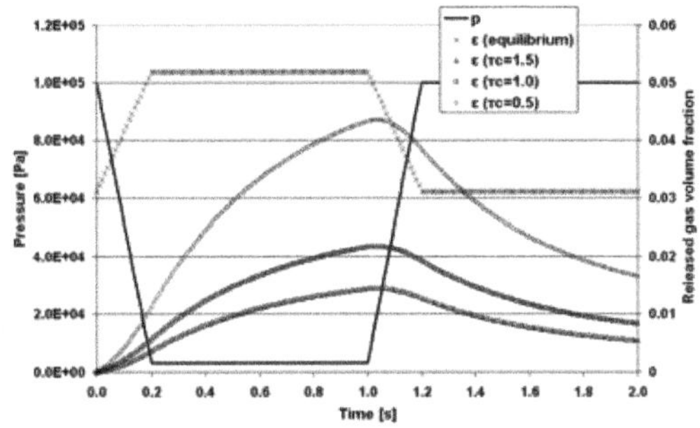

Fig. 10. Released gas volume fraction during an expansion and compression cycle

Different time constants are compared, and it is possible to notice that the released gas volume fraction is far from the equilibrium value both for the expansion and the compression processes. Thus, the released gas is not dissolved immediately and it lasts for a time period comparable with the process itself. The CFD analysis is carried out by means of the open source computational fluid dynamics code OpenFOAM. Bounded central differencing is used for the discretization of the momentum, second-order upwind for subgrid kinetic energy, and the mixture fraction. Pressure–velocity coupling is achieved via a Pressure Implicit Splitting of Operator algorithm similar procedure. The second-order implicit method is used for time integration scheme. The cavitation model originally implemented in OpenFOAM code does not account for the effects of turbulence. Therefore, the code is modified in order to account for the turbulence in the compressible flow equations. The approach to turbulence modelling is simplified since the effects of the liquid – gas/vapour interface are neglected, and this point remains an area for further investigation. In all the simulations carried out for the present work the turbulence is modelled by means of the two zonal version of the k-ω model, known as the shear stress transport model (Menter, 1993). The grids used in the simulations are constructed paying careful attention to comply with the requirements on the y^+ values for the k- ω SST formulation. Finally, the liquid and gaseous phase is tracked by using the volume of fluid (VOF) method where both fluid phases are compressible. The numerical approach is preliminarily validated by simulating a test case available in literature (Ramamurthi & Nandakumar, 1999) and comparing the numerical results with the experimental measurements. The selected geometry is a sharp-edged cylindrical orifice and the working fluid is water. The properties of water used in the simulations are the standard values at the reference temperature of 293 K. The dissolved gas fraction is assumed equal to the average value for tap water at ambient pressure, i.e. 0.01%. In the experiments the effects of varying the diameters of the sharp-edge orifice and the ratio of the inlet diameter (d) and the orifice length (l) are investigated for different Reynolds numbers of operation. The modelling of the experimental device focuses only on the orifice geometry, and particular care is paid to assure that the flow at the orifice extremes is not influenced by the inlet and outlet of the CFD domain. To do this, two plenums are considered at the orifice inlet and outlet, each one

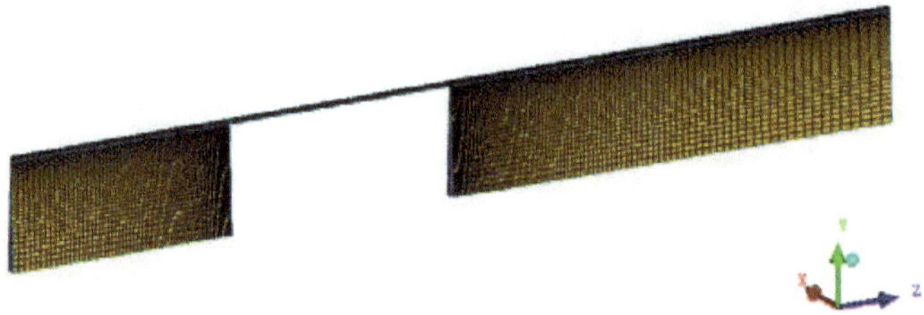

Fig. 11. CFD domain for the small sharp-edged cylindrical orifice case (d=1 and l/d = 20)

Section diameter	l/d	Reynolds number
0.5 mm	20	4.5x103; 1.1x104; 1.8x104
0.5 mm	50	5.9x103;8.5x104; 1.2x104
1.0 mm	20	1.1x104;1.8x104; 3.4x104
1.0 mm	50	8.8x104; 12.6x104; 3.3x104
2.0 mm	20	3.4x104; 4.9x104; 7.2x104
2.0 mm	50	1.2x104; 3.4x104; 5.6x104

Table 3. Test cases for the flow through a small sharp-edged cylindrical orifice

having a diameter and an axial length ten times larger than the orifice diameter. Thanks to the axial symmetry of the geometry and of the boundary conditions, it is possible to account only for a 5° sector of the whole domain, and to simulate the experiments as two dimensional cases. Fig. 11 shows an example of the CFD domain used. The present analysis focuses only on the geometrical configurations and operating conditions where cavitation was evidenced by the experiments in (Ramamurthi & Nandakumar, 1999). Table 3 details the test cases that are accounted for in the simulations. The numerical results are compared with measurements mainly in terms of the orifice discharge coefficients calculate as described in section 2.2. In order to evaluate the influence of cavitation and aeration modelling on the results' accuracy, the test cases are simulated by means of three different numerical approaches to cavitating flows analysis, i. e. a numerical approach which accounts for turbulent flows but does not include any cavitation model, a second approach which includes the cavitation model but assumes the flow as laminar and a third numerical approach with accounts both for turbulence and cavitation. Evidently, the computational cost increases from the first approach to the last one, and the accuracy of the results which can be obtained depends on the single problem that has to be simulated. Therefore, the predictive capabilities of the above mentioned numerical approaches are addressed when simulating cavitating flows. Fig. 12 shows the comparison between the experiments and calculations for the selected test case in terms of discharge coefficients. In the Figure the trend of the discharge coefficients in case of attached flow is also plotted (dotted line). This line represents the limit between the hydraulic flip cases and the reattached flow conditions. All the numerical approaches used in the simulation demonstrate to predict the discharge coefficient trend as a function of the Reynolds number. Nevertheless, the results' accuracy increases when adopting a cavitation model. In fact, when comparing Fig. 12 a) and c) and Fig. 12 b) and c) respectively, it can be noticed that better agreement between measurements

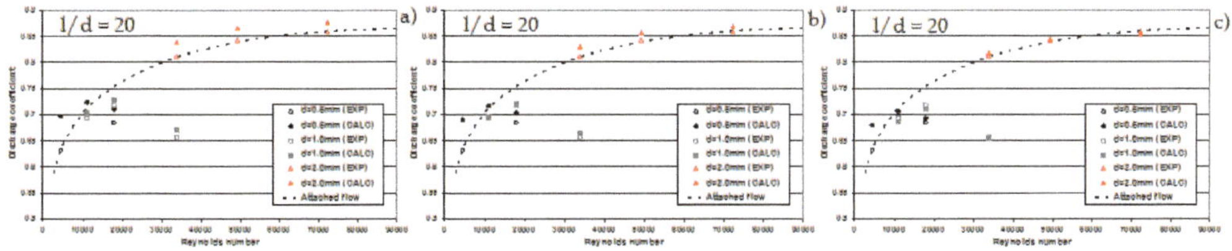

Fig. 12. Comparison between measured and calculated discharge coefficients for simulations a) without cavitation model and turbulent flow, b) with cavitation model and laminar assumption and c) with cavitation model and turbulent flow

and calculations can be achieved by the simulations accounting for cavitation and neglecting the turbulence effects. Thus, in order to improve the predictive capabilities of hydraulic components analysis, modelling the cavitating regions results to be more important than accounting for turbulence effects, when considering all the operating conditions used in the selected test cases,. This behaviour confirms what has been evidenced in (De Villiers, 2004) for the numerical analysis of the injector nozzle flow. In particular, in the simulations without any cavitation model, the low pressure regions immediately downstream the abrupt section change are underestimated (see Fig. 13). This behaviour can be explained by the fact that the velocity variation affects only pressure, while in the numerical analysis accounting for cavitation both pressure and density vary with velocity. As can be expected, the results' accuracy can be further improved by accounting both for the cavitation and the turbulence effects. Fig. 12 c) demonstrates that a better agreement between the measured and calculated discharge coefficients can be obtained by adopting the third numerical approach. When comparing the flow field obtained by the numerical approach with cavitation model and laminar assumption and the one with both cavitation model and turbulent flow no remarkable differences can be noticed within the orifice. The low pressure region downstream the abrupt section change results to be slightly larger for the former approach. In Fig14 the void fraction calculated by the numerical approaches without and with turbulence model is compared. It can be noticed that the gamma distribution is very

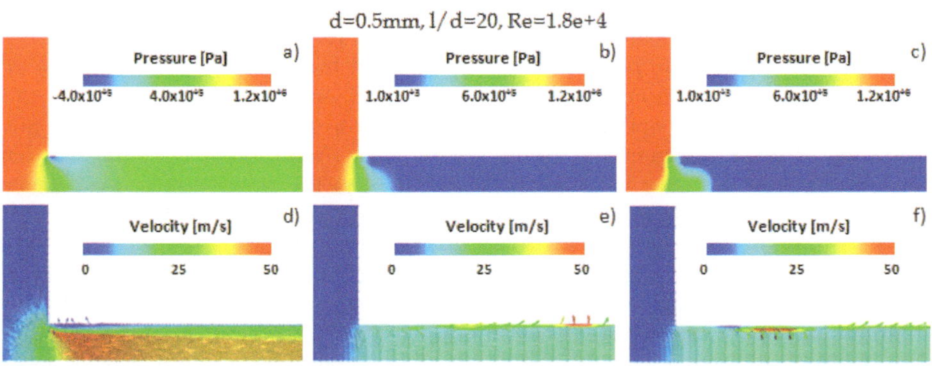

Fig. 13. Pressure and velocity distribution on a cut plane through the cylinder axis for different approaches: a) and d) no cavitation model and turbulent flow, b) and e) with cavitation model and laminar flow and c) and f) with cavitation model and turbulent flow

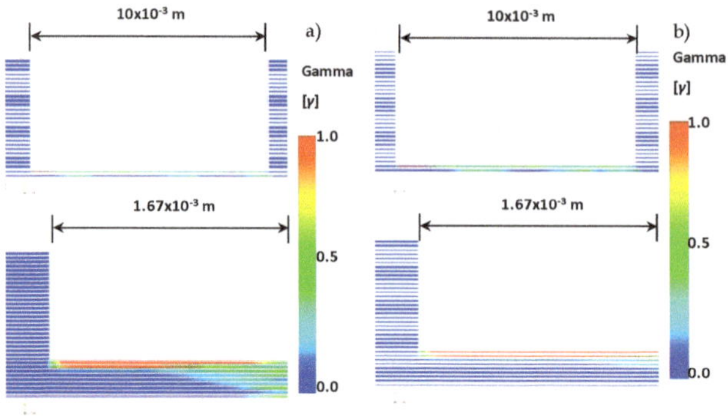

Fig. 14. Void fraction distribution on a cut section through the cylinder axis for a numerical approach with cavitation model: a) laminar assumption and b) turbulent flow

similar between the two models; nevertheless, when accounting for turbulence the gas/vapour phase is slightly more dispersed and it propagates closer to the cylinder axis. As can be seen in Fig. 12 for the case having orifice diameter equal to 0.5 mm, l/d = 50 and Reynolds Number of $1.8x10^4$ the flow does not reattach in the orifice. This behaviour is clearly evidenced by the gamma distribution in Fig. 14. Gas/vapour phase results to be distributed all through the orifice until the flow reaches the outlet plenum.

4. Steady state modelling of hydraulic components

The numerical models described in the previous sections are now adopted for the analysis of a specific hydraulic component: a closed centre electro-hydraulic load-sensing proportional control valve, usually adopted in multi-slice blocks to control parallel actuations of industrial, agricultural and earthmoving applications. In particular, the numerical simulation is used to evaluate the valve performance in terms of overall discharge coefficient, efflux angle, flow forces and pressure and velocity distributions in the critical region. In its simpler design, the proportional control valve for load-sensing applications is intended to directly react to a pressure signal (coming from the actuator) in order to maintain as constant as possible the pressure drop across its metering edges. This action is normally influenced by a local pressure compensator which could have either a single or a double stage configuration, and is usually placed upstream or downstream the control valve centre. Therefore, for a given operating position of the control valve spool, and for the flow-rate across the efflux area of metering orifices the pressure drop can be maintained constant independently by the actuator work-load. Fig. 15 depicts the closed centre load-sensing proportional control valve studied here. It is designed for operational field limits up to 100 l/min as maximum flow-rate and up to 350 bar as maximum pressure. As shown, the centre presents a Z connection between the high pressure port (P, in red) and the internal volume (P1, magenta), which is metered in all spool directions by a twice notched edge. At the same time, both connections between the internal volume and the actuator ports (A and B, green), and those between the actuators ports and the discharge line ports (T, blue) are metered by multiple notched edges. It is worth mentioning that the central volume indicated as P1 normally hosts the local pressure compensator (not included in the sketch in Fig. 15). The Figure presents also a view of the proportional control valve spool including a zoomed view

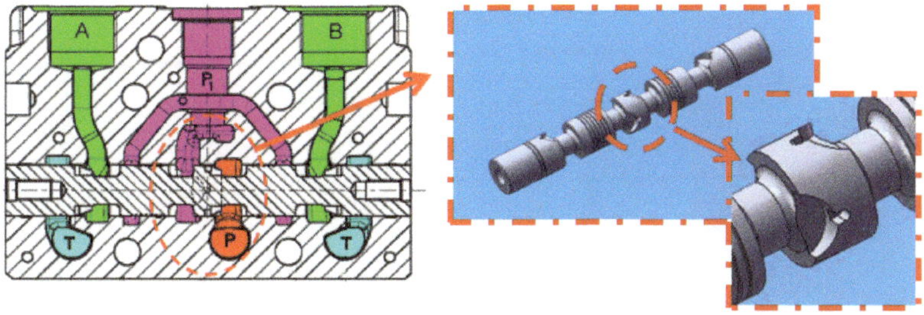

Fig. 15. Main section of the control valve and view of the spool's metering edges

of the design of the central edge. As shown, two equal notches located at 180° characterize each side of the central edge, and each couple of notches has a 90° rotation with respect to the other one. Moreover, the notches on the left side of the central edge meter the hydraulic power during the spool motion from the centre to the right (when P1 opens to A, and B opens to T), while the notches on the right side works during the motion from the centre to the left (when P1 opens to B, and A opens to T). In this way, the high pressure port P is connected to the control valve internal volume through an asymmetric path, which depends on the spool direction of motion. The CFD domain used in the simulations, see Fig. 16, includes the metering edges of the proportional control valve (upper, lower, front and back notch), the inlet duct and the outlet region up to the central volume where usually a local pressure compensator is positioned. The considered domain is marked with a dashed line in Fig. 15. The corresponding mesh is created using an unstructured grid paying particular care to the metering areas. Local refinements are used to obtain a large number of cells in the critical sections and at small opening positions. Moreover, wall cell layers are employed in order to have the proper wall cell height accordingly to the adopted turbulence model. The average mesh resolution is set to 0.3 mm corresponding to an overall cell number equal to 3 million elements. Fig. 16 b) shows the mesh of the zoomed views of a the metering edge. In the numerical analysis, several spool positions are considered in order to simulate different operating conditions of the control valve. In particular five spool displacements are investigated when the inlet and outlet regions are connected by means of the upper and lower notches (i.e. direct flow through the notches) and the corresponding five spool displacements in which the front and back notches connect the inlet and outlet of the valve (i.e. inverse flow through the notches). The former spool positions are considered as positive while the latter ones are assumed as negative even though the opening area is the same, since the spool displacement is symmetric with respect to the valve centre. Table 4 details all the operating conditions used in the analysis. As can be noticed from the Reynolds number listed in Table 4, all the cases are fully turbulent. Therefore, in the simulations carried out for the present work the turbulence is modelled by means of the two zonal version of the k-ω model, known as the shear stress transport model. Furthermore, careful attention is paid in order to comply with the requirements of the turbulence model adopted. The results of the numerical analysis of the load-sensing proportional control valve are discussed in terms of discharge coefficients, pressure and velocity fields, flow forces and flow acceleration angles. In particular the effect of the direct and inverse flow through the notches is highlighted. Fig. 17 a) depicts the comparison between the reference and calculated inlet pressures. In the simulations the outlet pressure is assumed as a constant boundary condition, while the inlet

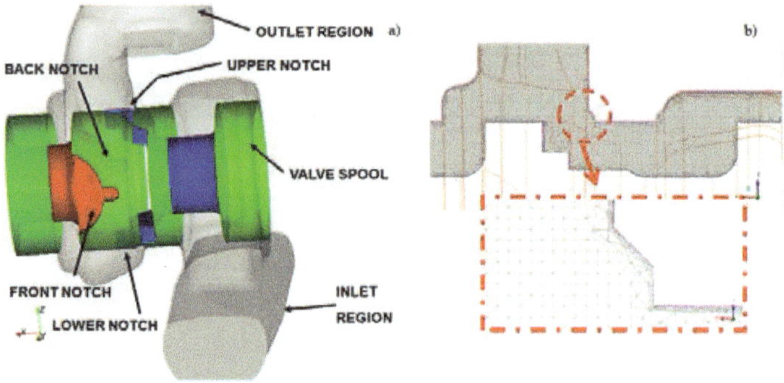

Fig. 16. a) The geometry of the control valve included in the CFD domain and b) detail of the mesh in the notch region for a mid range opening position of the spool

	Spool displacement (x/xmax)	Reynolds number
Inverse flow through the front and back notches	-1.00	3573
	-0.78	4209
	-0.65	3289
	-0.50	2174
	-0.35	1422
Direct flow through the upper and lower notches	0.35	1423
	0.50	2176
	0.65	3290
	0.78	4210
	1.00	3561

Table 4. Spool displacements and operating conditions used in the simulations

pressure is calculated as a consequence of the fluid - dynamics losses through the metering edges. As a first result, it can be noticed that even though the boundary conditions are symmetrical with respect to the control valve centre, the predicted inlet pressure is slightly different when comparing the direct and inverse flow. This behaviour is more evident when comparing the discharge coefficients, see Fig. 17 b). The discrepancy can be attributed to two main reasons. First, in the direct flow (i.e. positive spool displacements) the effective area decreases progressively as the liquid flows through the notches up to the vena contracta section close to the notch outlet; conversely, in the inverse flow, the area decreases abruptly at the notch entrance. Second, the front and back notches have very similar geometrical boundaries both at the inlet and at the outlet, while the geometrical volumes downstream the upper and lower notches differ significantly. In fact the liquid flowing out from the upper notches is heading for the outlet boundary, whilst the flow out from the lower notch finds a confined region causing further pressure losses. These reasons have opposite effects on the discharge coefficient; in fact the first one would advantage the direct flow while the second reason would advantage the inverse flow. The numerical simulations show that the trade-off between these two contrasting trends results in a better discharge coefficient for the inverse flow. This behaviour is also confirmed when comparing the flow

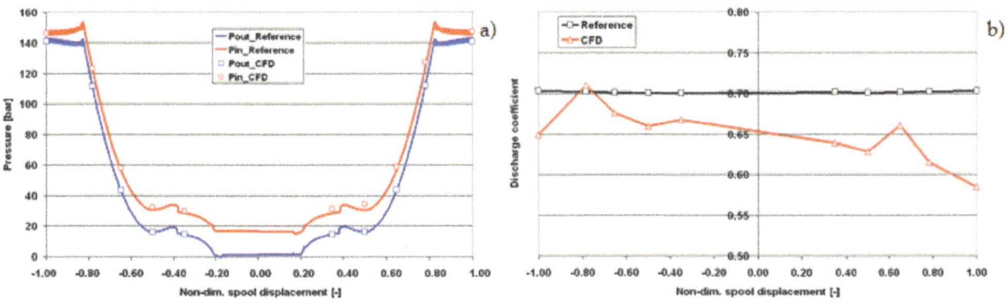

Fig. 17. Comparison a) between reference and calculated pressures at the valve inlet and outlet and b) between reference and calculated discharge coefficients

fields obtained for the different operating conditions. As an example, Fig. 18 plots the pressure and velocity fields on a cut section through the symmetry plane of the front and back notches (inverse flow) and the upper and lower notches (direct flow) for a small displacement case. While the flow through the front and back notches look very similar, the pressure and velocity fields are significantly different for the upper notch and the lower notch. In particular, the flow exiting the lower edge hits the outlet chamber wall and bounces back into the spool volume and then it is redirected to the outlet boundary. This

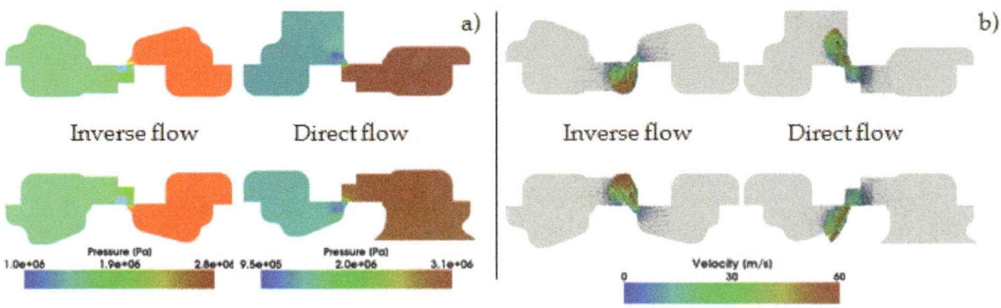

Fig. 18. a) Pressure and b) velocity distributions on a cut section through the symmetry plane of the metering section for the inverse and direct flow (x/x_{max}=0.35).

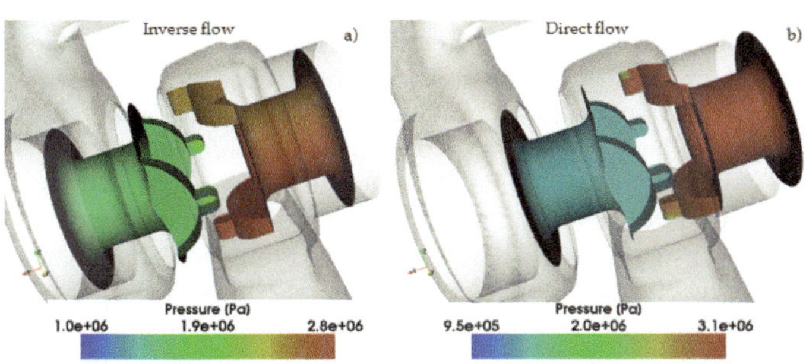

Fig. 19. Pressure distributions on the spool chambers used for the calculation of the flow forces for a) the inverse flow and b) the direct flow)

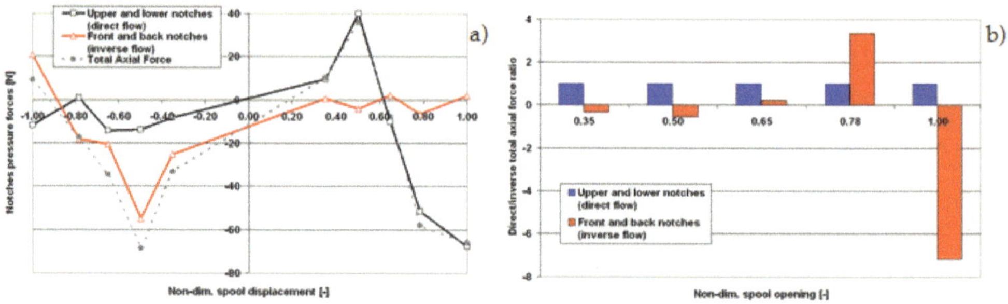

Fig. 20. a) Flow forces exerted on the valve spool in the axial direction and b) ratio between the total axial flow forces in case of direct and inverse flow

more complicated flow pattern increases the fluid - dynamics losses and thus determines a smaller discharge coefficient. By integrating the pressure distribution over the surface of the spool chambers' walls, it is also possible to calculate the axial forces exerted on the spool by the flow. Fig. 19 depicts the pressure contours on the walls of the spool chambers for all notches both in case of direct and inverse flow (as an example only the case with x/x_{max}=0.35 is plotted). It can be noticed that the pressure contour is quite uniform for the non operational notches, while the pressure varies significantly throughout the walls of the notches through which the oil flows. This behaviour results in a very low integrated value of the flow induced forces on the non operational chamber of the spool, as can be seen in Fig. 20. When comparing the integrated values of the forces for the direct flow (upper and lower notches) and the inverse flow (front and back notches), it is possible to outline that the behaviour is clearly not symmetrical with respect to the control valve centre, see Fig. 20 a). In fact, even though the trend looks similar and in both cases the sign of the forces reverses moving from the small to the large displacements, the magnitude is rather different. Fig. 20 b) details the ratio between the total axial forces obtained in case of direct and inverse flow. The forces have the same sign only at mid range displacements and the inverse flow induced forces are initially lower than the direct flow ones but approaching the maximum displacement they become much larger. Finally, by using the average velocity vector at the exit of the four notches it is possible to estimate the flow acceleration angle for each metering edge. Fig. 21 shows the convention used in this analysis for the calculation of the efflux angle with respect to the coordinate system axes. When values larger than 90 degrees are found, the negative supplementary angle is considered. Fig. 22 a) shows the efflux angles for the four metering sections as a function of the spool displacement. The angles are calculated for each notch only in its operational spool opening range. It is interesting to notice that each couple of notches has a quite similar behaviour. While this result could be expected for the front and back notches due to their similar geometrical boundaries, it is less evident in the case of the direct flow, where the downstream region is very different for the upper and lower notches. Furthermore, when comparing the efflux angles obtained for the direct and inverse flows, it can be remarked that the trend looks similar. In fact, smaller angles are characterizing the small spool displacements, as a consequence of the Coanda effect, whereas they are increasing moving towards the maximum opening. Nevertheless, in the average the direct flow results in larger flow acceleration angles; this behaviour is likely due to the fact that the main flow direction change happens close to the notch exit, while in the inverse flow the stream bends abruptly just entering the notch. By knowing the flow acceleration angles it is then possible to calculate the theoretical values of the axial forces, F_{th}, through the von Mises theory:

$$F_{th} = -2 \cdot C_d \cdot C_v \cdot \cos\varphi \cdot \Delta p \cdot A \tag{7}$$

where C_d and C_v are the discharge coefficient and the velocity coefficient (in the following assumed equal to 0.98), Δp is the pressure drop across the metering edge, A is its geometrical area and φ is the efflux angle. Fig. 22 b) depicts the comparison between the theoretical flow force and the one predicted by using the numerical simulation (grey lines). In the CFD results, the contributions of each metering edge are also plotted (red and black lines). The theoretical calculations and the numerical predictions are significantly different, and in particular by using the von Mises theory it is not possible to determine the change of direction moving from low to high spool displacements both for the inverse and direct flow.

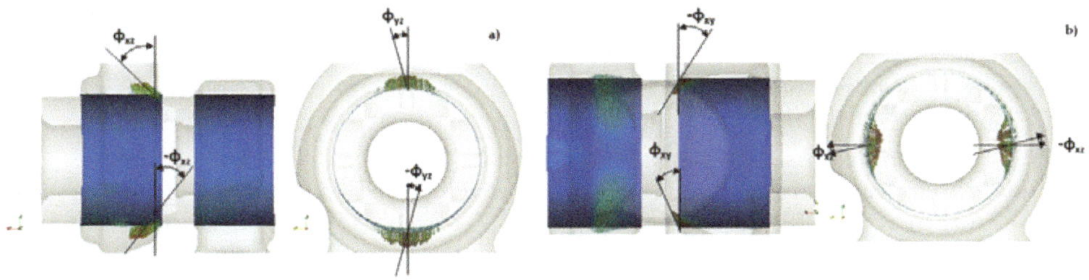

Fig. 21. Convention used for the calculation of the efflux angles

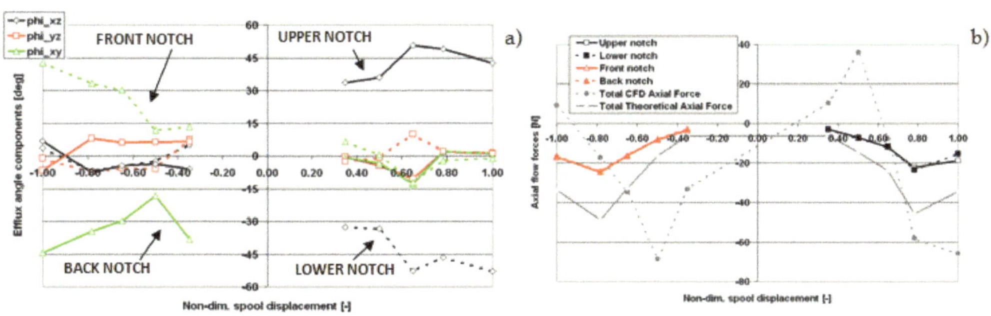

Fig. 22. a) Efflux angles for the different notches as a function of the spool displacement and b) comparison between theoretical and predicted axial forces

5. Fully transient modelling of hydraulic components

The numerical analysis of the hydraulic valve addressed in the previous section is further deepened by adopting a fully transient numerical approach. The main issue in this type of analysis is the moving mesh methodology employed to simulate the spool displacement profile versus time. To accomplish this task, the mesh changer libraries of the OpenFOAM code are modified to account for the linear motion of a portion of the geometry sliding over a second portion. The mesh motion is resolved by using a Generalized Grid Interface (GGI)

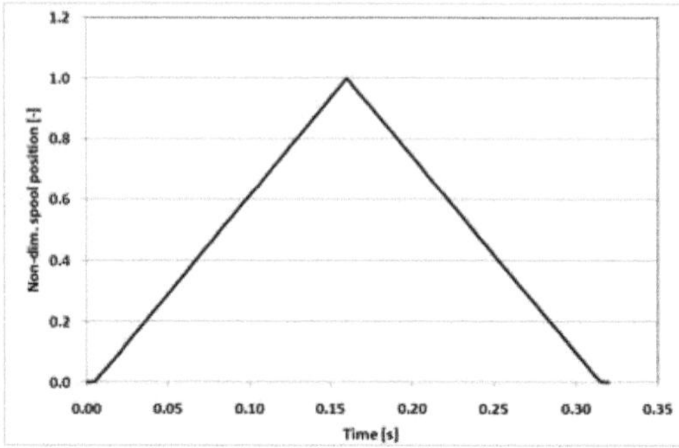

Fig. 23. PWM control applied to the spool motion

approach (Beaudoin & Jasak, 2008), originally developed for turbomachinery applications and modified to include not only rotational motion of the moving grid but also the linear displacement of a valve spool. The spool is moved accordingly to the PWM control profile depicted in Fig. 23. The other models adopted in the analysis are the same as the ones detailed in the previous sections. The valve is tested under several operating conditions. In particular, three constant flow rates at the inlet are considered, while a forth case is accounted for, in which the pressure across the valve is held constant. The four cases are first simulated by using a steady state approach and then a full transient approach. For the steady state simulations, four spool displacements are studied spanning from an almost closed position to mid range and maximum displacement. In case of steady state analysis there is no difference between the opening and the closing travels; on the contrary for the transient approach different inertia phenomena may arise. Table 5 lists the simulated cases, detailing the steady states and transient simulations. The comparison between the results obtained by the steady state calculations and the fully transient ones are discussed in terms of total pressure drop across the component, discharge coefficients, pressure and velocity fields, and flow acceleration angles. In Fig. 24 a) the total pressure drop across the valve is depicted for Cases # 1 to 3 and Cases # 5 to 7. The values predicted by the steady state calculations and the transient ones do not differ remarkably; particularly in the transient simulations the pressure variation during the opening and closing travels are quite similar. A large discrepancy can be seen when considering constant pressure at the boundaries. Fig. 24 b) plots the flow rate through the valve when holding constant the pressure drop across the component, i.e. 20 bar. The values predicted by the steady state simulations at small spool displacements are very close to the ones calculated during the opening phase with the transient approach. At large spool displacement the saturation flow rate resulted to be lower for the transient approach than for the steady state simulation. It is interesting to notice the difference between the flow rate calculated during the opening and the closure of the spool. The inertia of the flow through the metering edge of the valve causes a small fluctuation as the spool reverses it direction, while at mid range displacements the difference is even more evident. During this part of the metering curve, the flow inertia is being opposing to the spool motion and the total effect is a larger fluid dynamics loss. Similar behaviour can be seen also when considering the discharge coefficients, see Figs. 25 and 26. When the flow rate is held constant, the values during the first part of the opening travel are slightly larger than the ones during

	Inlet condition	Outlet condition	Spool positions	Numerical approach
Case #1	Constant flow rate (Q=12 l/min)	Constant pressure (pout=50 bar)	0.038; 0.35; 0.68; 1.00	Steady state simulations
Case #2	Constant flow rate (Q=30 l/min)	Constant pressure (pout=50 bar)	0.038; 0.35; 0.68; 1.00	Steady state simulations
Case #3	Constant flow rate (Q=100 l/min)	Constant pressure (pout=50 bar)	0.038; 0.35; 0.68; 1.00	Steady state simulations
Case #4	Constant pressure (pin=70 bar)	Constant pressure (pout=50 bar)	0.038; 0.35; 0.68; 1.00	Steady state simulations
Case #5	Constant flow rate (Q=12 l/min)	Constant pressure (pout=50 bar)	Function of time (see Fig. 25)	Transient simulation
Case #6	Constant flow rate (Q=30 l/min)	Constant pressure (pout=50 bar)	Function of time (see Fig. 25)	Transient simulation
Case #7	Constant flow rate (Q=100 l/min)	Constant pressure (pout=50 bar)	Function of time (see Fig. 25)	Transient simulation
Case #8	Constant pressure (pin=70 bar)	Constant pressure (pout=50 bar)	Function of time (see Fig. 25)	Transient simulation

Table 5. Boundary conditions used for the different simulated cases

closure, while as the direction reverses the discharge coefficients are larger due to the inertia, but decreases remarkably as the metering edge reduces as an effect of the larger losses. When comparing the discharge coefficients calculated by the steady state and transient approach, the values are quite close, particularly at large spool displacement and for the opening travel. Conversely, if the values during the closure travel are considered, the transient simulations predict lower discharge coefficients as a results of the behaviour mentioned before. Figs. 25 and 26 plot also the efflux angle for the different cases. The efflux angle is calculated accordingly to the reference system mentioned in section 4. A common trend of the efflux angle is predicted for all the transient simulations. The angle tends to increase rapidly during the small spool displacements reaching a local maximum between 0.5 and 1.5 mm, ranging from small to large flow rate, while it decreases during the mid range displacements and finally it starts increasing again at high spool axial positions. The efflux angle results to be slightly larger for the closure travel all through the displacements; the flow exiting the metering section during the closure phase results to be more disperse when compared to the flow during the opening travel. This behaviour is clearly visible when considering the flow field during the opening and closure travels, see Fig. 27. The efflux angle trend described above for the transient simulations is definitively different from the behaviour predicted by the steady state calculations; in fact, with the steady state approach the efflux angle is increasing as the spool displacements increases. Furthermore, the efflux angle values calculated by this approach resulted to be larger than the ones obtained with the transient simulations. This difference has a remarkable importance if the flow forces are accounted for, since they are significantly affected by the vena contracta angle.

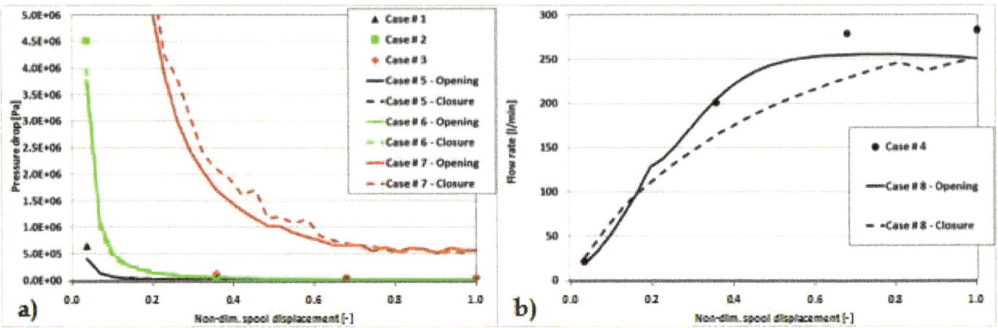

Fig. 24. Comparison between the a) pressure drop and b) flow rate through the metering area predicted by the steady state and transient simulations.

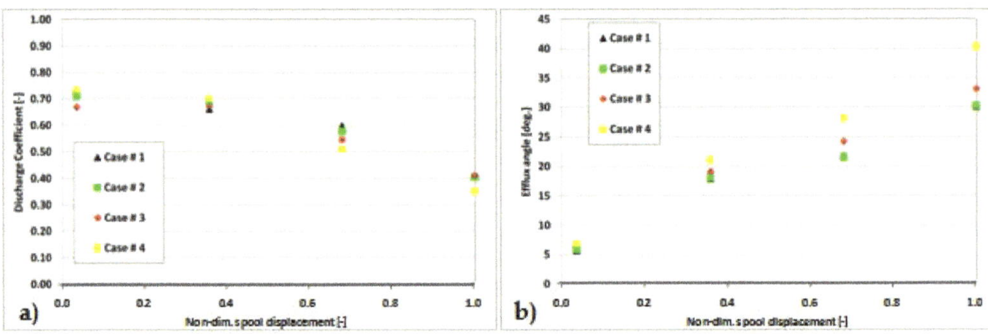

Fig. 25. a) Discharge coefficients and b) efflux angle for the steady state simulations

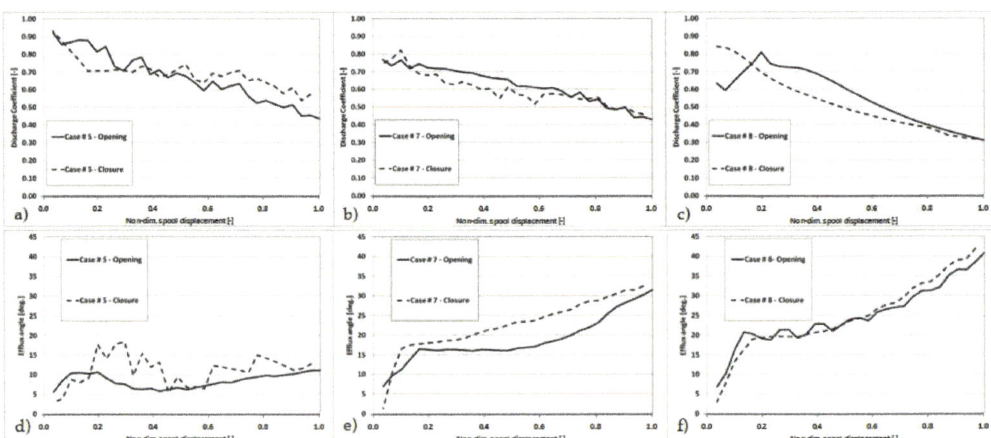

Fig. 26. a), b) and c) Discharge coefficient and d), e) and f) efflux angle for the transient cases

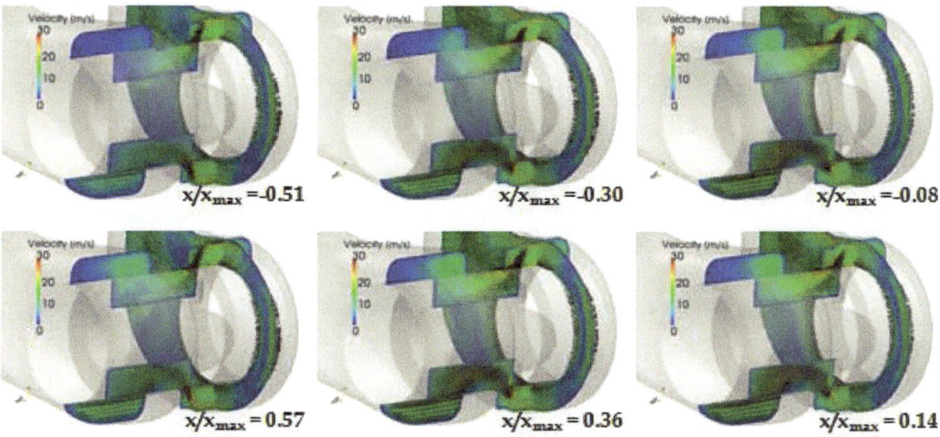

Fig. 27. Velocity distribution for different non-dim. spool displacements (Case # 7)

6. Modelling approach for a multi-fuel injection system

The numerical approach described in the previous section for the design and performance prediction of hydraulic components is employed for the analysis of a low pressure, common rail, multi-fuel injection system. In particular, system is working with a mean pressure close to 3.5 bar and every injector is driven by a modified PWM control characterized by 3rd order function opening and closing ramps and characteristic times adapted to the engine rotational speed through an ECU correction. In the numerical analysis of this hydraulic system, the turbulence as well as the cavitation models are used to address the flow within the injector in order to estimate the permeability characteristics of the injector when operating with different fuels. In the analysis, the operating conditions for the injectors are preliminarily evaluated by lumped and distributed parameter approach. Thus, the system behaviour and the injection profiles for several fuel blends are calculated. The general reliability of the one-dimension approach is defined with respect to experimental data mainly in terms of injected mass per stroke. Fig. 28 shows the injection system model including only the rail and the injectors. The connection to the supply is simplified by means of a flow-rate source. The Figure shows also the injector lift and the modified PWM wave; the total time, t_{inj}, is adjusted accordingly to the measurements. The CFD analysis focuses only on the injector geometry since it represents the main component influencing the permeability of the hydraulic system. The injector is characterized by six nozzles: the first one is in-line with the injector axis, while the other five are located circularly every 72 degrees. Therefore, the injector geometry and the boundary conditions are cyclic symmetrical allowing the simulations to be carried out only over a 72 degree sector mesh (see Fig. 29). In the simulations the pressure at the nozzles' exit is kept constant equal to 1 bar, while the inlet pressure is varied from 3.5, 5 and 10 bar. In fact, the lowest inlet pressure is the current injection pressure for this type of multi - fuel injection system, but the tendency is to increase it in order to comply with the mass flow rate required by new fuels and new injection strategies. Finally three different fuels are accounted for in the numerical analysis. Simulations with varying needle lifts and injection pressures are carried out for pure ethanol, pure gasoline and a fuel mixture corresponding to 50% gasoline and 50%

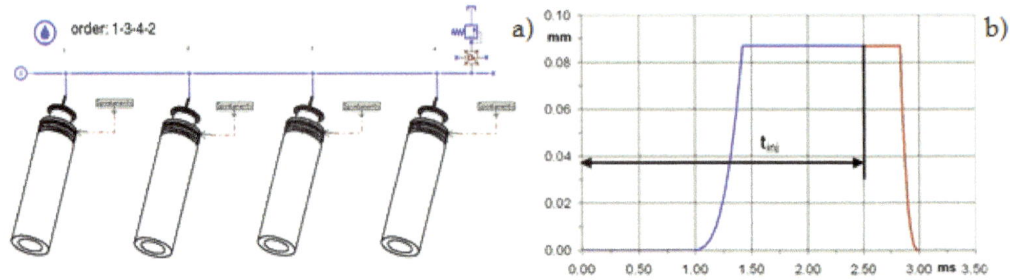

Fig. 28. a) Layout oft he lumped and distributed parameter model and b) modified PWM control applied to the injector lift

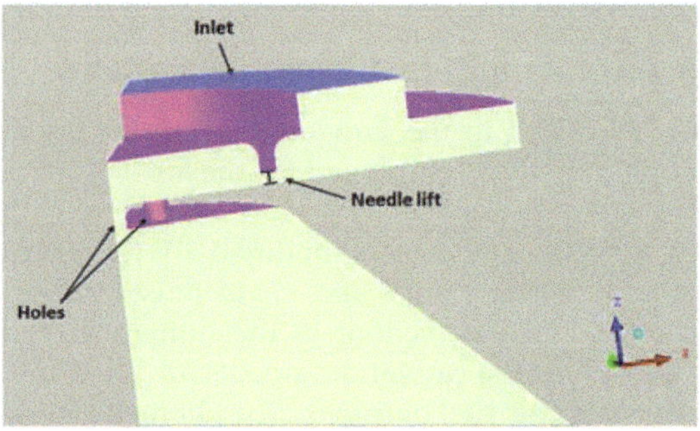

Fig. 29. Geometry of the 6 nozzle injector tip used in the simulation (72 degree sector)

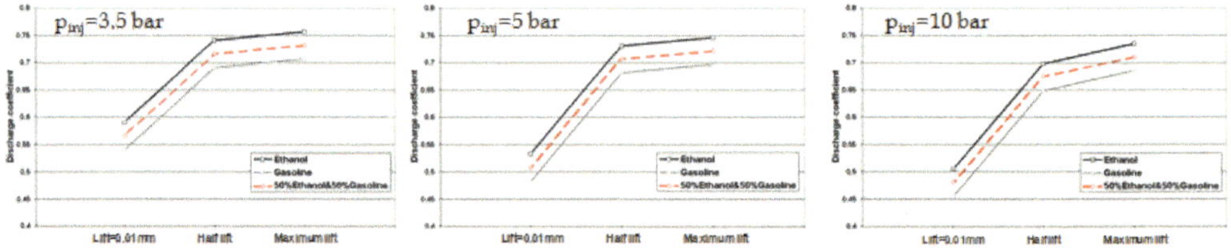

Fig. 30. Discharge coefficient vs. needle lift for different fuels

ethanol in terms of mass fraction. Fig. 30 shows the performance of the injection system in terms of discharge coefficients as a function of the lift, injection pressure and operating fuel. In all simulations the back pressure at the outlet was assumed to be constant at 1 bar. When comparing the results holding constant the injection pressure, it is possible to highlight that, as expected, the discharge coefficients for the maximum and half lifts are very similar and only for the almost closed position it reduces drastically. When considering the influence of the working fuel a significant difference can be seen for all the injection pressures and needle lifts. In fact, the discharge coefficients tend to decrease remarkably when injecting gasoline. The reduction reaches values up to 7% when compared to ethanol injection and the trend is rather homogeneous for all the considered cases. This result can be attributed to the different vapour pressure for ethanol and gasoline; the higher vapour pressure allows the fuel to vaporize earlier causing larger low density regions within the nozzle. This effect can be seen as a reduction of the effective area of the nozzle throat section. This behaviour can be clearly evidenced when considering the flow field within the injector nozzle in terms of pressure and void fraction distributions (see Figs. 31 and 32). By comparing the pressure and void fraction contour plots on a cut plane through the nozzles' axes holding constant the working fuel and varying the injection pressure and needle lift, see Fig 31, no relevant differences can be observed in the pressure field when halving the needle lift with the same injection pressure, while a larger low pressure region in both nozzle can be evidenced for the almost closed needle position. Same observation can be made when addressing the void fraction distribution; only at the highest injection pressure a remarkable high vapour fraction region appears. Fig. 32 depicts the results obtained by varying the injected fuel and needle lift while holding constant the injection pressure (i.e. 3.5 bar) in terms of pressure and vapour fraction distributions. As anticipated when describing the results in terms of discharge coefficients, in case of gasoline injection the low pressure regions within the nozzle are wider than in case of ethanol or ethanol/gasoline fuel mixture. Consequently, larger high vapour fraction regions form in the nozzle and close to the nozzle outlet, thus reducing the effective area.

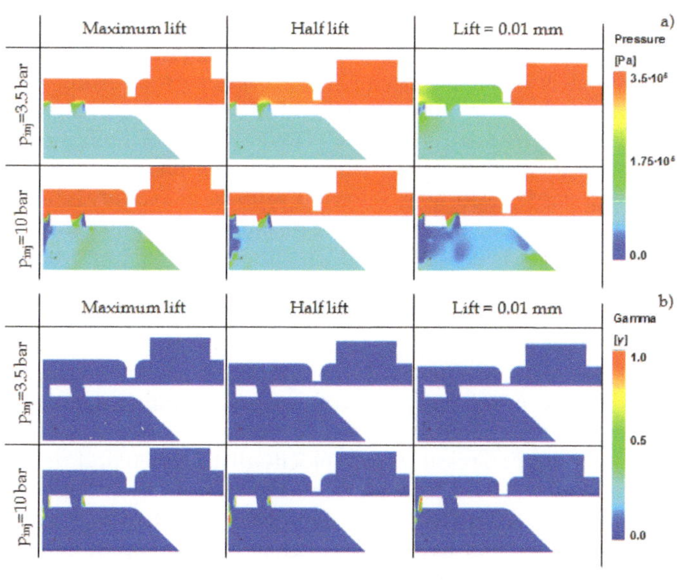

Fig. 31. Comparison of a) the pressure and b) the void fraction (gamma) distributions on a cut section through the nozzles' axes for different injection pressures and needle lifts (injected fuel: ethanol)

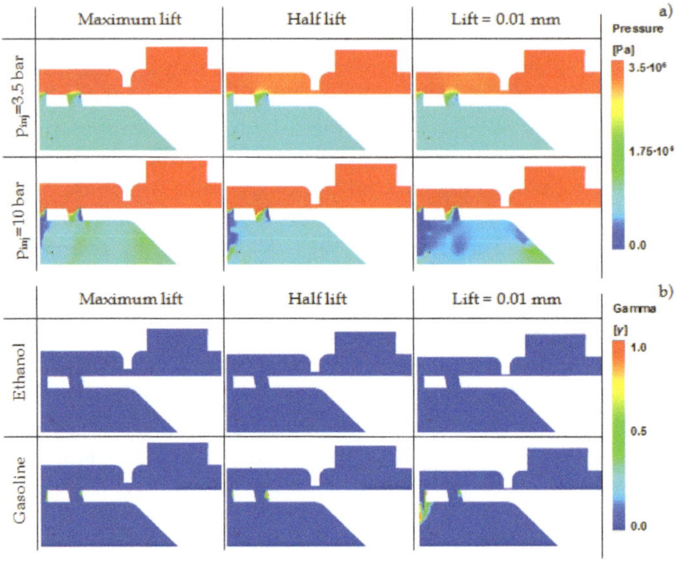

Fig. 32. Comparison of a) the pressure and b) the void fraction (gamma) distributions on a cut section through the nozzles' axes for different injected fuels and needle lifts (injection pressure: 3.5 bar)

7. Conclusion

In this chapter, a numerical approach for the multidimensional design of hydraulic components and systems has been shown. The procedure for setting up the numerical analysis have been highlighted in order to provide the designer with the guidelines for addressing the major issues that arise in the CFD simulation of hydraulic components. In particular, the approach to modelling the turbulent flows has been investigated and the results that can be obtained by adopting different numerical models have been compared. Furthermore, the influence of the boundary layer resolution as well as the cell dimension on the predictive capabilities has been addressed. Additionally, the importance of accounting for cavitation and aeration in the numerical modelling has been evidenced and the effects on the results' accuracy have been remarked. Finally, the steady state approach and the fully transient simulation have been confronted in terms of predictive capabilities of the behaviour of internal flow field of a hydraulic component, complexity and required time for the case set up as well as the computational effort and simulation duration. The results that can be obtained by means of the multidimensional numerical analysis of hydraulic systems have been discussed in terms of pressure and velocity distribution and recirculating regions within the components and of overall performance parameters, such as the discharge coefficient, the flow acceleration angle and the flow forces. In the chapter, the methodology for the multidimensional design of hydraulic systems has been applied to the analysis of a closed centre electro-hydraulic load-sensing proportional control valve, usually adopted in multi-slice blocks to control parallel actuations of industrial, agricultural and earthmoving machines. Examples of results that can be obtained by the numerical simulation have been provided and discussed in the terms of component performance and influence on the system operations. Furthermore, the same approach has been employed in the investigation of a low pressure, common rail, multi-fuel injection system. The behaviour of the injection system and in particular of the injector has been addressed under different operating conditions, and its performance has been also evaluated when extending the working range to higher pressure and to different fuel mixtures. The multidimensional CFD approach demonstrated to be a valuable tool in the design of hydraulic components and systems, since it provides reliable and accurate results on the flow behaviour that can help the designer understand their operations and improve the performance.

8. References

Barman, P. (2005). Computational fluid dynamic (CFD) analysis to optimize the pump suction line configuration of a hydraulic control system, *SAE 2005-01-3633* , ISSN 0148 - 7191

Beaudoin, M. & Jasak, H. . (2008). Development of a Generalized Grid Interface for Turbomachinery simulations with OpenFOAM. *Proceeding of the 2008 2nd Open Source CFD International Conference*, Berlin, Germany, December 4-5, 2008

De Villiers, E., Gosman, D. & Weller, H. (2004). Detailed Investigation of Diesel Spray Atomisation Using Quasi-Direct CFD Simulation, *Proceedings of the JSME 2004 6th International Symposium on Diagnostics and Modeling of Combustion in Internal Combustion Engines*, Vol.2004, No.6, pp. 295-302, Yokohama, Japan, August 2-5, 2004

Launder, B.E. & Sharma, B.I. (1974) Application of the Energy Dissipation Model of Turbulence to the Calculation of Flow Near a Spinning Disk, *Letters in Heat and Mass Transfer*, Vol.1, No.2, pp. 131-137, ISSN 0735-1933

Menter, F.R. (1993). Zonal Two Equation k-ω Turbulence Models for Aerodynamic Flows, *Proceedings of AIAA 1993 23rd Fluid Dynamics, Plasmadynamics, and Lasers Conference*, pp. 1-21, ISSN 0001-1452, Orlando, Florida, USA, July 6-9, 1993

Menter, F.R. (1994) Two-Equation Eddy-Viscosity Turbulence Models for Engineering Applications, *AIAA Journal*, Vol.32, No.8, pp. 1598-1605, ISSN 0001-1452

Ohm, T. R., Senser, D. W. & Lefebvre, H. (1991). Geometrical effects on discharge coefficients for plain-orifice atomizers, *Atomization and Sprays*, Vol.1, No.2, pp. 137-153, ISSN: 1044-5110

OpenCFD (2010). *OpenFOAM 1.7.1*, August 26 2010, Available from: http://www.openfoam.co.uk

OpenFOAM®-Extend Project (2010). *OpenFOAM 1.6-ext*, November 26 2010, Available from: http://www.extend-project.de

Oshima S. & Ichikawa T (1985). Cavitation Phenomena and Performance of Oil Hydraulic Poppet Valve, *Bulletin of JSME*, Vol.28 No.244, pp. 2264-2271, ISSN: 1881-1426

Oshima S. & Ichikawa T. (1986). Cavitation Phenomena and Performance of Oil Hydraulic Poppet Valve, *Bulletin of JSME*, Vol.29 No.251, pp. 1427-1433, ISSN: 1881-1426

Oshima, S., Leino, T., Linjama, M., Koskinen, K. T. & Vilenius, M. J. (2001). Effect of Caviation in Water Hydraulic Poppet Valves, *International Journal of Fluid Power*, Vol.2, No.3, pp 5 -13, ISSN 1439-9776

Pao, R. H. F. (1967) *Fluid Mechanics*, John Wiley & Sons, Inc, Publishers. ISBN 61-11499, New York, USA

Payri, R., Margot, X. & Salvador, F. (2002). A NumericalStudy of the Influence of Diesel Nozzle Geometry on the Inner Cavitating Flow, *SAE 2002-01-0215*, ISSN 0148 - 7191

Ramamurthi, K. & Nandakumar, K. (1999). Characteristics of flow through small sharp-edged cylindrical orifices, *Flow Measurement and Instrumentation,* Vol.10, No.3, pp. 133–143, ISSN 0955-5986

Weiss, R.F. (1970). The solubility of nitrogen, oxygen and argon in water and seawater, *Deep Sea Research and Oceanographic Abstracts*, Vol.17, No.4, pp. 721-735, ISSN 0967-0637

Wilcox, D.C. (1988) Reassessment of the Scale-Determining Equation for Advanced Turbulence Models, *AIAA Journal,* Vol.26, No.11, pp. 1299-1310, ISSN 0001-1452

Yadigaroglu, G. & Lahey, R. T. Jr. (1976). On the Various Forms of the Conservation Equations in two Phase Flow, *International Journal of Multiphase Flow,* Vol. 2, No. 5-6, pp. 477-494, ISSN 0301-9322

Yang, R. (2002). Simulations of Oil Flow and Flow-Induced Forces Inside Hydraulic Valves, *SAE 2002-01-1376* , ISSN 0148 - 7191

Yang, R. (2005). Predicting Hydraulic Valve Pressure Drop Using CFD, *SAE 2005-01-3635,* ISSN 0148 - 7191

Zarotti, G. (1998). *Fluidi Oleodinamici – nozioni e lineamenti introduttivi,* Vol.2. CEMOTER-CNR

13

Use of Fluid Dynamic Simulation to Improve the Design of Spouted Beds

Rodrigo Béttega[1], Ronaldo Guimarães Corrêa[2] and José Teixeira Freire[2]
[1]*Federal University of Uberlândia, School of Chemical Engineering*
[2]*Federal University of São Carlos, Department of Chemical Engineering*
Brazil

1. Introduction

The use of spouted beds in numerous chemical and pharmaceutical industrial processes and in the agricultural industry has become popular principally because they enable high rates of heat and mass transfer to take place between system phases. However, despite the advances achieved through research into spouted beds over the more than 50 years since their invention, it continues to be difficult to obtain detailed information regarding the hydrodynamics of this equipment. The constant movement of the particles and the geometric characteristics of the equipment are the principal factors complicating this evaluation. More specific information is still required about scaling up spouted beds and using data from half-column beds to infer data for full column beds.

The half-column technique applied for hydrodynamic evaluation of spouted beds is very old. Nevertheless, many authors have stated that great care must be taken in their application to obtain experimental data for use in representing full-column spouted bed phenomena. The use of optical fiber sensors is currently one of the most commonly recommended techniques for this type of measurement. However, measuring the particle dynamics by inserting optical fiber sensors may produce inaccurate results, as calibration is extremely difficult and the sensors may cause perturbations within the bed.

With regard to scaling up, larger spouted beds tend to be less stable and consequently harder to control and operate. This issue is difficult to study using experimental techniques because it relates directly to constructing beds on different scales. The flexibility of computational fluid dynamics, allied to the reliability of the simulated results that have already been proven by a range of studies, can contribute to this area. It enables the analysis of scaling up data in different configurations and scales and using fewer experimental modules, making it possible to conduct studies that remain difficult in the laboratory.

With regard to the study of fluid movement, the most recent studies of fluid mechanics are still based on the same Navier-Stokes equations. Although they were proposed approximately 200 years ago, analytical solutions for the Navier-Stokes equations are known for just a few cases due to the non-linear nature of the partial differential equations (PDAs). There is still little knowledge of analytical solutions for this mathematical class, making it impossible to obtain a solution for arbitrary regions and general boundary conditions. These are the principal reasons why analytical evaluation of the movement of the fluids is very difficult. In light of this, treatments using experimental tests are

predominant in this study. The costs involved in certain experiments and limited representative data make it necessary to improve the mathematical modeling of the problem.
The advent of computers in the '80s and their improved storage and processing capabilities have led to accelerated growth in the numerical simulation of fluid mechanics and heat transfer (Computational Fluid Dynamics - CFD) and its use in different areas. Computational fluid dynamics can be defined as the set of computational methods used to solve physical and physico-chemical problems occurring in dynamic fluids by applying sophisticated numerical and computational tools. By using these powerful numerical simulation tools, it is now possible to obtain detailed and accurate results for flows with the most diverse characteristics, among which multiphase gas-solid flow stands out. The speed of evolution of computational codes, allied to the consequent improvement in the simulated results, is making CFD an increasingly important analysis tool, both in terms of process design at laboratory or industrial scale.
This chapter describes new advances achieved using CFD numerical modeling in studies of cylindrical and semi-cylindrical spouted beds. These studies were conducted with the aim of evaluating the mathematical model parameters, as well as exploring numerical simulation applied to semi-cylindrical spouted bed in the assessment of hydrodynamic data pertaining to cylindrical beds. Recent research involving CFD and spouted beds is highlighted and discussed throughout this text with the aim of presenting recent advances in research in this area.

2. Mathematical modelling

2.1 Mathematical modeling of general multiphase flow

The basic axioms of fluid dynamics are the conservation laws, specifically, conservation of mass, conservation of linear momentum, and conservation of energy. The combination of these mathematical equations provides a fluid dynamic description of single-phase systems well defined and reliable. However, the development of a mathematical model to represent a general multiphase flow is still a great challenge to researchers due to the complexity of phenomenona involved on it.
Currently in gas-solid flows, although the equations are not yet clearly defined, there are several approaches available, especially for engineering problems. In gas-solid flow modeling there exist the Eulerian approach, which is based on the hypothesis of a continuous medium for the description of the individual behavior of each phase (fluid or solid phase), and the Lagrangian approach, which is based on the individual movement of each particle in the flow.
The Lagrangian approach gives a direct description of the particulate phase movement by mapping the path of each individual particle present in the phase. In this mathematical model, the movement of each particle is described by ordinary differential equations in Lagrangian coordinates. Consequently, the velocity of each particle, as well as its trajectory, can be determined from the integration of the Lagrangian equations of the particle. The Lagrangian approximation is especially useful to describe the transient nature of the particle movement.
To solve a Lagrangian equation for a given particle, the behavior of the gas phase and of the neighboring particles must be known beforehand. In some cases, the behavior of the fluid phase can be determined using an Eulerian approach. As the gas and the particulate phases are linked by gas-solid effects, a large number of interactions are necessary to obtain the convergence of the solution for all phases.

Lagrangian modeling is limited in its application by the large computational requirements for its solution. Due to its high processing cost, this approach is normally applied in cases in which the particulate phase can be considered diluted or in cases of flow of isolated particles for which the multi-fluid approach (Euler-Euler) is not appropriate. In this chapter, mathematical modeling will be limited to the Eulerian approximation, which is commonly applied to the solution of fluidized bed and spouted bed flows, among others.

In the Eulerian approach, each phase is assumed to obey the continuum assumption. The continuum assumption considers solid phase as continuous, rather than discrete. Consequently, properties such as density, pressure, temperature, and velocity are taken to be well-defined at infinitesimal volumes, and are assumed to vary continuously from one volume to another. This approach is basically an extension of the mathematical modeling of a single-phase flow to a multiphase flow. However, since the fluid phase and the particulate phase actually are not continuous media at any moment, forms of constructing a phase continuum should be established. The transported magnitudes of each pseudo continuous phase, or even of the turbulence model of each phase (in the case of turbulent flow), must be determined. Furthermore, the phase interactions must be expressed in a continuous form.

A fundamental step in the construction of a continuous model is the correct determination of the "fluid element" of the continuous medium. This element or control volume must be determined so that the particles present are sufficient to ensure that this volume has macroscopic properties, such as temperature or velocity. At the same time, the size of this element must be sufficiently small in comparison to the system dimensions from a macroscopic viewpoint to allow the use of the continuous phase concept.

Based on this, we present below a mathematical modeling of multiphase flow using the Euler approach. The manual of the software Fluent® was used as a reference to describe this modeling (Fluent, 2006).

In the description of multiphase flow as a continuous interpenetrant, the concept of phase volumetric fraction α_q is introduced. This fraction represents the space occupied by a certain phase in the global flow; the momentum and mass conservation laws will be applied to each of the phases. In this way, the volume V_q of a phase (either solid or fluid) is defined by:

$$V_q = \int_V \alpha_q dV \tag{1}$$

where:

$$\sum_{q=1}^{n} \alpha_q = 1 \tag{2}$$

The effective density of a phase q is:

$$\hat{\rho}_q = \alpha_q \rho_q \tag{3}$$

where ρ_q is the specific mass of phase q.

The continuity equation of phase q is determined from:

$$\frac{\partial}{\partial t}\left(\alpha_q \rho_q\right) + \nabla \cdot \left(\alpha_q \rho_q \vec{v}_q\right) = \sum_{p=1}^{n}\left(\dot{m}_{pq} - \dot{m}_{qp}\right) + S_q \tag{4}$$

where:
$\vec{v}_q$ - velocity of phase; $\dot{m}_{pq}$ - mass transfer of a p phase to a q phase; $\dot{m}_{qp}$ - mass transfer of a q phase to a p phase; S_q - generation of mass for the q phase; n - number of phases of the flow.
The mechanisms of mass transfer between phases, as well as the mass generation term, can be defined in different ways. However, these terms are commonly considered null in common gas-solid flow problems of particulate systems. The solution of Eq. 4 for all the secondary phases present in the flow together with Eq. 2 gives the volumetric fraction of the primary phase of the flow. This mathematical treatment is common to fluid-fluid and granular flows.
Momentum conservation balance for phase q is given by the following equation:

$$\frac{\partial}{\partial t}\left(\alpha_q\rho_q\vec{v}_q\right)+\nabla\cdot\left(\alpha_q\rho_q\vec{v}_q\vec{v}_q\right)=-\alpha_q\nabla p+\nabla\cdot\overline{\overline{\tau}}_q+\alpha_q\rho_q\vec{g}+ \\ +\sum_{p=1}^{n}\left(\vec{R}_{pq}+\dot{m}_{pq}\vec{v}_{pq}-\dot{m}_{qp}\vec{v}_{qp}\right)+\left(\vec{F}_q+\vec{F}_{lift,q}+\vec{F}_{vm,q}\right) \tag{5}$$

where: $\overline{\overline{\tau}}_q$ – tensor of viscous stresses of phase q; $\vec{v}_{pq}$ – phase relative velocity; $\vec{R}_{pq}$ – force of interaction between phases; $\vec{F}_q$ – external body force of phase q; $\vec{F}_{lift,q}$ – lift force; $\vec{F}_{vm,q}$ – virtual mass force; p – pressure shared by the phases; p_q – pressure of phase q; g - gravitational acceleration.
The viscous stress tensor $\overline{\overline{\tau}}_q$ of phase q is obtained by:

$$\overline{\overline{\tau}}_q=\alpha_q\mu_q\left(\nabla\vec{v}_q+\nabla\vec{v}_q^T\right)+\alpha_q\left(\lambda_q-\frac{2}{3}\mu_q\right)\nabla\cdot\vec{v}_q\overline{\overline{I}} \tag{6}$$

where μ_q represents the shear viscosity of phase q and λ_q represents the bulk viscosity of phase q. Relative velocity of the phases is defined as follows:
For $\dot{m}_{pq}>0 \xrightarrow{\text{mass transfer to phase } q} \vec{v}_{pq}=\vec{v}_p;\ \vec{v}_{qp}=\vec{v}_q$;

For $\dot{m}_{pq}<0 \xrightarrow{\text{mass transfer to phase } p} \vec{v}_{pq}=\vec{v}_q;\ \vec{v}_{qp}=\vec{v}_p$.

The force of interaction between the phases depends on friction and pressure, among other factors. This force must respect the conditions $\vec{R}_{pq}=-\vec{R}_{qp}$ and $\vec{R}_{qq}=0$. To describe this interaction, it is possible to solve this force by:

$$\sum_{p=1}^{n}\vec{R}_{pq}=\sum_{p=1}^{n}K_{pq}\left(\vec{v}_p-\vec{v}_q\right) \tag{7}$$

where $K_{pq}=K_{qp}$ is the coefficient of exchange of momentum between phases.
In multiphase flows, it is possible to include the effects of the lift force acting on the secondary phase. The lift forces act on a particle due mainly to the velocity gradients in the primary phase flow, and it is more important in flows with large particles and low packing. The lift force action on a secondary phase p in a primary phase q is calculated as:

$$\vec{F}_{lift}=-0.5\rho_q\alpha_p\left(\vec{v}_p-\vec{v}_q\right)\times\left(\nabla\times\vec{v}_q\right) \tag{8}$$

In most cases, the lift force is negligible when compared to the drag force. If its effects are truly significant, the terms of the force of ascension must be included in the equations of conservation of momentum for both phases considered:

$$\vec{F}_{lift,q} = -\vec{F}_{lift,p} \tag{9}$$

The effect of virtual mass force is important when the secondary phase p accelerates in relation to the primary phase q, and more significant when the secondary phase density is much smaller than the primary phase density. In the Eq. 10 is expressed this effect:

$$\vec{F}_{vm} = 0.5\rho_q \alpha_p \left(\frac{d_q \vec{v}_q}{dt} - \frac{d_p \vec{v}_p}{dt} \right) \tag{10}$$

The inertia of the primary phase that is found by the accelerating particles exerts a "virtual mass force" effect on the particles. The term $\frac{d_q}{dt}$ denotes the material derivative in phase q, which by definition is:

$$\frac{d_q(\phi)}{dt} = \frac{\partial(\phi)}{\partial t} + \left(\vec{v}_q \cdot \nabla\right)\phi \tag{11}$$

The force of the virtual mass is added to the right side of the momentum equation for both phases: $\vec{F}_{vm,q} = \vec{F}_{vm,p}$.

The conservation of energy in the Eulerian modeling is described by a separate equation for the enthalpy of each phase:

$$\frac{\partial}{\partial t}\left(\alpha_q \rho_q h_q\right) + \nabla \cdot \left(\alpha_q \rho_q \vec{u}_q h_q\right) = -\alpha_q \frac{\partial p_q}{\partial t} + \overline{\overline{\tau}}_q : \nabla \vec{u}_q - \nabla \cdot \vec{q}_q + S_q + \sum_{p=1}^{n}\left(\vec{Q}_{pq} + \dot{m}_{pq} h_{pq} - \dot{m}_{qp} h_{qp}\right) \tag{12}$$

Where: h_q - specific enthalpy of phase q; $\vec{q}_q$ - heat flow; S_q - source term that includes the enthalpy sources (due to chemical reaction or radiation); Q_{pq} - enthalpy exchange between the phases. The heat exchange between the phases must be calculated using the local balance conditions: $Q_{pq} = -Q_{qp}$ and $Q_{qq} = 0$.

The previously presented equations represent the global model for the solution of the velocity field of an arbitrary multiphase flow using the Euler-Euler approach. The next section will present the models used by the software Fluent® to solve a solid-fluid flow with more details of the adopted steps and equations involved.

2.2 Mathematical modeling of spouted bed systems

To describe the behavior of granular solid-fluid multiphase flow, the software Fluent® uses multi-fluid modeling, as previously described. The shear stress in the solid phase is obtained from an analogy between particles in random movement in the solid phase, and from the molecules moving in a certain gas phase. The result of the shock between the solid phase particles is treated as being analogous to the shock between gas molecules, taking into account the lack of shock elasticity in the solid phase. As in a gas, the intensity of the fluctuations determines the stresses, the viscosity, and the pressure of the solid phase; the kinetic energy associated with the velocity fluctuations is described by a pseudothermal

temperature called granular temperature, which is proportional to the square of the random movement of the particles.
The mathematical model is presented next for the case of two-phase solid-fluid flow, where p represents the particulate solid phase (secondary phase) and q is the fluid phase (primary phase). The mass transfer between the phases is neglected in this modeling aiming to obtain a simplified fluid dynamic model, but its effects can be included, as previously discussed.

2.2.1 Continuity equation

Considering the mass transfer between phase null, the continuity equation gives the phase volumetric fraction:

$$\frac{\partial}{\partial t}\left(\alpha_q \rho_q\right)+\nabla\cdot\left(\alpha_q \rho_q \vec{v}_q\right)=0 \quad (\textit{fluid phase}) \tag{13}$$

$$\frac{\partial}{\partial t}\left(\alpha_p \rho_p\right)+\nabla\cdot\left(\alpha_p \rho_p \vec{v}_p\right)=0 \quad (\textit{solid phase}) \tag{14}$$

2.2.2 Momentum conservation

From Eq. 5, including the indicated simplifications, and changing the indexes, we obtain the momentum conservation equation for the fluid and the solid phases:

$$\begin{aligned}&\frac{\partial}{\partial t}\left(\alpha_q \rho_q \vec{v}_q\right)+\nabla\cdot\left(\alpha_q \rho_q \vec{v}_q \vec{v}_q\right)=-\alpha_q \nabla p+\nabla\cdot\overset{=}{\tau}_q+\alpha_q \rho_q \vec{g}+\\&+\sum_{p=1}^{n}\left(K_{pq}\left(\vec{v}_p-\vec{v}_q\right)\right)+\left(\vec{F}_q+\vec{F}_{lift,q}+\vec{F}_{vm,q}\right) \quad (\textit{fluid phase})\end{aligned} \tag{15}$$

$$\begin{aligned}&\frac{\partial}{\partial t}\left(\alpha_p \rho_p \vec{v}_p\right)+\nabla\cdot\left(\alpha_p \rho_p \vec{v}_p \vec{v}_p\right)=-\alpha_p \nabla p-\nabla p_p+\nabla\cdot\overset{=}{\tau}_p+\alpha_p \rho_p \vec{g}+\\&+\sum_{q=1}^{n}\left(K_{qp}\left(\vec{v}_q-\vec{v}_p\right)\right)+\left(\vec{F}_p+\vec{F}_{lift,p}+\vec{F}_{vm,p}\right) \quad (\textit{solid phase})\end{aligned} \tag{16}$$

where p_p is the pressure of the granular solid phase and n is the number of phases that constitute the mixture, here n = 2. Having presented the simplified equations (Eq. 15-16), next we discuss the pressure and viscosity of the solid phase, as well as the drag between phases, which are critical variables in the modeling of this type of flow.
The spouted bed has regions of high and low volumetric fraction of solids (annulus and spout, respectively). Gidaspow's model (Gidaspow et al., 1992) is commonly used to represent the momentum transfer coefficient between phases in spouted beds since it results from the Wen & Yu (1966) model for the diluted phase region and the Ergun (1952) model for the dense phase. Thus, the Gidaspow's model (Gidaspow et al., 1992) is described as below:

$$K_{pq}=K_{qp}=\frac{3}{4}C_D\frac{\alpha_p\alpha_q\rho_q\left|\vec{v}_p-\vec{v}_q\right|}{d_p}\alpha_q^{-2.65}, \quad \alpha_q>0.8 \tag{17}$$

$$K_{pq} = K_{qp} = 150\frac{\alpha_p(1-\alpha_q)\mu_q}{\alpha_q d_p^2} + 1.75\frac{\alpha_p \rho_q |\vec{v}_p - \vec{v}_q|}{d_p}, \quad \alpha_q \leq 0.8 \tag{18}$$

The drag coefficient (C_D) is given by:

$$C_D = \frac{24}{\alpha_q \mathrm{Re}_p}\left[1 + 0.15(\alpha_q \mathrm{Re}_p)^{0.687}\right]. \tag{19}$$

Relative Reynolds coefficient (Re_p) is calculated with the following equation:

$$\mathrm{Re}_p = \frac{\rho_q d_p |\vec{v}_p - \vec{v}_q|}{\mu_q} \tag{20}$$

For granular flows in compressible regime (i.e., when the volumetric fraction of the solids is smaller than the maximum allowed value), the solid phase pressure must be included in the equation of conservation of momentum for the granular phase. Due to the distribution of the velocities for the particles, a variable called granular temperature, which appears in the pressure and viscosity equations of the solid phase, is added to the model. The granular temperature is a measure of the variation of kinetic energy of the particles caused by their shock. One of the possible formulas for the pressure of solids is given by Lun et al. (1984), in which the pressure equation is constituted of a kinetic contribution and a contribution of the collision of the particles:

$$p_p = \alpha_p \rho_p \theta_p + 2\rho_p(1 + e_{pp})\alpha_p^2 g_{0,pp}\theta_p \tag{21}$$

In Eq. 21, e_{pp} represents the restitution coefficient for the collision of the particles and $g_{0,pp}$ is the radial distribution function. The restitution coefficient must be altered, depending on the characteristics of the particulate solid.
The radial distribution function is a correction factor that alters the probability of collision between the particles when the granular phase is dense. This function can be interpreted as a non-dimensional distance between spheres:

$$g_0 = \frac{s + d_p}{s} \tag{22}$$

where s is the distance between the particles. It is possible to observe that for a diluted solid phase $s \to \infty$, $g_0 \to 1$. On the other hand, when the solid phase gets close to the packing limit $s \to 0$ and $g_0 \to \infty$. The radial distribution function is closely related to the X factor of the theory of non-uniformity of gases (Chapman & Cowling, 1991). X is equal to the unit for rarefied gases and increases, tending to infinite, when the molecules are so close that movement is not possible.
The literature has different formulas for the radial distribution function, which is given by Ogawa et al. (1980):

$$g_{0,pp} = \left[1 - \left(\frac{\alpha_p}{\alpha_{p,\max}}\right)^{1/3}\right]^{-1} \tag{23}$$

Where α_p is the volume fraction of solid phase and $\alpha_{p,\max}$ is maximum packing limit.
The tensor parameters are determined by the granular kinetic theory. The viscous stress tensor comprises stresses due to the shearing viscosity and the bulk viscosity, resulting from the exchange of quantity of movement due to the movement of the particles and their collision. A component that results from the friction between the particles can be included to calculate the effects that occur when the solid phase reaches maximum volumetric fraction. The collisional, kinetic, and frictional effects are added to give the shearing viscosity of the solid phase (μ_p):

$$\mu_p = \mu_{p,col} + \mu_{p,kin} + \mu_{p,fr} \tag{24}$$

Collisional effect (Eq. 25) and kinetic contribution (Eq. 26) are described by Gidaspow et al. (1992):

$$\mu_{p,col} = \frac{4}{5}\alpha_p \rho_p d_p g_{0,pp}\left(1+e_{pp}\right)\left(\frac{\theta_p}{\pi}\right)^{1/2} \tag{25}$$

$$\mu_{p,kin} = \frac{10\rho_p d_p \sqrt{\theta_p \pi}}{96\alpha_p\left(1+e_{pp}\right)g_{0,pp}}\left[1+\frac{4}{5}g_{0,pp}\alpha_p\left(1+e_{pp}\right)\right]^2 \tag{26}$$

The bulk viscosity (λ_s) comprises the resistance of the particles of the granular phase to compression and expansion. Equation (27) can be used for this viscosity, according to Lun et al. (1984):

$$\lambda_s = \frac{4}{3}\alpha_p \rho_p d_p g_{0,pp}\left(1+e_{pp}\right)\left(\frac{\theta_p}{\pi}\right)^{1/2} \tag{27}$$

In dense low velocity solid phase flows, in which the solid fraction is close to the maximum packing limit, the generation of stresses results mainly from the friction between the particles. One of the forms to calculate the effect of friction in the stresses is using Eq. (28) (Schaeffer, 1987), where ϕ is the internal friction angle and I_{2D} is the second invariant of the stress tensor.

$$\mu_{p,fr} = \frac{p_p \sin\phi}{2\sqrt{I_{2D}}} \tag{28}$$

In granular flows with high volumetric fractions of solids, the effects of collisions between particles become less important. The application of the granular kinetic theory in such cases would, therefore, be less relevant, as the effect of friction between the particles must be taken into account. To overcome this problem, the software Fluent® used a formula of the effect of friction extended for the combined application with the granular flow kinetic theory.
Stresses due to friction between particles are normally written in Newtonian form following Eq. 29, where $\tau_{friction}$ represents the stress due to frictional effects.

$$\tau_{friction} = -P_{friction}\overline{\overline{I}} + \mu_{friction}\left(\nabla\vec{u}_p + \left(\nabla\vec{u}_p\right)^T\right) \tag{29}$$

The resulting stress of the friction effects is added to the stress derived from the kinetic theory when the volumetric fraction of the solids exceeds a certain critical value. This value is normally considered as 0.5 when the flow is three-dimensional and the value of the packing limit of the solid phase is 0.63, thus:

$$P_p = P_{kinetic} + P_{friction} \tag{30}$$

$$\mu_p = \mu_{kinetic} + \mu_{friction} \tag{31}$$

The pressure due to friction is obtained mainly by semi-empirical form. The viscosity can be obtained by applying the modified Coulomb law, which gives the expression:

$$\mu_{friction} = \frac{P_{friction} sen\varphi}{2\sqrt{I_{2D}}} \tag{32}$$

The friction pressure can be determined with the model described by (Johnson & Jackson, 1987):

$$P_{friction} = Fr\frac{\left(\alpha_p - \alpha_{p,\min}\right)^j}{\left(\alpha_{p,\max} - \alpha_p\right)^k} \tag{33}$$

where the coefficients are j=2 and k=3 (Ocone et al., 1993). The friction coefficient is assumed to be a function of the volumetric fraction of the solids (Eq. 34) and the viscosity of friction is thus given by Eq. 35.

$$Fr = 0,1\alpha_p \tag{34}$$

$$\mu_{friction} = P_{friction}\sin\phi \tag{35}$$

Granular temperature of the solid phase (θ_p) is proportional to the kinetic energy produced by the random movement of the particles. This effect can be represented by:

$$\frac{3}{2}\left[\frac{\partial}{\partial t}\left(\rho_p\alpha_p\theta_p\right) + \nabla\cdot\left(\rho_p\alpha_p\vec{v}_p\theta_p\right)\right] = \left(-p_p\bar{\bar{I}} + \bar{\bar{\tau}}_p\right):\nabla\vec{v}_p + \nabla\cdot\left(k_{\theta_p}\nabla\theta_p\right) - \gamma_{\theta_p} + \phi_{qp} \tag{36}$$

where $\left(-p_p\bar{\bar{I}} + \bar{\bar{\tau}}_p\right):\nabla\vec{v}_p$ represents the generation of the energy by the stress tensor of the solid phase; $k_{\theta_p}\nabla\theta_p$ represents the energy diffusion (k_{θ_p} is the diffusion coefficient); γ_{θ_p} represents the energy dissipation produced by collisions and ϕ_{qp} represents the energy exchange between the solid and the fluid phases.

The diffusion coefficient is given by Gidaspow et al. (1992):

$$k_{\theta_p} = \frac{150\rho_p d_p\sqrt{\theta_p\pi}}{384\left(1+e_{pp}\right)g_{0,pp}}\left[1 + \frac{6}{5}\alpha_p g_{0,pp}\left(1+e_{pp}\right)\right]^2 + 2\alpha_p^2\rho_p d_p\left(1+e_{pp}\right)g_{0,pp}\sqrt{\frac{\theta_s}{\pi}} \tag{37}$$

Dissipation of energy due to the collisions γ_{θ_p} can be described by the expression from Lun et al. (1984):

$$\gamma_{\theta_p} = \frac{12\left(1-e_{pp}^2\right)g_{0,pp}}{d_p\sqrt{\pi}}\rho_p\alpha_p^2\theta_p^{3/2} \tag{38}$$

The energy exchange between the solid and the fluid phases due to the kinetic energy of the random movement of the particles ϕ_{qp} is given by Gidaspow et al. (1992):

$$\phi_{qp} = -3K_{qp}\theta_p \tag{39}$$

To solve the equation of granular temperature conservation using the software Fluent®, three methods are possible, besides the addition of a user function (UDF):

- Algebraic: obtained by leaving out the convection and diffusion terms of Eq. 36;
- Solution of the partial differential equation: obtained by solving Eq. 36 with all terms, and
- Constant granular temperature: useful in dense phase situations in which the particle fluctuation is small.

2.2.3 Boundary conditions

For the granular phase p, it is possible to define the shearing stress on the wall as follows:

$$\vec{\tau}_p = -\frac{\pi}{6}\sqrt{3}\phi\frac{\alpha_p}{\alpha_{p,\max}}\rho_p g_0\sqrt{\theta_p}\vec{U}_{s,w} \tag{40}$$

where: $\vec{U}_{s,w}$ - velocity of the particle moving parallel to the wall; ϕ - specularity coefficient. For the granular temperature, the general contour condition is given by (Johnson & Jackson, 1987):

$$q_s = \frac{\pi}{6}\sqrt{3}\phi\frac{\alpha_p}{\alpha_{p,\max}}\rho_p g_0\sqrt{\theta_p}\vec{U}_{s,w}\cdot\frac{\pi}{4}\sqrt{3}\frac{\alpha_p}{\alpha_{p,\max}}\left(1-e_{pw}^2\right)\rho_p g_0\theta_p^{3/2} \tag{41}$$

With the model developed here, it is possible to simulate the fluid dynamic behavior of several gas-solid flow systems, especially dense phase systems, for which the Eulerian approach is most used. The next section presents some of the most relevant current work in literature on the application of CFD to multiphase flow problems with the Eulerian approach.

3. Numerical simulation of the semi-cylindrical spouted bed

It is impossible to observe the spout channel in a cylindrical bed as the channel is surrounded by the (particle dense) annular zone. Consequently, the semi-cylindrical spouted bed arose as an alternative for obtaining experimental measurements for cylindrical spouted beds. This old technique was first used in a study by Mathur & Gishler (1955). This bed should not be classified as unconventional equipment as its purpose is to obtain experimental data relating to a full bed with similar geometric characteristics.

As shown in the diagram in Fig. 1, it is possible to obtain a semi-cylindrical bed by placing a transparent wall in a full column bed. The great advantage of constructing a semi-cylindrical spouted bed is that it makes it possible to view the internal behavior of the particles in the vessel as the spouted channel is in direct contact with the flat wall. From the supposition that the hydrodynamic behavior of the semi-cylindrical bed is similar to that of a cylindrical bed (for systems with similar geometric characteristics), one can infer experimental data for cylindrical beds from the data obtained for the half column (velocity of solids from image analysis, radius of the spout channel, fountain shape). After its first use, even with the caution recommended by its creators, some later studies, notably by Lefroy & Davidson (1969), suggested that the semi-cylinder did not show significant differences for the most general designs to which it should be applied.

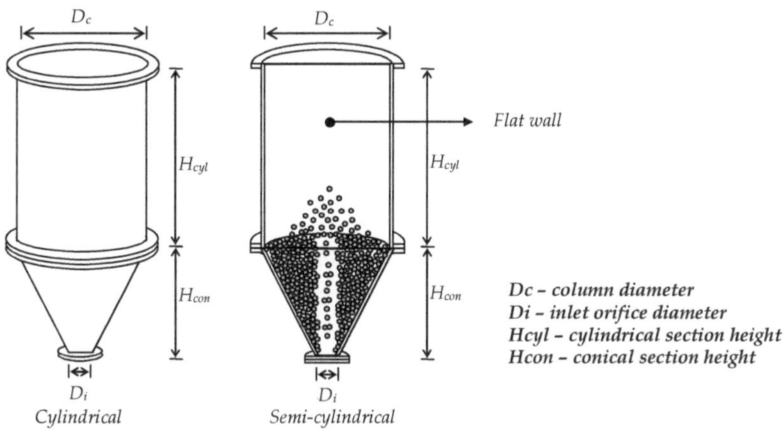

Fig. 1. Diagram of a semi-cylindrical (half-column) spouted bed

It is acknowledged that the insertion of a flat wall into a spouted bed causes significant modifications to the geometry of the equipment. He et al. (1994a) experimentally evaluated the hydrodynamic behavior of cylindrical and semi-cylindrical spouted beds with similar geometry. They observed that inserting the flat wall into the bed changed the velocity profile of the particles due to the additional system friction caused by the wall. In addition to the above study, He et al. (1994b) studied the behavior of the spouted bed porosity profile, comparing the results obtained for cylindrical and semi-cylindrical beds. It was verified that inserting the flat wall did not significantly change the porosity behavior of the bed in comparison with the change caused by the wall on the velocity profile of the solid.
These results, along with others in the literature, raise a question about the semi-cylindrical spouted bed: are the measurements obtained using the half-bed technique truly representative for inferring the hydrodynamic data for cylindrical beds and, if so, for which variables? The studies by He et al. (1994a, 1994b) are conclusive regarding the velocity of the solid phase and the porosity of the bed, indicating that special care is necessary when applying this technique. Despite the need for special care, visual information related to the hydrodynamic behavior of the cylindrical bed can only be obtained using the half-column technique. To analyze these effects, some numerical studies were carried out and the results are discussed in the following chapters.

3.1 Numerical evaluation of the specularity coefficient

Various authors have described the importance of friction between the particulate phase and the flat wall present in the half-column spouted bed. In view of this effect and the possibility of evaluating it numerically using CFD numerical simulation, this chapter presents the results of a numerical study involving the specularity coefficient (a parameter added to the Eulerian model of the particulate phase to represent the friction between this phase and the glass wall in the spouted bed). Experimental results were used to verify the simulation results in one of the cases. In addition to the results, the characteristics of the computational mesh adopted are presented together with the most relevant aspects of the numerical solution procedure.

3.1.1 Experimental data

The experimental data presented in this study were obtained for a spouted bed operating with air spheres at a controlled temperature. Table 1 presents the geometric and operational parameters for the spouted beds, together with the physical characteristics of the phases present in the equipment.

Semi-cylindrical Conical Spouted Bed		
D_c (m)	Column diameter	0.30
D_i (m)	Inlet orifice diameter	0.05
H_{con} (m)	Conical section height	0.23
H_{cyl} (m)	Cylindrical section height	0.80
H_e (m)	Static bed height	0.23
Φ (°)	Conical section angle	60.0
d_p (m)	Particle diameter	2.18
u_i (m/s)	Velocity inlet	18.0
T_{air} (°C)	Air temperature	50.0
ρ_p (kg/m3)	Solid density	2,512

Table 1. Parameters and geometric characteristics of the equipment used in this study

The experimental evaluation of the hydrodynamics of the spouted beds involved determining experimental data for a semi-cylindrical spouted bed operating under the conditions described in Table 1. The velocity of the particles in the annular, spout and fountain zones were obtained from videos filmed through the glass wall at different axial and radial positions of the half-column conical spouted bed. To be able to make measurements using the image analysis software, the software was first calibrated using standard scales. After being calibrated, the local velocity of the solid was calculated by analysis of the distance covered by the particle and the time taken to cover the distance.

A static pressure probe was fixed to the upper part of the bed with its point inside. By varying the position of the probe, the inner pressure of the bed was mapped. Note that the radial pressure data was obtained in the direction perpendicular to the glass wall in the semi-cylindrical spouted bed.

The height of the fountain was obtained from the images of the half-column conical spouted bed. These images were produced using a digital stills camera. Image software was also used to process the images and obtain the shape of the spout channel and the height of the fountain.

3.2 Computational mesh and numerical procedure

3.2.1 Computational mesh

The simulations were conducted using a three-dimensional mesh containing a symmetry plane to divide the domain. GAMBIT 2.13 software was used to generate the mesh. The spacing between nodes adopted was 5.0 mm, increasing gradually towards the outlet zone, where a spacing of 1.0 cm was adopted. The mesh used tetrahedral cells, reaching a mesh with 128,830 cells. Fig. 2 is a diagram of the mesh used in this part of the study.

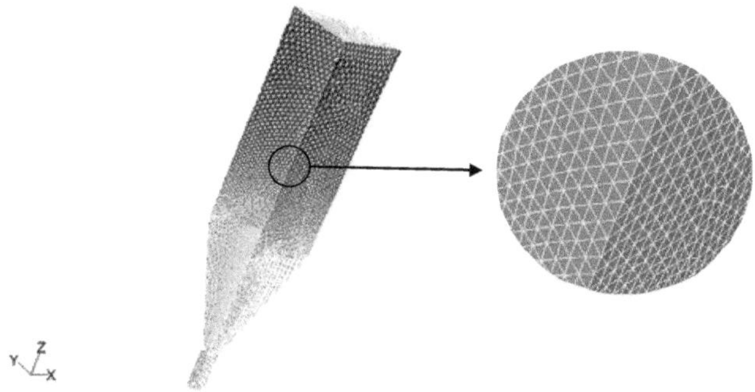

Fig. 2. Computational mesh

Table 2 gives the general characteristics adopted for the construction of the computational mesh illustrated in Figure 2. Further information on cell geometry and meshing schemes can be found in Fluent (2006).

Computational mesh caracteristics	
Minimum cell volume	$3.67x10^{-9}$ m^3
Maximum cell volume	$1.04x10^{-6}$ m^3
Total number of cells	128,830
Number of nodes:	25,596
Face meshing scheme	Inlet tube faces: TRI- PAVE[1] Cone faces: TRI- PAVE Cylinder faces: TRI- PAVE Intersection faces: TRI- PAVE
Volume meshing scheme:	Tube: TET/HYBRID- TGRID[2] Cone: TET/HYBRID- TGRID Cylinder: TET/HYBRID- TGRID
Cells with high asymmetry (>0.97)	0

Table 2. Characteristics of the computational mesh

[1] TRI-PAVE: triangular elements / PAVE meshing scheme.

[2] TET/HYBRID-TGRID: mainly tetrahedral elements; however, elements of other geometries may be used if required / TGRID meshing scheme.

3.2.2 Boundary conditions

The following boundary conditions were applied in this study:

- Fluid is injected at the system inlet in the axial direction at constant velocity. The condition of the inlet turbulence was obtained by the intensity and length scale method, using equations from the literature to calculate these parameters (I=4.36%; L=0.0022 m);
- The outlet pressure is defined as being atmospheric pressure. For the outlet turbulence conditions, the turbulence intensity and length scale method was again employed, using equations in the literature to calculate these parameters (I=5.4%; L=0.012 m);
- The yz plane was defined as the symmetry plane for both of the simulated systems.
- In the xz plane (flat wall), the non-slip condition was used for the fluid mechanics of the fluid phase of the system (air). For the particulate phase, the specularity coefficient was defined (the effect of this parameter on the hydrodynamics of the system is tested in this part of the study);
- At the bed walls, the non-slip condition was adopted for the fluid phase and the partial-slip condition for the solid phase (by means of the specularity coefficient).

Phase parameters used in the simulations are given in Table 3:

Parameter	Description	Value
α_s	Porosity of the static bed	0.41
e_{ss}	Restitution coefficient	0.9
ρ_s (kg/m^3)	Density of the solid	2,512
ρ_f (kg/m^3)	Density of the fluid (air)	1.0
μ (Pa.s)	Air viscosity	2.09x10^{-5}
d_s (m)	Diameter of the solid	2.18x10^{-3}
Θ	Sphericity	1.0

Table 3. Parameters applied to the simulation

3.2.3 Numerical solution

The numerical solution was obtained using the Fluent® 6.3.26 software. The SIMPLE algorithm for the coupled phase was applied to solve the equations of the discretized model. The second-order upwind discretization scheme was applied for the *momentum* conservation equations. Relaxation factors were selected with values between 0 and 0.4. As a transient regime simulation was used, the system was considered permanent after 8 s of simulated time, using the integration steps between 1.0x10^{-4} and 5.0x10^{-4} s.

3.3 Evaluation of the specularity coefficient

The effect of the specularity coefficient on the simulation results of the hydrodynamics of the semi-cylindrical spouted bed was analyzed with four different coefficient values. Specularity value was changed for wall positions between 0.0, 0.05, 0.1 and 0.2; the simulated results in response to these changes were analyzed. Simulated results for static pressure were then compared with experimental data. In addition to the static pressure results, the solid volume fractions and the solid and fluid velocity profiles were also evaluated numerically and discussed in the context of the numerical simulation.

3.3.1 Static pressure

In Fig. 3 are presented the simulated static pressure results for the different axial positions along the semi-cylindrical spouted bed. These results are compared with the experimental data obtained in the laboratory. One can see that the simulated static pressure curves in the radial direction for each of the different axial positions highlighted, Fig. 3, are not significantly influenced by changes in the value of the specularity coefficient. It shows that the simulated and experimental pressure results present qualitatively similar behavior for each of the axial positions. In the results for z=0.115m, the largest pressure values occurred in the curve for the value of φ=0.00, while with z=0.165m, φ=0.05 was responsible for the largest values. Hence, stressing the previous observation, the most pronounced deviations in these pressure results can be attributed to the numerical approximation.

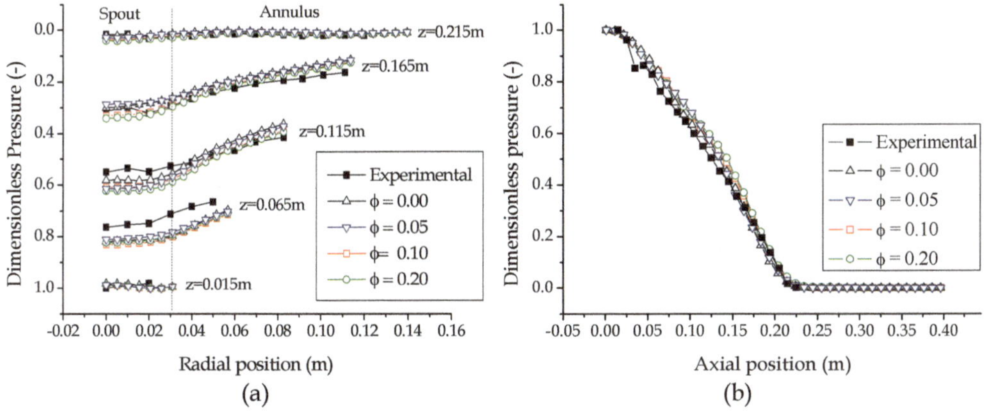

Fig. 3. Dimensionless static pressure for different specularity conditions: (a) radial profile; (b) axial profile with r=0.0m

Results presented in Fig. 3a indicate that the change in the specularity coefficient, and as a consequence the change in friction between the particulate phase and the flat wall, did not lead to significant changes in the static pressure in the radial direction. The experimental axial profile for the static pressure at the center of the spout channel is given in Fig. 3b, together with the simulated curves for the different specularity coefficients tested. Results in Fig. 3b reaffirm the qualitatively similar behavior of the static pressure in comparison with the experimental results. In this case as well, small deviations between the simulated responses are observed, demonstrating the small influence of specularity on the pressure behavior, as discussed above for the radial pressure curves given in Fig. 3a.

Due to the absence of significant changes in the static pressure results for the different specularity coefficients evaluated, it is possible to state that this coefficient did not have a significant effect on the behavior of this variable. Hence, it can be concluded from the simulated results that the magnitude of friction between the particulate phase and the flat wall did not significantly affect the behavior of static pressure in the semi-cylindrical spouted bed.

3.3.2 Solid volume fraction

Figure 4 shows the simulated contours of the solid volume fraction, for the different specularity coefficients evaluated, in the planes representing the glass wall and the symmetry plane of the bed.

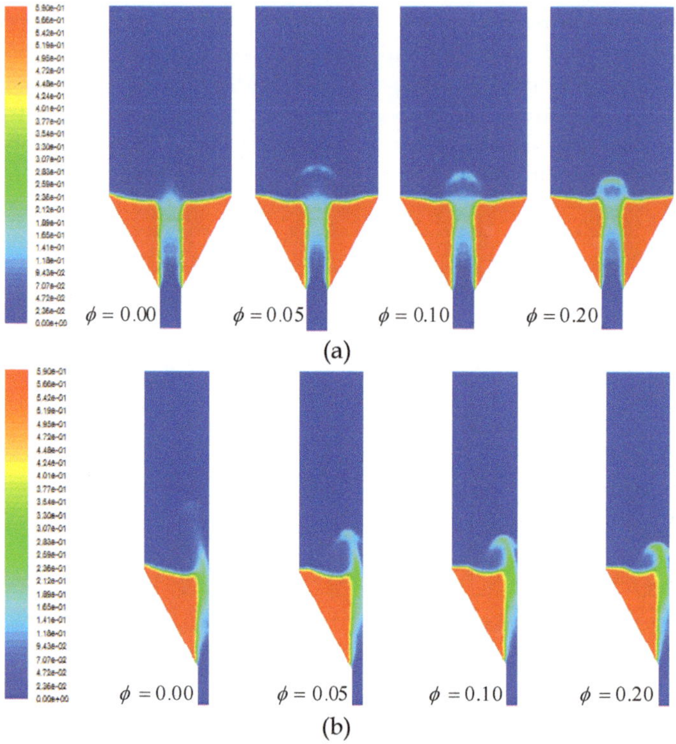

Fig. 4. Simulated contours for the solid volume fraction: (a) glass wall, (b) symmetry plane

It is possible to verify from these results that modifications to the specularity coefficient have a qualitative effect on the behavior of the solid fraction in the system. However, the large differences in behavior occur in the fountain zone of the system, while qualitatively similar behavior in the spouted channel zone was verified for all of the simulated cases.

The specific fountain shapes for the different conditions are directly related to the increase or reduction in friction between the wall and the particulate solid due to changes in the specularity coefficient. Changing the effectiveness of solid-wall friction changes the particulate solid velocity profile, thus modifying the maximum velocity reached by the solid in the spout channel, as well as the height that it reaches after its vertical launch at the channel outlet.

On the basis of the numerical results presented, it is possible to conclude that the solid fraction in the spouted bed is affected by the specularity coefficient, particularly in the fountain zone. In a real situation, it is expected from the physics perspective that an increase in the friction between the flat wall and the solid will cause the height of the fountain to decrease. Hence, in addition to showing the importance of this parameter for an accurate representation of the behavior of the solid fraction in the bed, the results presented here prove, qualitatively, the effectiveness of including the specularity coefficient in the mathematical model for the hydrodynamics of the spouted bed.

3.3.3 Velocity of the solid phase and fluid phase

Figure 5 shows the simulated velocity results for the particulate phase at the center of the spout channel along the axial direction. The qualitative similarity between the curves is verified by this figure, despite their quantitatively different behavior.

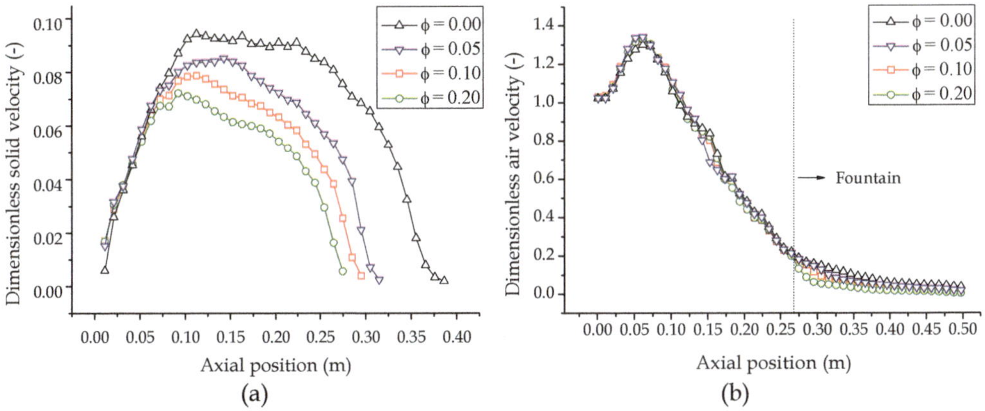

Fig. 5. Dimensionless velocity at the center of the spout channel in the axial direction: (a) particulate phase, (b) fluid phase.

Analysis of the shape of the curves given in Figure 5 shows that the maximum velocity achieved by the solid was inversely proportional to the increase in the value of the specularity coefficient. As previously discussed, the increase in the specularity coefficient reflects an increase in the effectiveness of the friction between the particulate phase and the flat wall, which in turn tends to reduce the velocity of the contacting solid. Comparing the curve for φ=0.00 with the other curves, it is possible to see that it has a region of practically constant velocity for intermediate spout channel positions. When the specularity coefficient was included, even at its smallest value, this tendency was not observed. It should be noted that this region of velocity with a flat profile is not found in experimental results present in the literature, which is a good indication of the need to include the specularity coefficient in the system model.

Figure 5b shows the simulated results for the velocity of the fluid phase at the center of the spout channel along the axial direction for different specularity coefficients. These results show that the specularity coefficient does not exert a significant influence on the behavior of the velocity of the fluid phase. Figure 5b also shows that the velocity of the fluid phase was not directly affected by the change in friction between the particulate phase and the flat wall. Hence, lower velocities in the particulate phase in the wall zone do not cause significant changes in the behavior of the velocity of the fluid phase. In the fountain zone, greater differences were seen between the curves for each of the situations evaluated. These differences must be due to the specific shape of the fountain that was investigated for each condition as it interfered with the distribution of air in this zone and caused the largest deviations found.

From the numerical study presented it is possible to observe the importance of the effect of friction between the flat wall and the particulate phase. Here, this effect was incorpored by the inclusion of the specularity coefficient in the hydrodynamic model for the spouted bed. As the specularity coefficient is one way to represent this effect, its careful evaluation and determination is fundamental to obtain quantitatively representative results.

4. Advances in CFD applied to spouted beds

A large number of studies about the movement of gases and solids in spouted bed can be found in the literature. However, it is still a difficult task to obtain precise information about the behavior of flow, temperature, mass transfer and particle distribution in fluidized and spouted systems using the classic models. The complexity of these systems and the interactions between particles and gas are important issues in designing spouted beds. The commonly used models are limited in their scope as they do not allow more in-depth analysis of the behavior.

Computational Fluid Dynamics (CFD) is one of the most productive numerical methods at present and has been widely applied to solving multiphase flow problems. In the case of the spouted bed, the use of this tool is still in its initial stage, but some studies can already be found in the literature.

Seeking to numerically evaluate the semi-cylindrical spouted bed, Béttega et al. (2010) present a study comparing the hydrodynamic results obtained experimentally and numerically for cylindrical and semi-cylindrical spouted beds. Fig. 6 shows the results for static pressure in the radial direction of the bed for different axial positions in the two systems evaluated. It is possible to observe the representative qualitative behavior between the experimental and numerical results in both of the systems covered by the authors, as shown by the results for the central position of the spout channel.

From Fig. 6a it can be seen that the experimental and simulated behavior of pressure in the radial direction was qualitatively similar. However, it also shows a greater divergence between the results in the region of particle acceleration. Moving in the radial direction in all of the curves in Fig. 6a, in the annular zone of the bed (approximately from r=0.03 m), the quantitative behavior is representative for all of the axial positions evaluated, except at z=0.065m, whose results diverge the most. The authors attribute errors in the experimental measurement at this position to the high circulation of particles and the resulting medium disturbances. The authors also stress that at z=0.065m, experimental measurement is even more difficult when compared to the other axial positions due to the effective presence of solids circulating at high velocity both in the annular zone and the spout channel.

The characteristics of the results found for the cylindrical bed are also observed for the semi-cylindrical bed, as can be seen in Fig. 6b. The differences between the numerical and experimental results seen at the center of the spout channel, apart from the greater divergence between the results at z=0.065m, are also more evident in these results. These occur because of the same problems highlighted for the cylindrical bed. The similar behavior of the numerical and experimental response for both beds is another indication that the numerical and experimental procedure adopted is representative.

Figure 7 shows the simulated results obtained for the velocity of the particulate phase in the radial direction for the cylindrical and semi-cylindrical spouted beds as obtained by Béttega et al (2010). Clear differences can be seen in the behavior of this variable for the two systems in this figure.

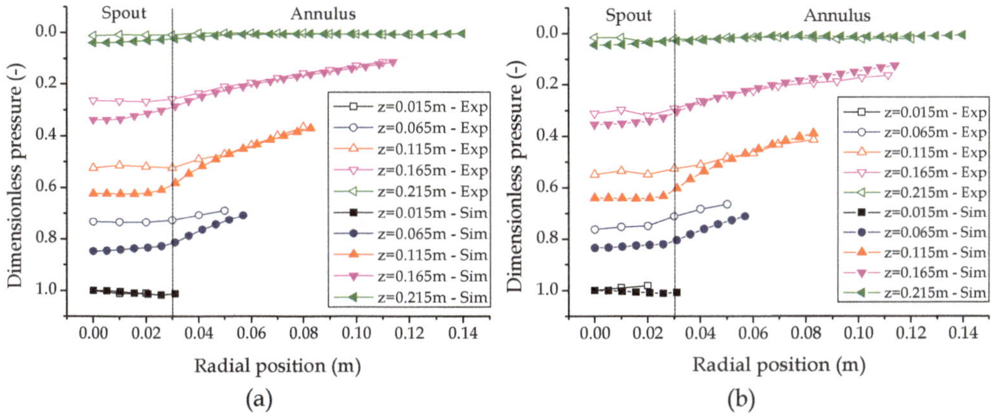

Fig. 6. Static pressure in the radial direction: (a) cylindrical bed, (b) semi-cylindrical bed

Quantitative and qualitative differences between the simulation results of both systems are evident in Fig. 7, especially in the spout channel, whose velocity of the solid is greater. These results indicate that the flat wall placed into the conventional spouted bed to produce the semi-cylindrical configuration changes the behavior of the particulate phase close to the flat wall significantly. From these results, using the simulated data, it is possible to reaffirm the observation of He et al. (1994a). The authors observed experimentally that the flat wall placed in the spouted bed caused additional friction to the system, provoking changes in the behavior of the solid phase velocity. The qualitative comparison of the results for the velocity of the solid obtained by He et al. (1994a) in conical-cylindrical spouted beds with the numerical results obtained in this study indicates qualitatively similar responses and that the simulation is representative.

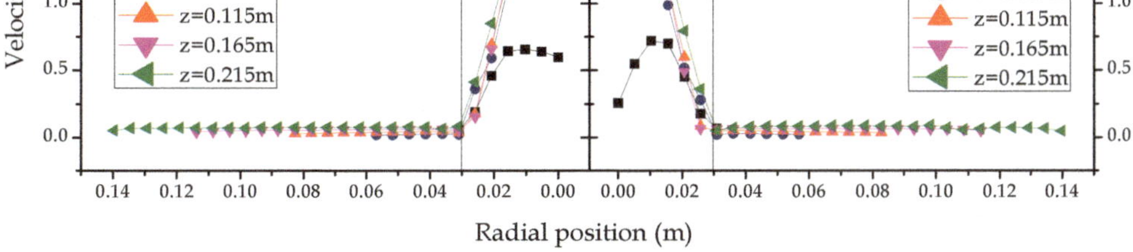

Fig. 7. Velocity of the simulated particulate phase along the radial direction

In Fig. 8 the simulated results for the volume fraction of solids achieved by Béttega et al. (2010) are compared with the image of the flat wall for the semi-cylindrical bed, making it possible to qualitatively evaluate parameters such as the spout channel shape and annular zone shape, both obtained numerically. From Fig. 8 it is possible to verify that the general behavior of the bed was correctly simulated, which can be demonstrated by comparing the value of the height of the fountain obtained numerically with its value obtained in the image and measured experimentally.

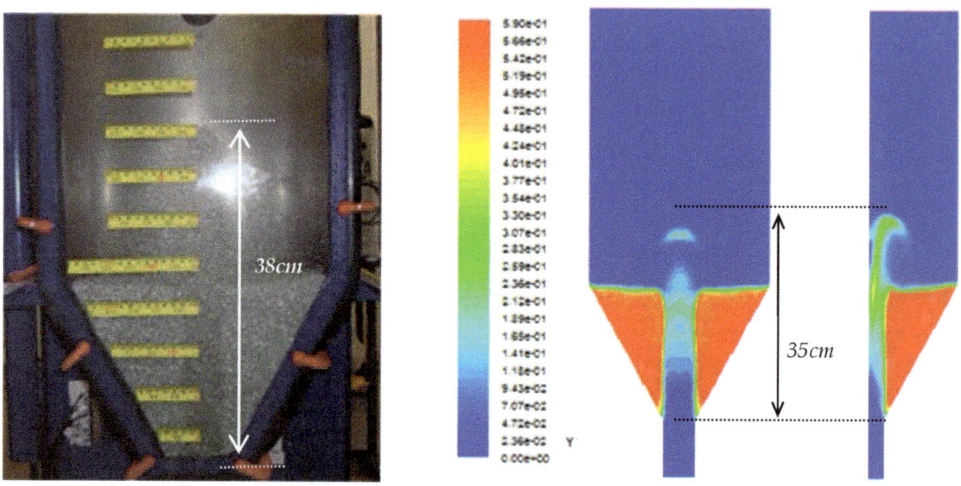

(a) (b)

Fig. 8. Semi-cylindrical spouted bed in permanent regime: (a) photograph of the equipment; (b) simulated contours of the solid volume fraction

According to the authors, the numerical results obtained using the Eulerian model, as well as the numerical procedure adopted in this study, are adequate for three-dimensional numerical simulation of semi-cylindrical and cylindrical spouted beds. Comparing the numerical results with the experimental data shows that the divergences between them are small, indicating that the model represents the system well.

In addition to the studies presented in detail above, other important recent advances on the subject are presented in the next paragraphs.

Szafran & Kmiec (2004) - the authors present a study on the application of computational fluid dynamics to the simulation of hydrodynamics and heat and mass transfer in a spouted bed grain dryer with a draft tube. The authors used the Fluent® software for resolving the hydrodynamics of the system. From the comparison of the results simulated using CFD and the results of correlations in the literature, the authors confirm that the simulated results predict the behavior of systems very well for mass transfer. However, the results show a significant divergence when heat transfer is evaluated.

Du et al. (2006a, 2006b) - they present a study that evaluates the use of different models for how the drag force influences the fluid dynamic behavior of a spouted bed. Using software Fluent®, the authors observed the capacity of some of the tested models to qualitatively represent the fluid dynamic behavior of a spouted bed. The Gidaspow model afforded the best results in relation to the experimental data. Continuing this work, the authors presented a new study evaluating the effect of drag, maximum packing, and restitution coefficient on the fluid dynamics of the spouted bed of Du et al. (2006b). These studies allowed proposing a new initial estimate of these important parameters to simulate a spouted bed, as well as knowing their qualitative effects on numerical simulation.

Zhonghua & Mujumdar (2008) - using the Fluent® 6.2 commercial package, the authors applied CFD to the numerical simulation of the behavior of spouted beds in two and three dimensions based on experimental systems in the literature. To verify the results in two dimensions, experimental data for particulate velocity presented in He et al. (1994a), obtained in a conical-cylindrical bed with optical fiber sensors, were used. For the three-dimensional bed, the parameters and experimental data presented by Link et al. (2005) were used. These authors used image analysis to obtain velocity and porosity data for a rectangular spouted bed. The behavior of the numerical results for both of the systems satisfactorily described the behavior of the two cases evaluated. In the three-dimensional case, the simulator was capable of capturing hydrodynamic perturbations in the equipment that appeared as instabilities in the bed. Simulated pressure curves were also presented in the study for the three-dimensional case analyzed. The authors verified that the pressure behavior was affected by the instabilities present in the equipment.

Duarte et al. (2009) - this work presented CFD simulations of conical and conical-cylindrical spouted beds using experimental data, from the literature and their own experimentation, to verify the numerical results. This study presented numerical results for the characteristic curve of the spouted bed and compared it directly to experimental results, thus confirming their agreement. The minimal spouted flow for the conical cylindrical bed was also numerically evaluated and compared with correlations from the literature, indicating the capacity of the numerical tool to obtain a good approximation for this variable. The authors also illustrated, with the contours of volumetric fraction of the solids, the evolution of the spouted bed from static bed to spouted condition in both evaluated cases.

Béttega et al. (2009a) - they evaluated the problem of increase in scale of spouted beds using numerical simulation. The software Fluent® was used to simulate the spouted beds in different scales, respecting the non-dimensional parameters obtained from the work of He et al. (1994a, 1994b) to ensure the conditions of geometric and dynamic similarity. They used simulation to verify the capacity of the numerical tool to reproduce the increase in the scale when the non-dimensional parameters are maintained. They numerically confirmed the experimental observation that when these parameters are satisfied, a similar fluid dynamic behavior of beds with different scales is ensured. They also evaluated the response of the spouted bed to the increase in scale for the usual grain drying conditions, in which the values of parameters such as density and viscosity of the fluid remained unchanged to satisfy the scale relationships. In some of the analyzed cases, the numerical response indicated the presence of instabilities for small scale alterations, as it is well known to occur in spouted beds - a great difficulty in increasing scale.

Béttega et al. (2009b) - this work presented a three-dimensional simulation of a half-column conical spouted bed. The authors compared the simulation results for parameters such as the spouted channel format, fountain format and height, as well the surface of the annulus, with information generated from images based on an experiment carried out under the same conditions as those of simulation. The simulated particle velocity profiles were also evaluated from the experimental measures performed using a high speed camera, revealing

a great coherence between both the numerical and experimental results. For the simulation, the authors neglected the friction between the wall and the particulate solid phase. The characteristic curve experimentally obtained is also presented. The numerical pressure was compared to the experimental value under the same conditions.
Rosa and Freire (2009) - in this work the fluid dynamic behavior of a spouted bed with a draft tube inside and continuously feeded with particle from the bottom was studied through experimental data of static pressure, fountain height, and fluid mass flow in the annulus, as well as using the computational fluid dynamics. The authors observed that the two-dimensional CFD model based on the Euler-Euler approach was successfully applied to describe the fluid dynamic behavior of this system. By comparing experimental and numerical data, the authors highlight the possibility of identifying the different regions of the equipment based on the fluid dynamic data.

5. Conclusions

From the numerical study presented in this chapter it was possible to identify and evaluate, in a qualitative way, the importance of the effect of friction between the flat wall and the particulate phase, represented by the inclusion of the specularity coefficient in the hydrodynamic model for the spouted bed. It was verified by these simulated results that this coefficient shows a strong influence on the behavior of the particulate phase in the semi-cylindrical spouted bed. However, its effect on the static pressure, the velocity of the fluid phase in the bed and on the solid fraction behavior in the spout channel is not very pronounced.
Hence, it is possible to conclude that correctly determining the friction between the flat wall and the particulate phase is extremely important for an effective simulation of the hydrodynamics of the semi-cylindrical spouted bed. As the specularity coefficient is one way to represent this effect, its careful evaluation and determination is fundamental to obtain quantitatively representative results.

6. References

Béttega, R., Corrêa, R. G. & Freire, J. T. (2009a). Scale-up Study of Spouted Beds Using Computational Fluid Dynamics. *Canadian Journal of Chemical Engineering*, Vol.87, No.2 (April 2009), pp. 193-203, ISSN 0008-4034.

Béttega, R., Almeida, A. R. F., Corrêa, R. G. & Freire, J. T. (2009b). CFD Evaluation of a Semi-cylindrical Spouted Bed: Numerical Simulation and Experimental Verification, *Canadian Journal of Chemical Engineering*, Vol.87, No.2 (April 2009), pp. 177-184, ISSN 0008.

Béttega, R., Corrêa, R. G. & Freire, J. T. (2010). Three-Dimensional Numerical Simulation of Semi-Cylindrical and Cylindrical Spouted Bed Hydrodynamics. *Drying Technology*, Vol.28, No.11, (November 2010), pp. 1266—1276, ISSN 0737-3937.

Chapman, S. & Cowling, T. G. (1991). *The Mathematical Theory of Non-Uniform Gases*, Cambridge University Press, ISBN 052140844X, New York, USA.

Du, W., Bao, X., Xu, J. & Wei, W. (2006a). Computational Fluid Dynamics (CFD) Modeling of Spouted Bed: Assessment of Drag Coefficient Correlations. *Chemical Engineering Science*, Vol.61, No.5, (March 2006), pp. 1401-1420, ISSN 0009-2509.

Du, W., Bao, X., Xu, J. & Wei, W. (2006b). Computational Fluid Dynamics (CFD) Modeling of Spouted Bed: Influence of Stress, Maximum Packing Limit and Coefficient of Restitution of Particles. *Chemical Engineering Science*, Vol.61, No.14, (July 2006), pp. 4558-4570, ISSN 0009-2509.

Duarte, C. R., Olazar, M., Murata, V.V. & Barrozo, M.A.S. (2009). Numerical Simulation and Experimental Study of Fluid–particle Flows in a Spouted Bed. *Powder Technology*, vol. 188, No.3, (January 2009), pp. 195–205, ISSN: 0032-5910.

Ergun, S. (1952). Fluid Flow Through Packed Columns. *Chemical Engineering Progress*, Vol.48, No.2, (February 1952), pp. 89–94.

Fluent, (2006). *Fluent 6.3 User's Guide*. Fluent Inc., Lebanon, New Hampshire, USA.

Gidaspow, D., Bezburuah, R. & Ding, J.(1992). Hydrodynamics of Circulating Fluidized beds, Kinetic Theory Approach. In: *Fluidization VII – Proceedings of the 7th Engineering Foundation Conference on Fluidization*, pp. 75-82, Brisbane, Australia.

He, Y. L., Lim, C. J., Grace, J. R., Zhu, J. X. & Qin, S. Z. (1994b). Measurements of Voidage Profiles in Spouted Beds. *Canadian Journal of Chemical Engineering*, Vol.72, No.2, (April 1994), pp. 229-234, ISSN 0008-4034.

He, Y. L., Qin, S. Z., Lim, C. J. & Grace, J. R. (1994a). Particle Velocity Profiles and Solid Flow Patterns in Spouted Beds. *Canadian Journal of Chemical Engineering*, Vol.72, No.4, (August 1994), pp. 561-568, ISSN 0008-4034.

Johnson, P.C & Jackson, R. (1987). Frictional-collisional Constitutive Relations for Granular Materials with Application to Plane Shearing. *Journal of Fluid Mechanics*, Vol.176, (March 1987), pp. 67-93, ISSN 0022-1120.

Lefroy, G. A. & Davidson J. F. (1969). The Mechanics of Spouted Beds. *Trans. Instn. Chem Eng.*, Vol.47, pp. 120-128.

Link, J. M., Cuypers, L. A., Deen, N. G. & Kuipers, J. A. M. (2005). Flow Regimes in a Spout-Fluid Bed: A Combinated Experimental and Simulation Study. *Chemical Engineering Science*, Vol.60, No.13, (July 2005), pp. 3425-3442, ISSN 0009-2509.

Lun, C. K. K., Savage, S. B & Jeffrey, D.J. (1984). Kinetics Theories for Granular Flow: Inelastic Particles in Flow and Slightly Inelastic Particles in General Flow Field. *Journal of Fluid Mechanics*, Vol.140, (March 1984), pp. 223-256, ISSN 0022-1120.

Mathur, K. B. & Gishler, P. E. (1955). A Technique for Contacting Gases with Coarse Solid Particles. *AIChE Journal*, Vol.1, No.2, (June 1955), pp. 157-168, ISSN 1547-5905.

Ocone, R., Sundaresan, S. & Jackson, R. (1993). Gas-particle Flow in a Duct of Arbitrary Inclination with Particle-particle Interaction. *AIChE Journal*, Vol.39, No.8, (August 1993), pp. 1261-1271, ISSN 1547-5905.

Ogawa, S., Umemura, A. & Oshima, N. (1980). On the Equation of Fully Fluidized Granular Materials. *Journal of Applied Mathematics and Physics*, Vol.31, No. 4, (July 1980), pp. 483-493, ISSN 0044-2275.

Rosa, C.A. & Freire, J.T. (2009). Fluid Dynamics Analysis of a Draft-tube Continuous Spouted Bed with Particles Bottom Feed Using CFD. *Industrial &. Engineering Chemistry. Research*, Vol.48, No.16, (August 2009), pp. 7813-7820, ISSN 0888-5885.

Schaeffer, D.G. (1987). Instability in the Evolution Equations Describing Incompressible Granular Flow. *Journal of Differencial Equations*, Vol.66, (January 1987), pp. 19-50, ISSN: 0022-0396.

Szafran, R. G. & Kmiec, A. (2004). CFD Modeling of Heat and Mass Transfer in a Spouted Bed Dryer. *Industrial & Engineering Chemstry Research*, Vol.43, No.4, (January 2004), pp. 1113-1124, ISSN 0888-5885.

Wen, C.Y. & Yu, Y.H. (1966). Mechanics of Fluidization. *Chem. Eng. Prog. Symp. Series*, Vol. 62, pp. 100-111.

Zhonghua, W. & Mujumdar, A. S. (2008). CFD Modeling of the Gas-Particle Flow Behavior in Spouted Beds. *Powder Technology*, Vol.183, No.2, (April 2008), pp. 260-272, ISSN: 0032-5910.

14

Study of an Individual Air-Conditioning Energy-Saving Equipment

Nguyen Anh Tuan[1], Wu-Chieh Wu[1,2] and K-David Huang[1]
[1]*Institute of Mechanical and Electrical Engineering, National Taipei University of Technology, Taiwan*
[2]*Network Operations Laboratory, Chunghwa Telecommunication Laboratories, Taiwan*
Republic of China

1. Introduction

Room air-conditioning does not always meet the optimal desire for thermal comfort of occupants and energy-saving goals [1]. This is due to the following reasons:

It is well established that differences exist between individuals in required comfortable temperature for their own local environments. (ii) Traditional air conditioning airflow distribution can stagnate around furniture, lights, and partitions which mean that airflow is distributed without consideration for the occupants' comfort demands. (iii) air-conditioning airflow has been distributed without consideration of occupants' demands beyond setting temperature and fan speed to meet thermal comfort standards, such as Predicted Mean Votes (PMV) and Predicted Percentage of Dissatisfied (PPD) indices (ISO-7730 2005) [2]. It may be that the actual presence of occupants in the workplace is less than that was set up from the original air conditioning design. This causes serious waste of cooling load to the system through unnecessary runtime. (iv) The traditional air conditioning air supply unit is set high on the ceiling or wall, which means that much of the cool air proportionally much consumed by light fittings, walls, and other surrounding objects. Therefore, this increases the demand on energy supply. In response to these concerns, several personalized air-conditioning systems have been showed and it rightly remains a great attention in efforts to design more intelligent workplaces.

Cho and Kim [3] introduced a personal environment module (PEM), the PEM was studied as an alternative air-conditioning system for improving thermal comfort in workspaces. Compared to a typical under-floor air-conditioning system, the advantages of the PEM is more flexible and easier to control.

Zeng et al. [4] introduced personalized ventilation systems (PVSs), that allow each occupant to control their own thermal comfort provision, such as airflow direction and temperature. The perceived air quality of the PVS is superior to that of a conventional ventilation system with the same amount of supplied air.

Melikov [5] have systematic reviews on the performance of personalized ventilation (PV) and on human response to it. PV, in comparison with traditional ventilation, has two important advantages: (i) it has potential for improving the inhaled air quality, and (ii) each occupant is delegated the authority to optimise as well as control the temperature, the flow rate, and the direction of supplied air according to their own preferences and to thermal comfort requirements.

Nielsen et al. [6, 7] studied a chair with integrated personalized ventilation for minimizing cross infection. The idea behind this PV system is to utilize the fact that the head or the body in natural contact with surfaces as mattresses, pillows, neck support pillows, headrest, clothing, blankets, walls, and those surfaces are designed also to be a supply opening of fresh air. The chair with integrated diffuser has proven to be a very efficient system to protect people from cross infection because clean air is supplied direct to the breathing zone. Effectiveness up to 95% can be reached, and the different systems have in general effectives larger than 50% to 80%. The flow rate to the system should be at a level of 10 l/s in most cases.

Niu et al. [8] investigated a chair-based PVS that can potentially be applied in theatres, cinemas, lecture halls, aircraft, and even offices. Perceived air quality improved greatly by serving cool air directly to the breathing zone. Feelings of irritation and local drafts could be eliminated by proper designs. Personalized air with a temperature below that of room air was able to bring "a cool head" and increased thermal comfort in comparison with traditional ventilation.

Schiavon et al. [9–11] studied the energy consumption of a PVS and energy saving strategies which can be used to control a PVS and to develop and test an index for evaluating the cooling fan efficiency in a laboratory. The potential saving of cooling energy elevated air speed was quantified by means of simulations using EnergyPlus software. Fifty-four cases covering six cities (Helsinky, Berlin, Bordeaux, Rome, Jerusalem, and Athens), three indoor environment categories, and three air velocities (< 0.2, 0.5, and 0.8 m/s) were simulated. Cooling energy savings in the range of 17–48% and a reduction of the maximum cooling power in the range of 10–28% have been obtained. The PVS may reduce the energy consumption substantially (up to 51%) compared to mixing ventilation when the control strategies are applied.

Yang et al. [12] showed a new type of PV in which the PV air terminal device (ATD) is mounted on the ceiling is developed. Thus, the personalized airflow spreads and has a target area which covers larger area of the occupant body. Thirty-two subjects performed normal office work and could choose to be exposed to four different PV airflow rates (4, 8, 12, and 16 L/s). The results showed that if PV airflow rate was increased or temperature was reduced inhaled air was perceived cooler and perceived air quality and air freshness improved. In a following paper Yang et al. [13] evaluated energy-saving potential of the ceiling mounted personalized ventilation (PV) system in conjunction with back ground mixing ventilation compared with mixing ventilation system alone and with mixing ventilation system when occupants are provided with individually controlled desk fans for generating additional air movement at each desk. Comparing with mixing ventilation plus desk fans, ceiling mounted personalized ventilation cannot only realize better cooling effect but also decrease the total energy consumption.

Faulkner et al. [14] studied the ability of two task/ambient conditioning (TAC) systems using air supplied from desk-mounted outlets to efficiently ventilate the breathing zone of heated manikins seated at desks. The TAC provided 100% outside air at a flow rate of 7 to 10 L s^{-1} per occupant. A high value of air change effectiveness (~1.3 to 1.9) was presented and high values of pollutant-removal efficiency (~1.2 to 1.6) were achieved.

In recently published designs, Huang et al. [15–17] adopted computational fluid dynamics (CFD) simulations and experimental measurements in a room with an airflow management technique to control airflow in the room to meet the demand for a regional steady-state temperature. A regional air-conditioning mechanism (RACM) system was constructed with an air uptake and outlet that created airflow circulation cells in the seating area of an occupant. Those studies considered the change in distance between the inlet and outlet ports, port heights, and air-inlet velocity. Results showed that the RACM could successfully establish an individual thermal environment zone. Compared to traditional air-conditioning systems, the RACM is predicted to save energy.

Nomenclature

d	diameter of RACM pipe, m	q_{wall}	thermal conductivity, W/m.K
g	gravitational acceleration, m/s^2	V	air velocity, m/s
h_1	inlet port height, m	T	air temperature, ^oC
h_2	outlet port height, m	T_{in}	air inlet temperature, ^oC
k	turbulent kinetic energy, m^2/s^2	t	time, s
L	length of RACM pipe, m	x, y	Cartesian coordinates
L_1	distance between inlet port and floor surface, m	φ_1	angle of inlet port, ^{o}
L_2	distance between outlet port and floor surface, m	φ_2	angle of outlet port, ^{o}
P_{out}	outlet vacuum pressure, Pa	ε	turbulent dissipation, m^2/s^3
PMV	predicted mean vote	η	temperature gradient, K/m
PPD	predicted percentage dissatisfied, %	*Subscripts*	
Q_{in}	cooling air supply flow rate, m^3/h	*oz*	occupied zone

This paper presents ten cases divided into two groups in order to study the effect of two parameters on the creation of the airflow circulation cell by using the CFD simulation: the cool air supply flow rate (Q_{in}) and outlet port position (L_2). Experimental platforms were set up to test the validity of a simulation model, and the tested results showed good agreement with simulation. Our simulation results showed that the airflow circulation cell was quite sensitive to changes in Q_{in} and L_2. In comparison with the conventional air-conditioning, the RACM could create a desirable thermal comfort zone and reduce energy consumption by as much as 91.6 %. In addition, the experimental study shows that the airflow circulation cell could be managed easily within the occupied zone no matter with or without the manikin object. These results could be helpful for designers and consultants with needed knowledge for design of RACM systems.

2. System analysis

RACM system is a personalized air conditioning system, which is integrated by main cooling air supply system (CASS), a blower fan, a regional air conditioning system, a sucker fan and two dampers, as shown in Fig. 1. The CASS supplies cooling air to the RACM by the blower fan. During operating process, the cooling air will channel through the damper, regulate the cooling air and activate the occupied zone to generate a comfortable zone according to the opening angle of damper. The cooling air will direct to the RACM, where the cooling air will exchange the heat with occupied zone in the room, after that the cooling air was sucked back to the main air conditioning by the sucker fan through the outlet port with the aim of recycling the cooling air from the occupied zone. Therefore, the RACM can turn on an independent airflow circulation cell. It cannot only satisfy thermal comfort occupants' required, but also reduce much energy consumption for the cooling air supply system.

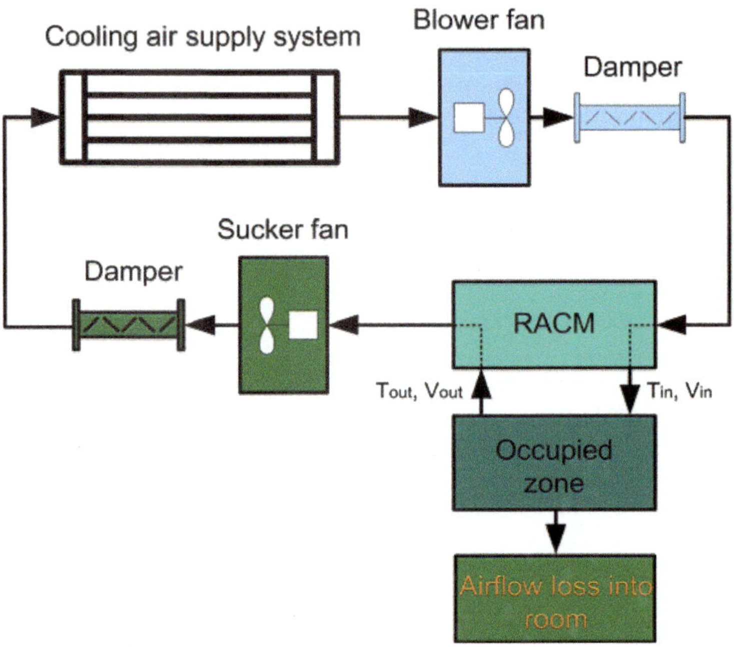

Fig. 1. Flow diagram of airflow in RACM system

3. Numerical method

3.1 Definition of model geometry and grid generation

An upper round part of the RACM has been designed for the air inlet and a lower round part has been designed for the air outlet on the main pipe of the RACM module as shown in Fig. 2. To improve the airflow distribution feature for space for human occupancy the adjustable louvers were fastened on inlet and outlet ports. The RACM enable the occupants to satisfy with respect to their own thermal comfort demands. The CFD code used in this study is FLUENT 6.3 [18]. The RACM was placed in the centre of the an empty workroom with no fixed furniture, nor occupants. Therefore, instead of using a 3D simulated model, a 2D one was applied in this study, as shown in Fig. 3. The 2D model workroom is the symmetrical two-dimensional (STD) model was described in detail in Huang's studies [15-17]. By using the solution-adaptive refinement such as gradient adaption of velocity, the initial STD model mesh can be refined by adding cells in occupied zone to enable the physical feature of the independent velocity field to be better resolved and coarsen in the rest of model room using GAMBIT software [19]. Therefore, the mesh quality of gradient adaption seems to better than initial mesh around occupied zone of model. About 11555 nodes of triangular mesh are used in the STD model, thus, the gradient adaption has been chosen for further simulation.

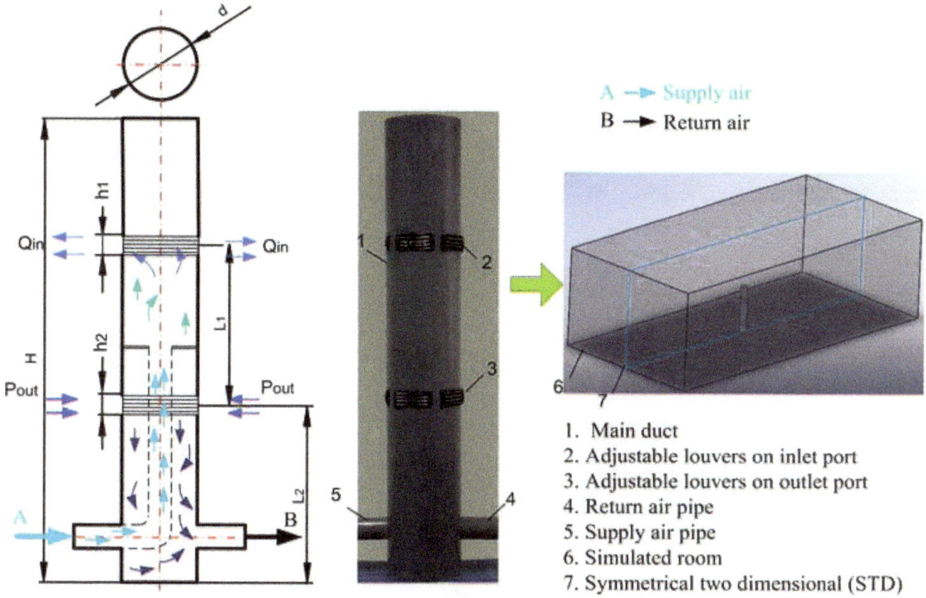

Fig. 2. Model of the regional air conditioning mechanism in the simulated room

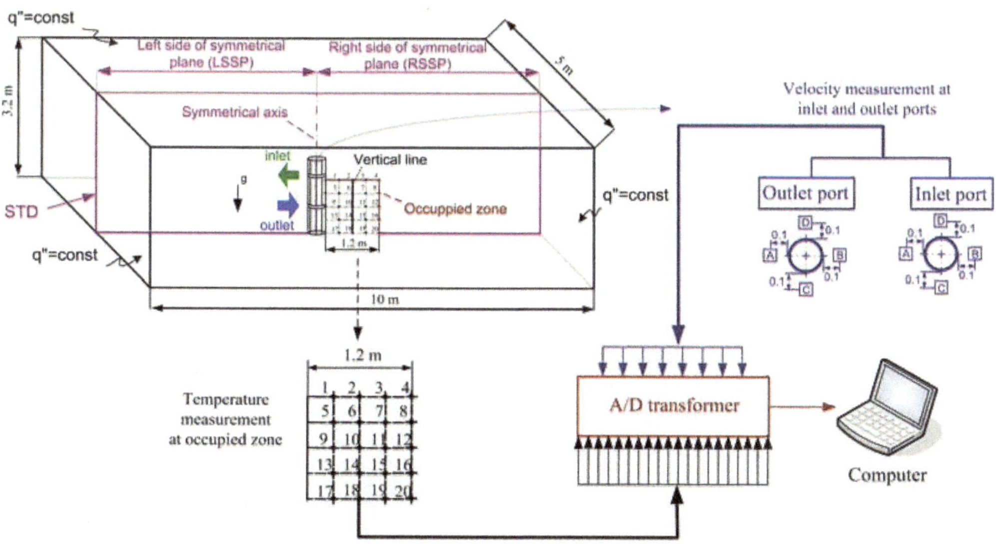

Fig. 3. Sketch layout of the RACM in the room

3.2 Simulation method

Assumptions are made with respect to the simulation of this study as follows:
The major assumptions of our analysis are as follows:

a. Newtonian fluid,
b. unsteady state,
c. effect of gravity and buoyancy are considered,
d. uniform air velocity at the inlet port,
e. standard $k-\varepsilon$ model is employed for modeling turbulent flow.

The following are the governing equations:
Continuity equation:
The mass conservation equation, or continuity equation, can be written as follows:

$$\frac{\partial p}{\partial t}+\frac{\partial}{\partial x_i}(\rho u_i)=S_m \tag{1}$$

Momentum equation:

$$\frac{\partial}{\partial t}(\rho\vec{\upsilon})+\nabla p+\rho\vec{f} \tag{2}$$

Energy equation:

$$\frac{\partial}{\partial t}\left[\rho(e+\frac{\upsilon^2}{2})\right]+\nabla.\left[\rho(e+\frac{\upsilon^2}{2})\vec{\upsilon}\right]=-\nabla.(p\vec{\upsilon})+\rho q+\rho(\vec{f}.\vec{\upsilon}) \tag{3}$$

Where, t is the time, ρ is the mass density, υ is the flow velocity, p is the fluid pressure. The source S_m is the mass added to the continuous phase from the dispersed second phase and any user-defined sources, q is the heat transfer rate per unit area, e is the internal energy, and f is the body force.
The turbulence equation, which was proposed by Launder and Spalding (1974) to model the turbulent kinematics viscosity, is primarily related to the eddy viscosity [20].

3.3 Boundary conditions

The airflow equations were solved under the conditions for air inlet/outlet ports, and wall boundaries. The air supply flow rate (Q_{in}) and distance between floor surface and outlet port (L_2) were chosen for all ten cases, as shown in Table 1. The outlet vacuum pressure and the

Names	Q_{inlet} (m^3/h)	L_2 (m)
Group 1	Changed from 20 to 110	0.5
Group 2	45	Changed from 0.1 to 0.9

Table 1. Calculation conditions

inlet air temperature for overall simulation were –2 Pa and 20 °C, respectively. Heat flux of vertical, ceiling, and a floor wall was 50 W m^{-2}. An adiabatic wall condition was chosen for the RACM walls and symmetrical axis. The surfaces of the walls of the room were assumed to have an emissivity ε of 0.85. The distance between inlet port and outlet port (L_1) was 0.6 m. RACM diameter (d) was 0.2 m. The inlet height port (h_1) was from 0.06 m to 0.1 m, outlet port height (h_2) was 0.06 m.

3.4 Validation of CFD simulations

The experiments were carried out in the Department of Mechanical and Electrical Engineering, National Taipei University of Technology. The schematic layout of the equipment and system used in this study is shown in Fig. 3. This system consists of a working room with the internal dimensions of $10m \times 5m \times 3.2m$ (ceiling height), a cooling air supply system, and an innovated RACM. The empty working room had an experimental temperature of 30 °C and a relative humidity of 50%. The study was conducted in Taipei, Taiwan during the summer, which is usually quite hot and humid. Detailed information about the equipment used in this study is shown in Table 2. Type-T thermocouple temperature sensors were set up at 20 points in the occupied zone to compare the experimental and simulation results, as shown in Fig. 3. A flow meter sensor was used for velocity measurement at four positions for both the inlet and outlet ports. Signals from temperature sensors and flow meter sensors were automatically transmitted to an A/D transformer and Testo−400 with recordings at 1 s intervals to ensure continuous reading by InstruNet and Comsoft-3 software running on a PC. All measurements were repeated at least three times at each test setting and the average experimental values of the three tests were considered to ensure stability and consistency.

Working room	$10m \times 5m \times 3.2m$
Cooling air supply system	2.7 kW–220 V
Diameter of RACM pipe	0.2 m
Total length of RACM pipe	1.7 m
Height of inlet and outlet ports	(0.06–0.1) m and 0.06 m
A/D transformer	InstruNet Model 100
Temperature sensor	(-40)–(+125) °C
Flow meter sensor	0–60 m/s
Computer	Joybook A52E

Table 2. Main component for the test RACM in the experimental setup.

3.4.1 Temperature at the occupied zone

The sensors at 20 points in the occupied zone were set up to compare in-room experimental and simulation results in the case of Q_{in} = 45 m³/h and L_2 = 0.5 m, as shown in Fig. 4. The simulation column shows trends similar to the experimental data column. The temperature difference between the simulation and experimental results was 0.18−1.38 °C in the occupied zone. It can be seen that the highest difference between the experimental and simulation results for temperature at the 20 points in the case of Q_{in} = 45 m³/h and L_2 = 0.5 m is about 4.7 %. Considering that the airflow is almost uncontrollable and natural, the simulation results of temperature in the occupied zone are in good agreement with the experimental data. Deviation at position 4 is larger than that of the other positions in the occupied zone presumably because the position is for the cooling air supply. On the other hand, part of the deviation may also be due to results from unsteady operation of the experiment.

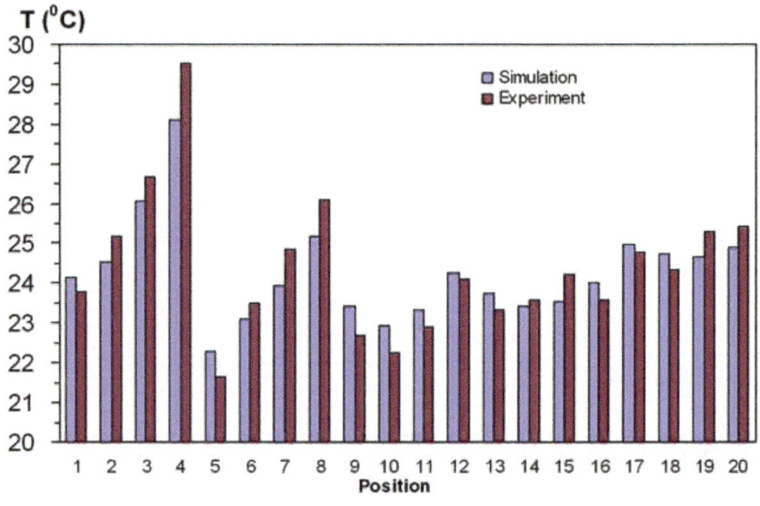

Fig. 4. Comparison of temperature between the experimental and simulation in the occupied zone of the room.

3.4.2 Velocity at inlet and outlet ports

Table 3 shows the comparisons of the experimental and simulation results of velocity for eight positions at the inlet and outlet ports under Q_{in} = 45 m³/h and L_2 = 0.5 m. The simulation and experimental results under the positions were good; however, it seems that the largest deviation is still large (~9.4 %).

In short, experimental and CFD data include the temperature and velocity in the occupied zone. The results of simulation are slightly different from the results of the experiments, around 9.4%. The fairly good agreement between experimental and simulation results indicates that the CFD simulation is possible to use in the investigation of in-room characteristics of the RACM.

Position	Experiment (m^3/h)	Simulation (m^3/h)
Air supply flow rate	44	45
Front inlet port		
Position A	0.212	0.234
Position B	0.25	0.248
Position C	0.231	0.245
Position D	0.213	0.225
Front outlet port		
Position A	0.172	0.181
Position B	0.161	0.168
Position C	0.172	0.181
Position D	0.152	0.165

Table 3. Comparison of velocities from experiment and simulation for the model room.

3.5 Calculus simulation

Based on the theory that the airflow convection effect is much stronger than the diffusion effect, this research aims to limit the air energy to a certain area by means of creating a circulation cell. The inlet airflow speed acts as the inertia, and the outlet vacuum pressure provides the centrifugal force required to form a circulation cell. In this manner, the RACM can create in the room a certain area that has the temperature different from the rest of the room. In this study, a total of two groups were simulated using FLUENT for a summer period. The room's internal dimensions were $10m \times 5m \times 3.2m$ as shown in Fig. 3. The left corner zone of the right side of symmetrical plane (RSSP) (length = 0.1–1.2 m; height = 0–1.4 m) corresponds to the occupied zone of the room as depicted in Fig. 3.

4. Results and discussion

4.1 CFD results

In this study, the RACM system was built based on the tested RACM in room with the length, diameter, and height of outlet port of pipe fixed, distance between inlet and outlet ports as constants. Nevertheless, the distance between floor surface and outlet port (ground effect distance) is adjusted with L_2 from 0.1 to 0.9 m. The height of inlet port of pipe are adjusted with h_1 from 0.06m to 0.1m.

4.1.1 Effect level of Q_{in} on establishing independent airflow cell process

The simulated results show the effect of the levels of Qin and L2 on the thermal comfort at occupied zone and energy saving potential. The RSSP is the important plane, since it represents the condition of the symmetrical plane well. The level effect of Q_{in} adjustment on establishing independent cell process in the case of L_2 = 0.5 m and the inlet velocity being adjusted from Q_{in} = 20 m^3/h to 110 m^3/h is presented in Fig. 5. Fig. 5 depicts the thermal environment on the RSSP after five hundred seconds of cooling. It is seen from Fig. 5 that the air supply flow rate acts as the inertia and the outlet vacuum pressure provides the centrifugal force to form an independent airflow circulation cell well. Energy and concentration are kept well inside the cell. It is also observed from Fig. 5 that it can form the coolest region at the left corner zone of the RSSP in some of the studied cases. The airflow loss increases as inlet velocity increases, there is airflow changing to horizontal direction and moving further away from occupied zone that is the cause of making whole cooled room and consume much cooling load. The average air temperature and velocity in these zones is in the range of 22.1–24.4 °C and 0.21–0.34 m/s, respectively. These are the comfortable temperature and velocity magnitude for the occupant in mainly sedentary, activity during summer conditions [21].

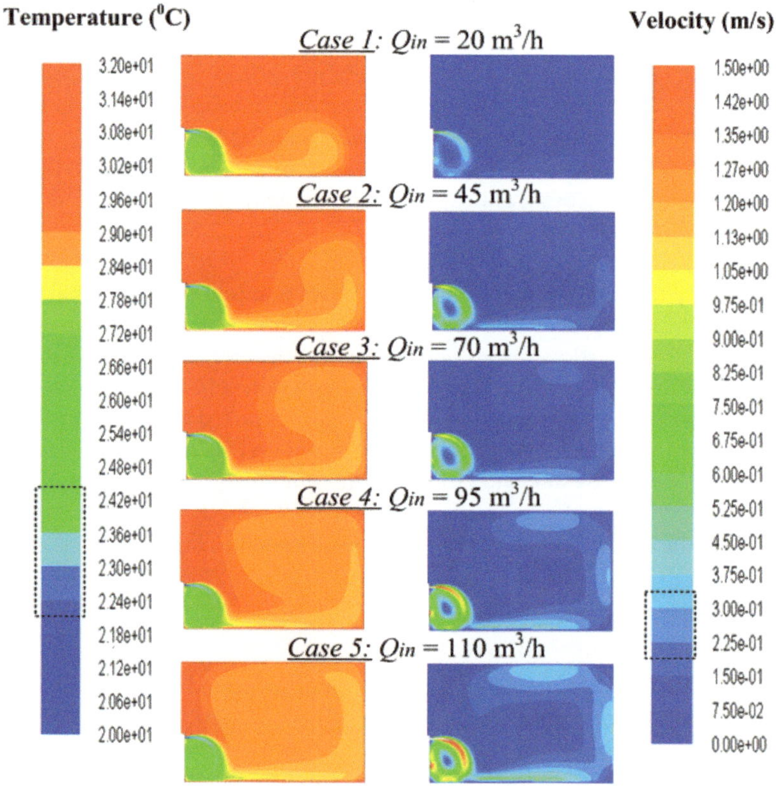

Fig. 5. Temperature and velocity distribution at the occupied zone of the room RSSP with different Q_{inlet} ($L_2 = 0.5$m).

4.1.2 Effect level of L_2 on establishing independent airflow cell process

The level of the ground effect adjustment on creating of the independent airflow circulation cell in the case under the Q_{in} = 45 m³/h and the ground effect distance being adjusted from the L_2 = 0.1 to 0.9 m is presented in Fig. 6. This figure shows clearly that the independent cell is significantly affected by the L_2 adjustment. With L_2 = 0.1 and 0.3 m, the independent cell at occupied zone is smaller than that of the case of L_2 = 0.5 m. The reason for this is that presumably the cooling airflow entering the room travels shorter distance possible than that of L_2 = 0.5 m case before leaving the room. For the case of L_2 = 0.5 m the relative distance between room ground and outlet port is long enough to form a good independent cell that fills almost all of the occupied zone with just a small airflow loss that is located at the bottom surface. Going through the two cases of L_2 = 0.7 and 0.9 m of Fig. 6, one can observe the changes in the flow field where the position of the inlet port is higher than that of the L_2 = 0.5 m case along the RACM wall. In these cases, the area occupied by the independent airflow cell diminishes and the temperature distribution will increase to a larger area of the room as L_2 is increased. This is because of the fact that as the distance between floor surface and outlet port increases the ground effect is not strong enough to aid this cooling air inlet flows back to the outlet port. The mean velocity of the occupied zone varies from 0.19 m/s to about 0.27 m/s. The temperature in the same group varies between 22 °C to about 25.1 °C in the occupied zone. This indicates a temperature rise about 3.3 °C for the varied relative distance between room ground and outlet port. Therefore, the distribution patterns in Figs. 5 and 6 show that an independent airflow cell can be successfully established. The suitable adjustment of the L_2 for a better creating of the independent airflow cell was achieved in the occupied zone as the L_2 is 0.5 m.

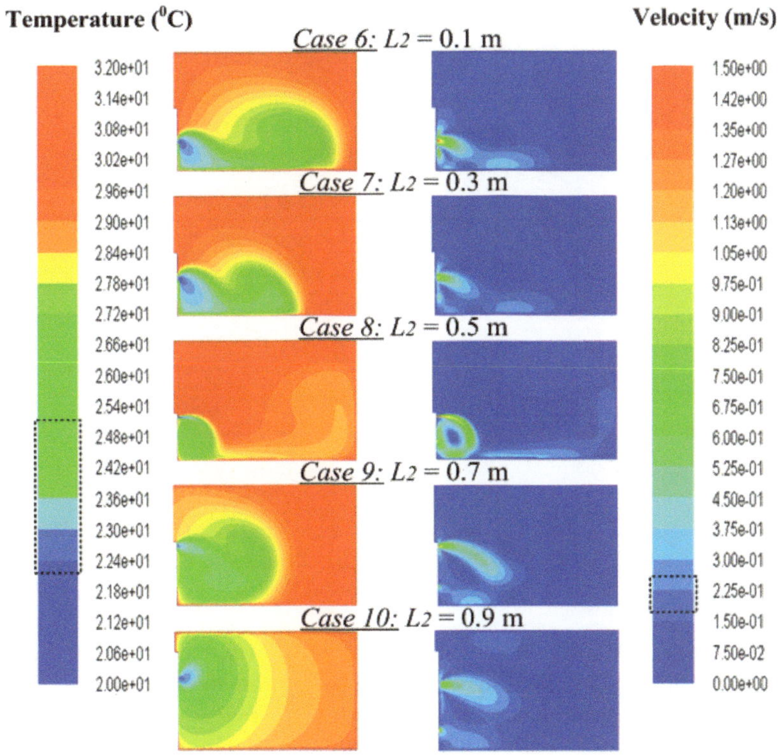

Fig. 6. Temperature and velocity distribution at occupied zone of room with different L_2 (Q_{inlet} = 45 m^3/h).

4.2 Thermal comfort indices

The thermal comfort of an occupied zone is an important issue. One method of assessing thermal comfort is to use the equations for predicted mean vote (PMV) proposed by Fanger [21, 22]. PMV is defined by six thermal variables of indoor air and conditions of human occupants: air velocity, air temperature, mean radiant temperature (MRT), relative humidity of the air, clothing, and physical activity. Figure 7 shows a combination of each thermal variable affecting PMV level. Fanger showed that values of PMV between –0.5 and +0.5 are in the range within which 90% of people are satisfied [21, 22]. The PMV corresponds to a prediction of the mean value of the votes of a large group of persons on the seven-point thermal sensation scale as shown in Fig.7.

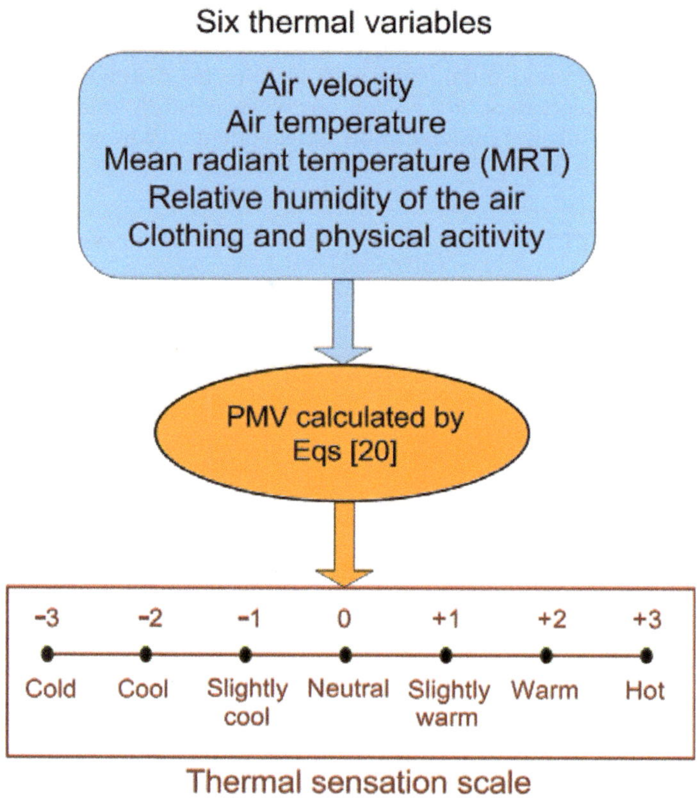

Fig. 7. Thermal comfort index.

The thermal comfort PMV index in this study was calculated by using parameters for cooling conditions in summer, mainly sedentary activity level –1.0 met and with a clothing thermal resistance of –0.5 clo. Relative humidity is assumed to be 50 % in the room, and MRT has been estimated using the P_1 radiation model [21, 22]. The overall thermal climate predicted by the PMV index can be seen in Fig. 8. It is seen from Fig. 8. that the PMV index is sensitive in air supply flow rate increase and various outlet positions. In group 1, case 5 has the worst PMV index among five cases, because the highest Q_{in} enters the room that causes the occupied zone to become less comfortable. The case 2 with Q_{in} = 45 m³/h seems to offer higher PMV potential among five cases in group 1, it was therefore chosen to be investigated the effect of L_2 in this study. Moving on to the group 2, the PMV index in case 8 gives higher potential with L_2 = 0.5 m. When L_2 is 0.1 m in case 6, the cooling flow entering the room travels the shortest distance between the floor surface and the outlet port before leaving the room, therefore the airflow circulation cell is smaller than that of the rest cases, it may cause a lower PMV value in case 6. When L_2 is 0.9 m in case ten, the airflow circulation cell diminishes that causes the PMV potential decrease in occupied zone in the room. Case 8 seems to offer higher PMV potential in group 2. Therefore, it may be seen that if the parameters are set for the RACM system as in case 8, it will be the one among ten simulated cases in this study which offer the most thermal comfort.

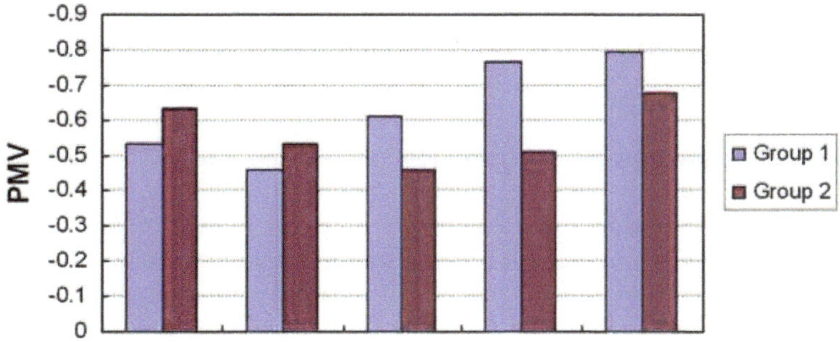

Fig. 8. PMV variation in the occupied zone

4.3 Vertical temperature profiles

The difference in the vertical air temperature between the feet and the head must be considered in determining conditions for acceptable thermal comfort. The vertical temperature profile is a function of sedentary occupant height. One measure indicates that the most comfortable thermal environment is achieved when the air temperature at head level is lower than the temperature of the floor surface [22]. It is worth noting that the difference of an allowable vertical temperature between 1.1 m and 0.1 m (head and ankle level) shall be less than 3 °C [22].

The predicted vertical temperatures were calculated for x is 0.6 m from the symmetrical axis as shown in Fig. 3 at different heights of 0, 0.1, 0.4, 0.6, 0.9 and 1.1 m from the floor surface level for all groups. These heights are assumed to represent the feet, ankle, knee, waist, shoulder, and the head of the occupant. It is obvious from Fig. 9 of group 1 that the difference for height from ankle (0.1 m) to the head level (1.1 m) is in the range of (–0.41 °C) – (–0.34 °C) the differences in vertical temperature are relatively low for all cases. Increasing Q_{in} leads relatively to the moving of air profiles to left hand side of Fig. 9. The reason for this is that, the inlet air flow rate increases, proportionally more flow blows into the room. For group 2, as the outlet port increases, the temperature profile up to head height is affected slightly in comparison to those in group 1 with the same physical parameter. The difference in temperature from the ankle to the head level is in the range of (–0.43 °C) – (–0.36 °C). It seen Fig. 9 that group 1 is the most sensitive from the point of view of varying air temperature profile along vertical line from surface to the head of the occupant as inlet air flow rate changes.

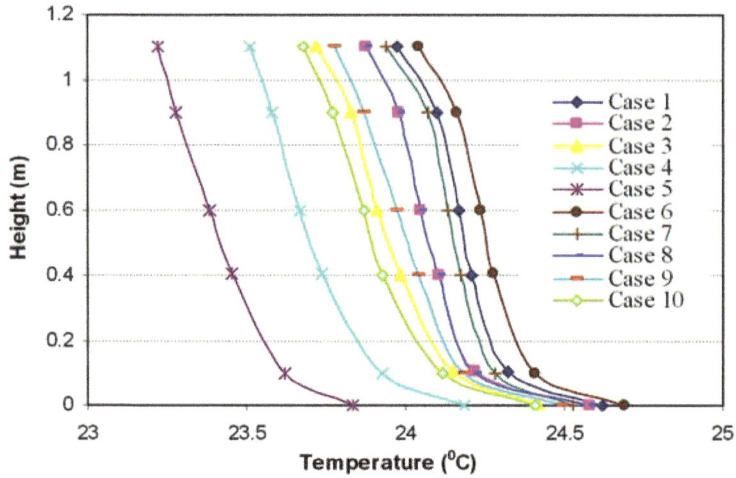

Fig. 9. Vertical temperature profile in the middle of the occupied zone

4.4 Energy-saving potential in a test case

In the case 2 study, Q_{in} = 45 m³/h. Inlet air temperature of 20 °C is blown from inlet port to the room. This supply continues for one hour, which makes 0.9 m³/h air supply rate for a 1 m² floor area. The difference between inlet and outlet temperatures, $\Delta T = 27 - 20 = 7$ K, is observed. The cooling capacity is then calculated as 2.1 W/m² using following formula:

$$\text{Cooling capacity} = (\dot{V}\rho)C_p\Delta T \tag{6}$$

Where $\dot{V}$ is volume flow rate of the air (2.5.10^{-4} m³ s^{-1}), ρ is density of air (1.2 kg m^{-3}), C_p is specific heat capacity of the air (1.10³ J kg^{-1}K^{-1}), and ΔT is the difference between the inlet and outlet air temperatures (7 K).

By means of a conventional air conditioning technique for a room three meters high, temperature difference $\Delta T = 9\,°C$, the design temperature gradient $T_{1.1} - T_{0.1}$ = 3 K m^{-1} with an occupancy design air temperature of $T_{1.1}$ = 23 °C. Temperature supply of T_s = 17 °C and the cooling capacity of 25 W m^{-2} are given by Levermore [23]. Compared to conventional air-conditioning, supplying air at 17 °C, the RACM, supplying at 20 °C, will save on chiller energy. In fact, the free cooling obtained by using cool fresh air will contribute a considerable amount to the cooling, especially, in summer conditions. Traditional air conditioning typically maintains the room at an air temperature of 23 °C whereas RACM has a typical occupied zone temperature of 23.8 °C as shown in case 2 of Fig.5. so this technique also saves on thermal energy whereas it does not reduce comfort [23]. But cooling differentials also have to be considered. For conventional cooling, the differential is 23–17 = 5 °C and the RACM the differential is 23.81–24.22 = –0.349 °C as shown in case 2 of Fig.6. For the same heat load, the conventional system will require higher flow rates than the RACM in order to maintain air distribution uniformly in the whole room while the RACM just requires this for the occupied zone. The RACM can cope with about 2.1 W/m² whereas a conventional system can cope with more than 25 W/m², proportionally energy efficiency reduce 91.6 %, so the RACM will use considerably less energy.

4.5 Experimental test

The regional air conditioning system is designed to create the independent airflow circulation cell based on convection effect being much greater than the diffusion effect. The room is divided into region with separate temperature section around the RACM as predicted in Figs. 5 and 6. This is to be expected because the airflow from the high pressure area to the low pressure area, plus the inlet air that blows straight down and is sucked back to outlet port by the outlet vacuum pressure.

4.5.1 Comparisons of the measured thermocouple data

Air temperature was monitored at six locations in the measurement room, then compared during a 500-second cool-down experiment. As shown in Fig. 10, the six locations were placed on the RSSP. The measured thermal data for occupied and non-occupied zone (the middle and right corner of the RSSP) are shown in Figs. 11 and 12.

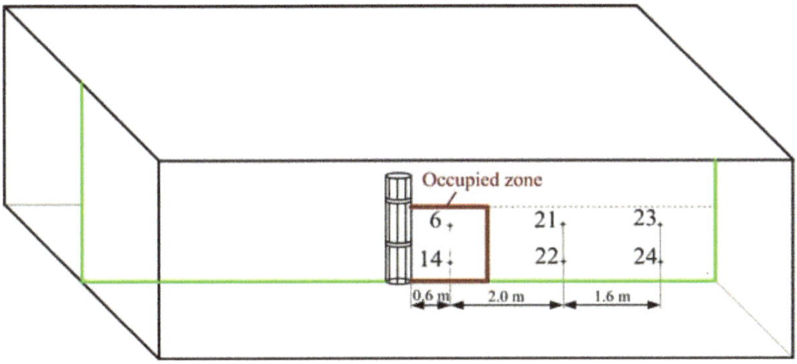

Fig. 10. The sketch of the eight monitored temperature locations in the RSSP

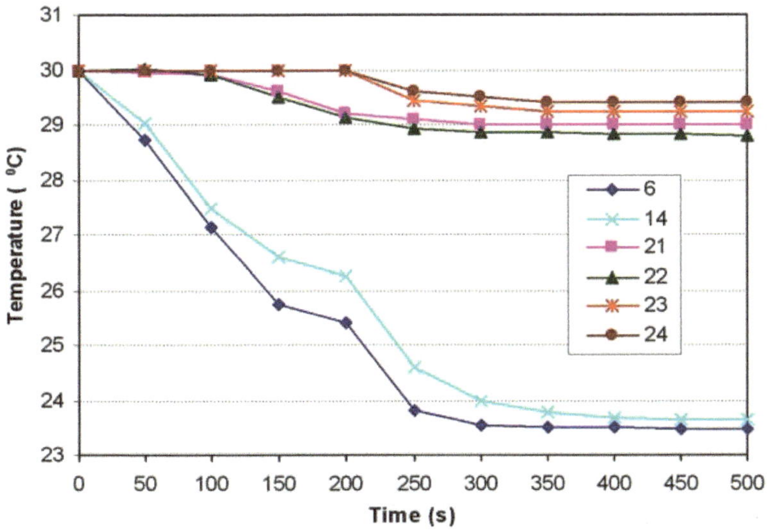

Fig. 11. Experimental temperature at six locations in the room (without a manikin)

4.5.2 Case I: Measurement without a manikin

Experiments were conducted in the initial room temperature of 30 °C under the Q_{in} of 176 m^3/h and the L_2 of 0.5 m. The temperature sensors were located at six positions as shown in Fig. 10, and then compared during a five hundred second cool-down experiment. Judging from the temperature field experiment as shown in Fig. 11, when the RACM is open, the system will run to maintain the occupied zone at a temperature around 23.5 °C, the middle of the RSSP around 29 °C, and the right corner of the RSSP around 29.5 °C. With the temperature different between the occupied zone and the middle of the RSSP are up to 6 °C and that at the right corner of RSSP are up to 5.5 °C. The occupant can feel different temperatures, which means that the occupied zone temperature distribution has met the demand of regional air-conditioning. Thus, the airflow circulation cell was successfully created in the occupied zone.

4.5.3 Case II: Measurement with a manikin

The influence of a manikin is shown in Fig. 12. A manikin was placed in the occupied zone and the RACM was turned on at an inlet temperature preset to 20 °C. The occupied zone temperature was from 24 to 24.5 °C, as shown in Fig. 12. This result shows that the airflow circulation cell is barely affected by a manikin and the target region is still accurately controlled to the preset temperature range. However, other regions are influenced by conduction and radiation, resulting in a drop of 0.5−1 °C to reach 29−30 °C. Compared with Case I, temperature distributions in the occupied zone still met the low temperature. Its goal can make the room divided into two regions with separate temperature sections without any partitions.

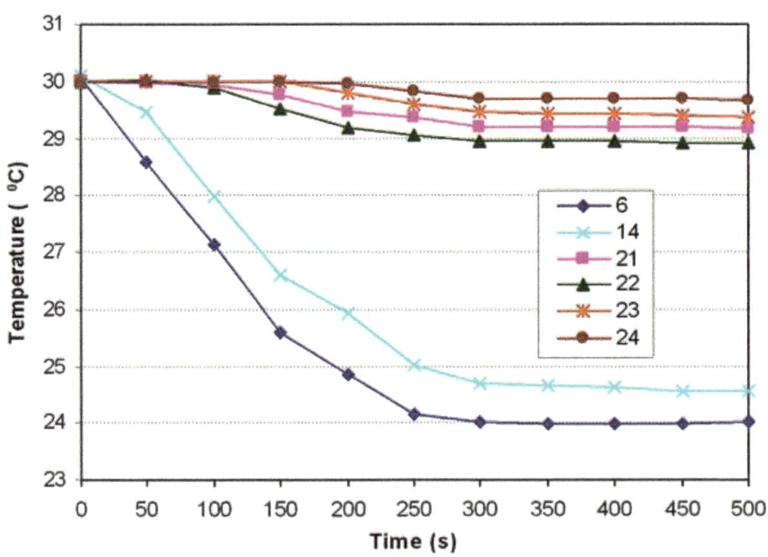

Fig. 12. Experimental temperature at six locations in the room (with manikin)

5. Conclusions

The present study introduces a new concept of energy saving mechanism design referred to as a regional air conditioning mechanism (RACM). Numerical predictions of distribution patterns in a room with this mechanism were conducted and reported. Experimental work was carried out to study the thermal regions of the RACM system. The results showed that:

i. Air convection is dominant inside the cell; inter cell only air diffusion operates and there is no mixing effect. Since the convection effect of airflow is much stronger than diffusion, the energy and concentration can basically be kept inside the cell. Therefore, the regional air-conditioning mechanism can be successfully established.
ii. The parameters of the RACM, including Q_{in} and L_2, influenced thermal comfort in the occupied zone. This result is very helpful in showing the important impacts, which must be considered when designing the RACM system.
iii. Under suitable adjustments of Q_{in} = 45 m^3/h and L_2 = 0.5 m, the highest level of thermal comfort demand can be achieved.
iv. The vertical temperature profiles in the middle of the occupied zone were almost uniform from floor to head level. In two group studies, group 1 with various values of Q_{in} is the most sensitive from the point of view of varying air temperature distribution along the vertical line from the foot to head level of a sedentary occupant in the room.
v. No energy saving index which can be generally accepted has been established so far. Compared to conventional air-conditioning, the RACM is predicted to be energy saving and to provide individual thermal comfort.
vi. It is possible to create two independent temperature regions in the experimental room.

6. Acknowledgement

This work is supported by Chunghwa Telecommunication Laboratories, Taiwan, ROC. We would like to give special thanks to Dr. Huy-Bich Nguyen, Nong Lam University, Vietnam, for optimizing our simulation model and we are also appreciative to Dr. Ngoc-Han Tran, National University of Singapore, Singapore and Dr. Anh-Dinh Le, University of Windsor, Canada for their suggestions of developing experimental work.

7. References

[1] Y. Murakami, M. Terano, K. Mizutani, Field experiments on energy consumption and thermal comfort in the office environment controlled by occupants' requirements from PC terminal, Build. Environ. 42 (2007) 17–27.

[2] ISO 7730 Moderate thermal environments - Determination of PMV and PPD indices and specification of the conditions for thermal comfort, International Organization for Standardization, Geneva. 1994.

[3] S.H. Cho, W.T. Kim, Thermal characteristics of a personal environment modules task air conditioning system: an experimental study, Energy Convers. Manag. 42 (2001) 1023–1031.

[4] Q. Zeng, J. Kaczmarczyk, A. Melikov, P.O. Fanger, Perceived air quality and thermal sensation with personalized ventilation system. Proceedings of 8th International Conference of Air Distribution in Rooms-Room Vent 2002, Copenhagen, Denmark 61–64.

[5] Melikov, A.K. Personalized ventilation, *Indoor Air,* 2004, 14, 157–167.
[6] Nielsen PV, Bartholomaeussen NM, Jakubowska E, Jiang H, Jonsson OT, Krawiecka K, Mierzejewski A, Thomas SJ, Trampczynska K, Polak M, Soennichsen M. Chair with integrated personalized ventilation for minimizing cross infection, Proceeding of Room Vent 2007, paper 1078 (2007)
[7] Nielsen PV, Barszcz E, Czarnota T, Dymalski DP, Jasiensky MA, Nowotka A, Mozer A, Wiankowska SM, Jensen RL. The influence of draught on a seat with integrated personalized ventilation, the 11th International Conference on Indoor Air Quality and Climate, Copenhagen, Denmark Paper ID: 247 (2008)
[8] J. Niu, N. Gao, M. Phoebe, H. Zuo, Experimental study on a chair-based personalized ventilation system, Build. Environ. 42 (2007) 913-925.
[9] S. Schiavon, A. Melikov, Energy saving and improved comfort by increasing air movement, Energy Build. 40 (2008) 1954–1960.
[10] S. Schiavon, A. Melikov, Energy strategies with personalized ventilation in cold climantes, Energy Build. 41 (2008) 543–550.
[11] S. Schiavon, A. Melilov, C. Sekhar, Energy analysis of the personalized ventilation system in hot and humid climates, Energy Build. 42 (2010) 699–707.
[12] B. Yang, A.K. Melikov, C. Sekhar, Ceiling mounted personalized ventilation system integrated with a secondary air distribution system - a human response study in hot and humid climate, Indoor Air 20 (4) (2010) 309 - 319.
[13] B. Yang, C. Sekhar, A.K. Melikov, Ceiling mounted personalized ventilation sytem in hot and humid climate - An energy analysis, Energy and Buildings 42 (2010) 2304 - 2308.
[14] D. Faulkner, W. J. Fisk, D.P. Sullivan, D.P. Wyon, Ventilation efficiencies of task/ambient conditioning systems with desk-mounted air supplies, Indoor Air. 9 (1999) 273–281.
[15] K.D. Huang, N.A. Tuan, K.T. Tseng, A numerical study of regional air-conditioning mechanism, Paper presented at 2008 IEEE International Conference on Sustainable Energy Technologies, 2008, Singapore.
[16] K.D. Huang, N.A. Tuan, Y.C. Shih, Energy-saving and thermal comfort studies of a regional air-conditioning mechanism. Proceedings of the Institution of Mechanical Engineers, Part A: J. Power Energy. 223 (2010) 12–33.
[17] K.D. Huang, N.A. Tuan, Numerical analysis of an air-conditioning energy-saving mechanism, Build. Simul.: Int. J. 3 (2010) 63–73.
[18] Fluent 6.3, Fluent Inc., Lebanon, NH 03766, USA, 2006.
[19] GAMbit 2.4. User guide, all volumes. Lebanon, NH 03766, USA: Fluent Inc. 2007.
[20] B.E. Launder, D.B. Spalding, The numerical computation of turbulent flows, Comput. Methods Appl. Mech. Eng. 3 (1974) 269–289.
[21] ASHRAE handbook fundamentals, ASHRAE, Inc., Atlanta, USA, Chapter 8, 2005, pp. 8.1–8.29.
[22] Thermal environmental conditions for human occupancy, ANSI/ASHRAE Standard 55–1992.
[23] Levermore, G.J. and Levermore, G. Building energy management systems: application to low-energy HVAC and natural ventilation control, E&FN Spon, London, 2000, Chapter 12, pp. 389–399.

15

Unsteady Differentiation of Aerodynamic Coefficients: Methodology and Application

Carlo Necci and Nicola Ceresola
Alenia Aeronautica SpA
Italy

1. Introduction

The evaluation of aerodynamic forces and of the relevant coefficients is a fundamental task in aircraft design. A number of numerical methods have been investigated and some of them, the most reliable and cost effective, are daily applied in the Aerospace field for design, development and research purposes. Conversely, not so many methods have been considered reliable enough for computing aerodynamic derivatives, which are fundamental to evaluate sensitivities to aerodynamic or shape parameters. Despite the efforts deployed by former researchers, the basic mathematical complexity to evaluate derivatives strongly limits the application of several methods. Numerical differentiation is surely a task widely studied and most of the schemes which have been developed are based on the finite differences method. Finite differences are flexible and easily applicable, but they are just an algebraic transposition of the Taylor series expansion, and as such they are an approximation. As a matter of fact, the attempt to apply the mathematical theory of differentiation to computer programs has to be handled differently. Furthermore, while steady derivatives allow quantifying different sensitivities when flying in steady condition, dynamic derivatives are supposed to provide useful information on what happens in unsteady condition, but this introduces the extra complexity of the time dependency. It is hence again evident that the need of a solid mathematical basis is crucial when dealing with these kinds of problems.
A possible approach is the experimental use of physical models in wind tunnels; data acquired while simulating different steady flight conditions can be interpolated to compute steady derivatives. While unsteady derivatives can be quantified by both mathematically handling steady values or moving the model in the wind tunnel and then managing the data streams that have been acquired. The drawback of this method is its high cost, especially when complex motions are tested. Furthermore, experimental work is complex, since setup and test implementation require a wide experience both in preparing the system setup and in understanding results. An extensive work has been done so far in this field; some references (Almosnino, 1994; Altun & Iyigun, 2004; Anderson & Newmann, 1999; Anderson et al., 1984; Guglieri & Quagliotti, 1993; Guglieri et al., 1993; Murphy & Klein, 2001) can give just an idea of the huge effort so far deployed.
But in the latest years numerical approaches have been extensively developed and employed to fulfil the same results in a cheaper and faster way. The simplest algebraic way

to evaluate derivatives is, as said, the use of finite differences (Anderson et al., 1984) which simply requires evaluating the function of interest at two or more nearby states. This method can also be used as a validation technique and it is attractive since it does not require any modifications to existing source codes but it has the drawback to be sensitive to cancellation errors. It is important using a small step size to approximate the mathematical derivative, but the issue of subtractive cancellation of the terms in the numerator will always be experienced. An alternative to this is the Complex Variable Differentiation method (CVD), suggested by Anderson and other Authors (Anderson et al., 1999). Conceptually, CVD uses a series expansion of functions and a complex perturbation of terms, so in effect splitting up real and imaginary parts of the relevant function. For the first derivative, this method does not require subtracting the values computed in different points, so avoiding errors connected to cancellation; it has been successfully tested in computation of aero-structure sensitivity derivatives (Newmann et al., 1998) and works adequately. But it has the drawback to double the memory required for computation, since for each function both the real and the corresponding complex component have to be managed; runtime cost increases by a factor close to three.

Another possibility consists in using codes based on panel method. This method bases its applicability on the conceptual simplicity of panels but it showed to be unreliable in some particular conditions (Almosnino, 1994). In 2004 Green published a good work (Green et al., 2004) focused on the possibility to computationally separate dynamic and stability derivatives, which during the experimental testing are measured in combination. Their study was furthermore interesting for the introduction of the automatic differentiation technique, shortly identified as AD. This was the beginning for a new branch of numerical research, which gave remarkable results. In Europe, the French researcher Laurent Hascoet carried out a study that finally was formalised in a public paper (Hascoet, 2005) focused on parallelisation and differentiation techniques based on a strong mathematical background and algebraic adherence to the rules of mathematical analysis. The work was then formalised in a more extended research context when the French 'Institute National de Recherche en Informatique et en Automatique' (INRIA) developed a tool (Tapenade) completely devoted to differentiation of Fortran source codes (Hascoet & Pascual, 2004).

Industrial applications of Tapenade were carried out across Europe by different Aerospace Companies; it was showed that the AD approach can be applied for both aircraft sensitivities studies and for shape optimisation (Selmin, 2004). It was proved that the AD method was effective for differentiating both the only CFD solver and the complete computation chain (shape parameters + solver). Comparisons were carried out at numerical level and results were than compared again with finite differences computations; the outcomes were more than satisfactory.

In 2009 another work was published by the Authors et Al. (Necci, et al., 2009); its goal was to demonstrate the practical applicability of AD to industrial problems dealing with complex configurations. It was shown that automatic differentiation could be successfully used not only for research purposes but also to carry out a design activity on complex industrial configurations. Progress has been made from that point on and AD capabilities have been extended to compute stability and control dynamic derivatives for civil and fighter aircrafts. This paper deals with this. A first conceptual comparison between AD and other methods will be provided in order to explain AD peculiarities; then a deeper study of the AD numerical technology will be carried out. Finally some results will be given.

2. Automatic differentiation concept

Automatic differentiation is based on concepts which are extensively described in Ref. [10] and [11]. To summarise, let $\mathbf{X} \in \Re^n$ be a vector argument and $\mathbf{Y} = \mathbf{F}(\mathbf{X}) \in \Re^m$ a corresponding vector function. Equation $F_m = F_m(x_1, x_2, ..., x_n)$ expresses the idea to have m functions depending on n variables. A computer program P able to evaluate Y can only evaluate simple functions; so we have to split up F_m into several sub-functions f_k, each one implemented by a corresponding instruction I_k. Function F will then be given by the composition $F = f_p \circ f_{p-1} \circ ... \circ f_1$ and the corresponding source code has to be a sequence like $P = \{I_1; I_2; ...; I_p\}$. As a mathematical limitation, all the functions which are implemented in the code functions have to be differentiable, despite some exceptions (e.g. the square root); however, in general, functions implemented by arithmetic operations are indeed differentiable. The automatic differentiation tool evaluates derivatives by using the chain rule, i.e. it applies the following mathematical concept:

$$F'(X) = \underbrace{f_p'(X_{p-1}) \times f_{p-1}'(X_{p-2}) \times ... \times f_1'(X_0)}_{J} \quad (1)$$

Here $X_k = f_k(X_{k-1})$ is the value of a generic sub-function and $X_0 = X$ is its first value. J is clearly the Jacobean of F. Using the concept expressed by equation (1), derivatives can be translated back into a sequence of instructions I_k', the instructions will be coded and inserted back into a copy of the control program P. This new set of instructions, added to P, yields to program P′, which now embeds instructions both for primitive and for derivative functions. Conceptually, the automatic differentiation works on some basic assumptions; firstly, P is considered simply as a run-time sequence of instructions and the AD tool differentiates this sequence. Secondly, each sequence is a composition of vector functions, each one assumed differentiable. Since each function is differentiable, we have the theoretical basis to differentiate according to the fundamental rule of mathematical analysis: we have to work not only on the sequence of functions but also on each single function. Furthermore one has to observe that the generic function $F_m = F_m(x_1, x_2, ..., x_n)$ depends on n terms, so differentiation of a single function imposes computing n derivatives. Applying this concept to a vector function it is evident that equation (1) is a Jacobean and, as such, it requires multiplying matrices with matrices. This opens another issue: what is the best strategy to reduce the computational cost? Direct (tangent) or backwards (reverse) modes stem from this issue.

Physically, often we just need evaluating the sensibility of a quantity over a design parameter; an example is the sensibility of lift coefficient with respect to the angle of attack $(C_{L\alpha})$ or lateral force sensitivity to sideslip angle $(C_{Y\beta})$. For an aerodynamic coefficient, this means taking into account just its component related to a given aerodynamic parameter. Mathematically, this requires projecting a function (a total derivative) onto a given direction (a design parameter), as conceptually shown in Fig. 1.

Since our basis has n components, a generic directional derivative is the projection of the total derivative $\mathbf{F}'(\mathbf{X})$ along one component of the basis. Formally, we have to evaluate:

$$\mathbf{F}'(\mathbf{X}) \cdot \dot{\mathbf{X}} = \underbrace{f_p'(X_{p-1}) \times f_{p-1}'(X_{p-2}) \times ... \times f_1'(X_0)}_{J} \cdot \dot{X} \quad (2)$$

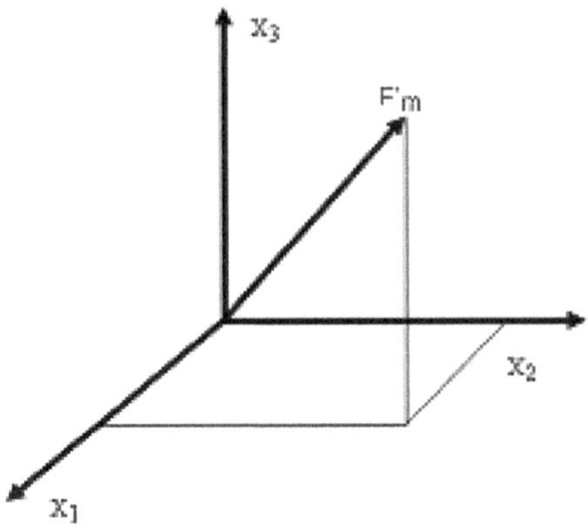

Fig. 1. Generic reference basis.

$\mathbf{F}'(\mathbf{X})$ is a $\{m \times n\}$ matrix and the most efficient way to evaluate the quantity given in (2) is from right to left; this requires multiplying a matrix by a vector. We have to start the differentiation from the first instruction $f_1'(X_0)$ in the source code and then we have to progress on, until the last instruction is reached. This approach makes things easier from the point of view of program coding, as it allows merging the original source code with the lines of differenced functions immediately after each corresponding primitive instruction. Such a concept is the basis of direct (or tangent) differentiation.

Conversely, in optimization processes or inverse problems, it is required minimising a generic cost function with respect to a number of design parameters; a physical example of this can be the minimisation of drag index with respect to the angle of attack, to Mach number and to a number of shape parameters. This imposes the differentiation of a function with respect to a number of parameters and then to weight each component with a dedicated coefficient. Mathematically, this implies transforming $\mathbf{Y} = \mathbf{F}(\underline{\mathbf{X}})$ vector into a row (using transposition) and then multiplying this row by a weight vector $\overline{\mathbf{F}}(m)$, which clearly has to be an input. Formally, the product becomes $\mathbf{F}^{\mathrm{T}}(\mathbf{X}) \cdot \overline{\mathbf{F}}$ and its gradient is $\mathbf{F}'^{\mathrm{T}}(\mathbf{X}) \cdot \overline{\mathbf{F}}$; transposing $\mathbf{F}'$ means transposing the product $F' = f_p' \times f_{p-1}' \times \ldots \times f_1'$ and, remembering that $(\mathbf{A} \times \mathbf{B})^T = \mathbf{B}^T \times \mathbf{A}^T$, we have:

$$\mathbf{F}'^T = \left(f_p' \times f_{p-1}' \times \ldots \times f_1'\right)^T = f_1'^T \times f_2'^T \times \ldots \times f_p'^T \tag{3}$$

from which:

$$\frac{\partial \mathbf{F}}{\partial \mathbf{X}} = \mathbf{F}'^T \times \overline{\mathbf{F}} = \underbrace{f_1'^T(X_0) \times f_2'^T(X_1) \times \ldots \times f_p'^T\left(X_{p-1}\right)}_{J^T} \times \overline{F} \tag{4}$$

Here $\mathbf{F}'^T\{n \times m\}$ is the transposed Jacobean. Even in this case computation is more efficient from right to left; we can progress using a matrix by vector product, computationally cheaper than a matrix by matrix computation. In this second case, computation starts differencing the last p^{th} function and then going back to first function. The definition of backwards (or inverse) differentiation is due to this.

A general rule, easy to remember when choosing the best differentiation method comes from the following observation; the less number of rows in the multiplying vectors ($\dot{X}$ or $\overline{F}$) the less will be the operations. For direct mode, short $\dot{X}$ corresponds to small n , thus direct mode can be computationally effective when differencing a number of functions with respect to a few parameters. Conversely, for backwards mode, short $\overline{F}$ correspond to small m , and this means a few functions to differentiate; it is attractive when differencing with respect to a number of parameters, i.e. in case of optimisation problems.

3. Numerical approach

By taking into account a generic solution depending on the position vector, normal vectors and boundary conditions, the sensitivity of a generic aerodynamic solution with respect to a parameter β can be expressed as (Ref. [13]):

$$\frac{\partial \mathbf{U}}{\partial \beta} = \frac{\partial \mathbf{U}}{\partial \mathbf{x}}\frac{\partial \mathbf{x}}{\partial \beta} + \frac{\partial \mathbf{U}}{\partial \boldsymbol{\eta}_{ij}}\frac{\partial \boldsymbol{\eta}_{ij}}{\partial \beta} + \frac{\partial \mathbf{U}}{\partial \boldsymbol{\eta}_i^b}\frac{\partial \boldsymbol{\eta}_i^b}{\partial \beta} \tag{5}$$

If the automatic differentiation tool is used in order to evaluate the quantities expressed in (5), it will generate the total derivative with respect to the specified design parameter. But we use the automatic differentiation to evaluate the sensitivity to just some parameters, so the tool has to be tuned according to real needs, i.e. it has to compute just some derivatives, not the total derivative. For this reason, the differentiation procedure has to take into account the following issues:

- when computing the sensitivity with respect to a single parameter, the relevant parameter β has to be set to 'one' before entering the differentiation routines and all other parameters have to be set to 'zero'; this can be done directly into the code or using an external input file;
- in order to reduce computational costs, differentiation should be applied just after the primitive solution has converged, so to avoid computing also the iterations related to derivatives. In this way the application of differenced functions to a converged solution will imply only one further computation run.

However, despite the conceptual simplicity and elegance of such an optimized approach, the real implementation of automatic differentiation is very challenging from the software engineering side, especially for complex industrial tools.

The solver that includes differentiated functions, here defined 'augmented', will be discussed preliminary for the static differentiation, so limiting the action to its steady loop; dynamic differentiation will follow and will extend the action to the transient loop. Three different structures of the augmented algorithm will be studied; all of them will include some preliminary steps, i.e. the acquisition of initial values, check of grid metrics and acquisition of initial solution. These steps will not be subject to any differentiation, since just the core of computation has to be differenced and not all the modules of the solver. After these three steps, some differences among the structures will be discussed.

In structure 1 (Fig. 2), after some initial checks, the parameter with respect to which we want to differentiate is set to unity. Then the augmented algorithm is engaged and both primitive and differenced parameters are computed. This configuration is structurally simple but it requires computing derivatives for all iterations while the primitive solution is still reaching its convergence.

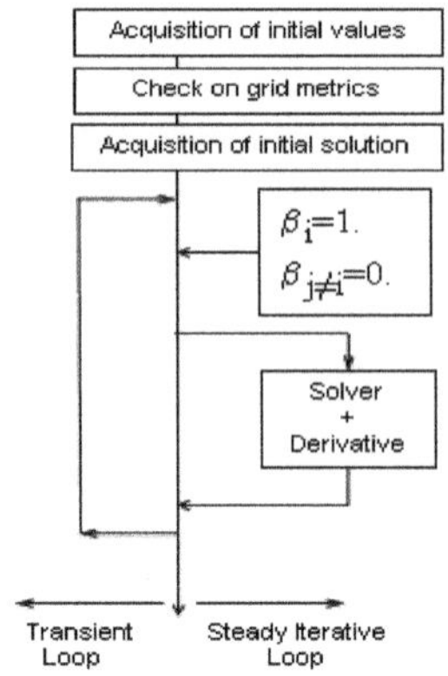

Fig. 2. Structure 1

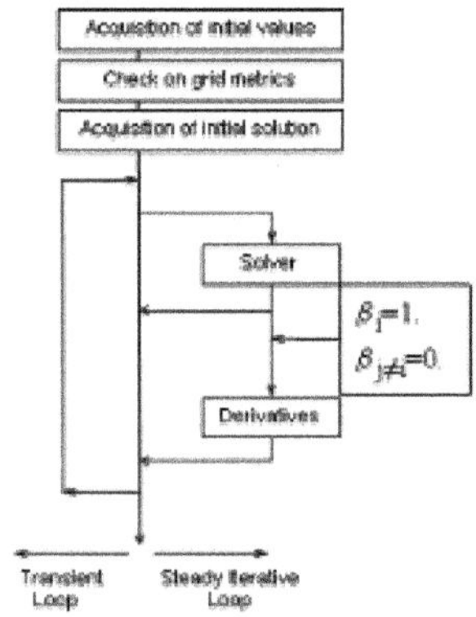

Fig. 3. Structure 2

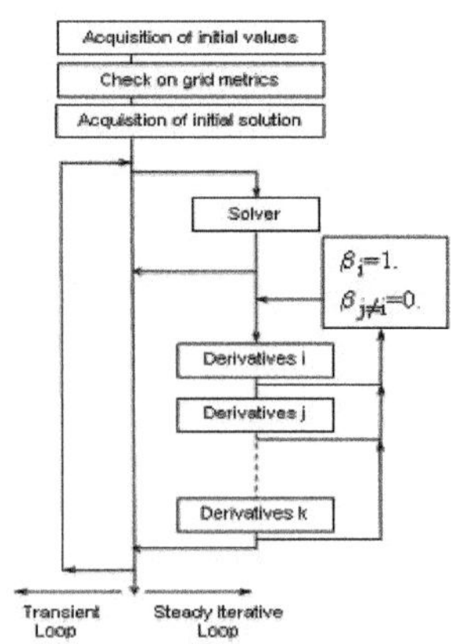

Fig. 4. Structure 3

This increases the runtime cost without any real benefit, unless the user is interested in studying the iterative evolution of derivatives. Structure 2 (Fig. 3) shows a variant; derivatives are computed just after the primitive solution has converged; only one extra run is required. Differentiation is paid with a minimal increase of computational cost and numerical efficiency has clearly increased. At this point, we can look at the problem from another point of view; we can remember that the AD tool provides the capability to differentiate a function with respect to one or more parameters. Surely, differencing with respect to several parameters involves a higher complexity of the augmented software because of the higher number of added instructions but, conceptually, structures 1 and 2 can be used indifferently to differentiate with respect to one or more parameters. An alternative structure of the augmented solver, suitable for differencing with respect to many parameters, is shown in structure 3 (Fig. 4). In this last case, a number of differentiation blocks can be cascaded, and each block computes the derivatives with respect to one only parameter. Then, iteratively, all the blocks can be engaged in sequence. It is necessary looping the cycle in such a way to reset to 'one' the correct β_i parameter before calling each block. The benefit comes from the capability to evaluate all the derivatives with respect to β_i with just one extra iteration for each parameter, like in Structure 2. Compared to structure 2, structure 3 has an advantage and a disadvantage. The advantage is that structure 2 manages a larger module involving several derivatives, and the use of only one block makes it easier linking such a block with the solver. The disadvantage is that if the user needs rearranging the differentiation with respect to one parameter, for any particular purpose, the use of structure 2 makes the work very hard. Structure 3 is simply a modular approach, which requires more initial work to integrate each differentiation module in the overall structure but that guarantees a higher flexibility in case of further reworking. It is evident that structure 3 has a practical difficulty; while in structures 1 and 2 the software has to be optimized for just one large block of augmented variables, in structure 3 more sub-blocks are involved, so optimization is longer and more complex. In this case, the modularity of the primitive code and software engineering skills of the user make the difference. Once the structure has been decided, a practical possibility from the software point of view is that all parameters that have to be set to 1. Then these can be read from an input file; then all different blocks, linked together with the relevant declaration files, are called by the program. These and a number of possible variants can be explored; it is just important to say that, whatever the user wants to do, he needs a complete control of the software. This means that the automatic differentiation technology cannot be applied to commercial tools, which are closed applications.

Moving now from static to dynamic differentiation, the only difference is in the type of parameters; static differentiation involves incidence angle, sideslip angle or geometric functions while dynamic differentiation involves roll, pitch and yaw velocities, i.e. it involves the time. Because of the dynamic nature of these parameters, AD impacts now the transient loop of the software. But in this case the good modularity of the solver that has been used in Alenia Aeronautica has shown its benefits, since the AD procedure has been used without any real problem and according to the rules described for the static differentiation.

4. Numerical applications and dynamic computing procedure

The results described in this section have been achieved using the tangent mode differentiation. Using the following expression:

$$H_d = \frac{\partial \mathbf{H}}{\partial \mathbf{B}} \tag{6}$$

we indicate with $\mathbf{H} = \{C_L, C_D, C_Y, C_l, C_m, C_n\}^T$ the aerodynamic coefficients vector and with $\mathbf{B} = (p, q, r)$ the design variables vector.

Because of the structure of the solver that has been used, the numerical sequence for computing a derivative requires three different steps:

1. the computation of a primitive steady solution (PS1);
2. the computation of a primitive unsteady solution (PS2);
3. the computation of the differentiated solution (DS).

Two tests cases have been investigated, a NACA0012 airfoil studied in viscid condition and a complete configuration, the DLR-F12, studied in inviscid condition. This second test case has been the real test bench, since the results that have been achieved have been compared with those collected by other Partners involved in the SIMSAC programme.

4.1 NACA0012 airfoil

NACA0012 has been studied for $M = 0.4$ and $\text{Re} = 6.2e + 06$. The grid is a C-type, 66880 nodes and 24924 elements, generated by an Alenia proprietary mesh generator. The skin has been meshed by placing 160 nodes on the upper and lower surfaces; the wake has been similarly meshed, by placing 50 nodes after the trailing edge. The skin mesh has then been extended in the outwards normal direction; the boundary layer has been filled up by building 57 parallel layers in the region between the skin and the area where the 99% of the freestream velocity is reached. Then, 200 more layers have been used in the farfield. Dimensions of the grid are as follows:

$$\Delta x = -0.15000e + 02; +0.16000e + 02$$

$$\Delta y = +0.0000e + 00; +0.10000e + 01$$

$$\Delta z = -0.15056e + 02; +0.15055e + 02$$

The mesh around both leading and trailing edges has been built by setting the initial spacing layer to 0.1e-02 chord units. Starting from the first layer, a geometrical progression has then been used; setting a growth factor 1.208, the mesh has been expanded in the normal outwards direction, until the final layer of the farfield has been reached. The airfoil chord length has been set to unity. The three-dimensions solver has been applied on this two-dimensions mesh by simply redefining the grid itself; the two-dimensions mesh has been generated on a vertical plane, which has then been doubled onto a second parallel plane. The two planes have then been joined together, so generating a three-dimensional structure. Fig. 5 shows how the two-dimensions mesh has been converted in three-dimensions.

As said earlier, the first step of the computation process has been the achievement of a stable primitive steady solution, for M=0.4 and α=2°; 15000 iterations have been used to completely minimise residuals and to reach an excellent solution. After this, two different dynamic analyses have been carried out, one involving some variable frequencies of oscillation and another involving some variable angles of attack. The laminar Prandtl number has been set to 0.72 and the turbulent to 0.90. CFL number has been set to 2.0 and no residual averaging has been necessary. The steady computation has required about 448 seconds on a parallel

Quadrics machine, while unsteady runs almost doubled this value. Thirty-two processors have been used, 28 for computing purposes and 4 for data traffic management.

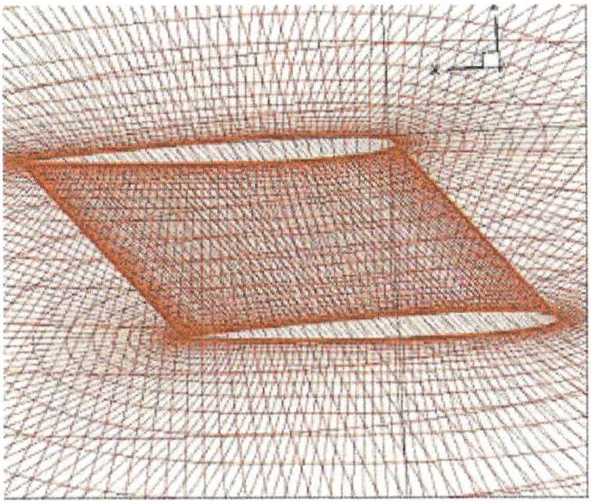

Fig. 5. Internal mesh - conversion to three-dimensions.

4.1.1 Variable frequency

The analysis for variable frequencies has required computing a number of primitive steady solutions, one for each frequency in the (used) range $0.025\,rad/s \le \omega \le 0.1\,rad/s$. For the sake of clarity, we can say that here we will just describe and discuss some details related to the numerical test case $M = 0.4, \alpha = 2°, \omega = 0.025\,rad/s$, remembering that the used process has been exactly the same for all the frequencies. After computing a stable primitive steady solution, the dynamic solver has been engaged. A pitch range $\Delta\alpha = \pm 1°$ has been imposed about the initial angle of attack and a forced oscillation has been applied. The rotation centre has been placed at $x = c/4$ and, in order to minimise the numerical transient due to the initial steady solution (PS1), three complete oscillations have been run. For each oscillation, 96 iterations have been computed for a total amount of 288 iterations; each one of these iterations required 200 sub-iterations to converge. Fig. 6 shows a graphical example.
Along the X axis is reported the number of iterations, while on y axis are shown both the corresponding values of the lift coefficient and of the angle of attack. It is evident that each time the local angle of attack increases, a corresponding increase of the lift coefficient is experienced as well. In order to show that each point in Fig. 6 corresponds to a converged value of the iteration process, Fig. 7 focuses the evolution of the solution for the lift coefficient during the very first step of the first oscillation. It is just the case to mention that this particular case has been chosen because the first unsteady iteration run after the initial (steady) solution is the most impacted from the numerical stability point of view. We can see that, when the sub-iteration process starts, the value of the lift coefficient is close to 0.24, which is the value provided by the steady solution PS1. After this, the unsteady computation begins and a slight fluctuation is experienced. Anyway, after about 150 sub-iterations, a convergence is reached. This last fact confirmed that the choice of 200 sub-iterations was successful, since the numerical fluctuation was almost completely eliminated. In order to prove this last sentence with numerical evidence, Table 1 provides maximum and minimum values for Cl during the forced three oscillations, as well as the relevant percentage fluctuations.

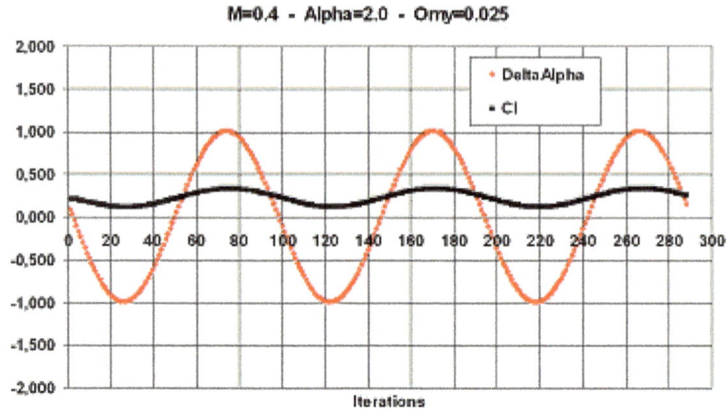

Fig. 6. NACA0012 oscillation evolution for $\left(M=0.4, \alpha=2°, \Delta\alpha=\pm1°, \omega_y=0.025\,rad/s\right)$.

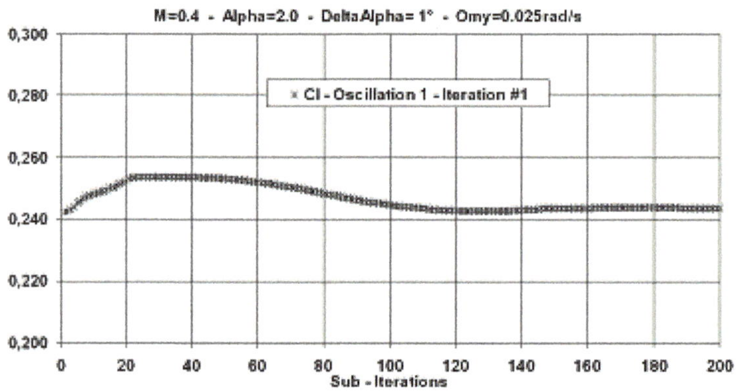

Fig. 7. NACA0012 - Lift coefficient evolution for unsteady computation $\left(M=0.4, \alpha=2°, \Delta\alpha=1°, \omega_y=0.025\,rad/s\right)$

	Cl min	Cl max	% Fluctuation
Oscillation 1 - Sub-Iteration 1-100	2.4283e-01	2.6130e-01	7.068%
Oscillation 1 - Sub-Iteration 101-200	2.4302e-01	2.4787e-01	1.956%
Oscillation 2 - Sub-Iteration 1-100	2.4213e-01	2.5384e-01	4.613%
Oscillation 2 - Sub-Iteration 101-200	2.4356e-01	2.4376e-01	0.082%
Oscillation 3 - Sub-Iteration 1-100	2.4212e-01	2.5382e-01	4.609%
Oscillation 3 - Sub-Iteration 101-200	2.4356e-01	2.4375e-01	0.077%

Table 1. Cl convergence - values and errors

The table is arranged in such a way to allow an analysis for groups of sub-iterations. Each one of the three oscillations is split up in two phases; the first one focuses on sub-iterations 1 to 100 while the second focuses on sub-iterations 101 to 200. During oscillation 1, Cl values in the first phase range from 2.4283e-01 to 2.6130e-01 and the relevant percentage fluctuation is 7.068%; values in the second phase range from 2.4302e-01 to 2.4787e-01 and the corresponding percentage fluctuation drops down to 1.956%. This means that, at the end of oscillation 1, the solution has almost converged. Going on and computing the same values for oscillations 2 and 3, all percentage variations in the first phases become even smaller (4.613% and 4.609%) and almost disappear during the second phases (0.082% and 0.077%). This proofs that the unsteady solution PS2 has completely converged at the end of the third cycle. Stated this, we can study the evolution of our differenced solutions; in particular, it is interesting comparing the results provided by the augmented solver with those computed by using the finite differences method. Fig. 8 deals with this and it still provides information for the case $M = 0.4, \alpha = 2°, \omega = 0.025$.

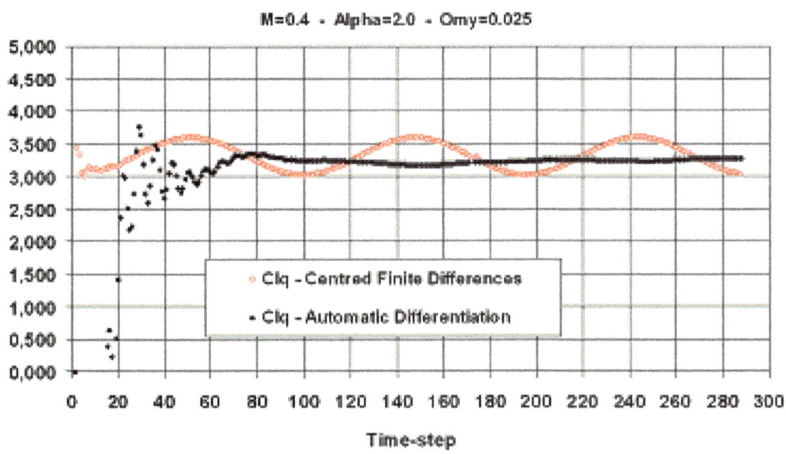

Fig. 8. Comparison between automatic differentiation and finite differences Clq derivatives $\left(M = 0.4, \alpha = 2°, \Delta\alpha = 1°, \omega_y = 0.025\, rad/s\right)$.

Finite differences have been computed by remembering that, from the analytical point of view, we have:

$$C_{l_{\dot{\alpha}}} = \partial C_l / \partial\left(\bar{c}\alpha/2U_\infty\right), \quad C_{m_q} = \partial C_m / \partial\left(\bar{c}q/2U_\infty\right) \tag{7}$$

Moving now from differential values to finite values, these expressions can be rewritten as:

$$C_{l_{\dot{\alpha}}} = \Delta C_l / \Delta\left(\bar{c}\alpha/2U_\infty\right), \quad C_{m_q} = \Delta C_m / \Delta\left(\bar{c}q/2U_\infty\right) \tag{8}$$

The curve related to finite differences fluctuates because of the fluctuating values of ΔC_l, computed by using the primitive solver. Conversely, the AD curve is much more stable; after an initial transient, the curve becomes almost steady and quickly converges. Table 2 shows both values and percentage errors between the two curves; it is evident that errors do vary mainly because of the finite differences fluctuations, while automatic differentiation numerical values show a good steadiness.

Time-step	Clq FD	Clq AD	Error %
0	3.43E+00	2.76E+01	-704.96%
20	3.18E+00	2.37E+00	25.52%
40	3.52E+00	2.81E+00	20.18%
60	3.53E+00	3.05E+00	13.61%
80	3.20E+00	3.33E+00	-3.97%
100	3.02E+00	3.25E+00	-7.59%
120	3.25E+00	3.23E+00	0.74%
140	3.56E+00	3.19E+00	10.51%
160	3.48E+00	3.19E+00	8.55%
180	3.13E+00	3.22E+00	-2.68%
200	3.03E+00	3.25E+00	-7.09%
220	3.26E+00	3.25E+00	0.21%
240	3.58E+00	3.24E+00	9.68%
260	3.42E+00	3.26E+00	4.77%
280	3.08E+00	3.27E+00	-6.04%

Table 2. Clq comparison - AD vs FD - values and errors M=0.4, α=2.0°, Δα=1°, ωy=0.025 rad/s

Similar considerations can be repeated for C_{m_q} (Fig. 2). In this case percentage errors are initially very high but convergence is fast and the error becomes smaller than 5%

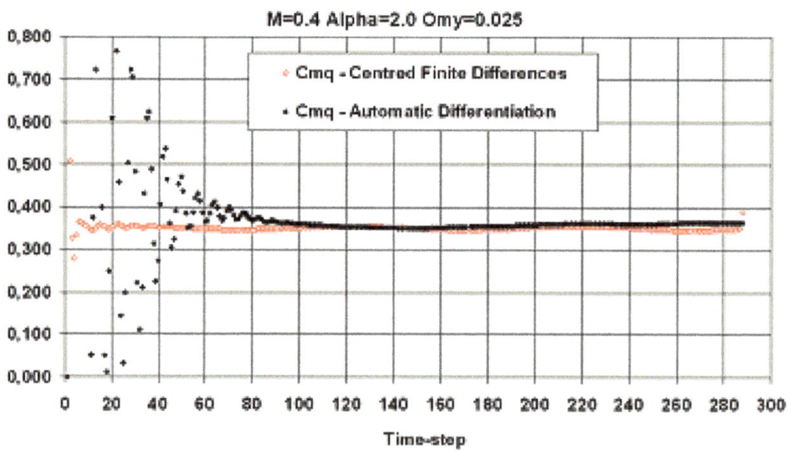

Fig. 9. Comparison between automatic differentiation and finite differences Cmq derivatives $\left(M = 0.4, \alpha = 2°, \Delta\alpha = 1°, \omega_y = 0.025 rad/s\right)$.

The following figures (Fig. 10 and 11), show the values of derivatives for the same flight condition, but with an oscillation frequency $\omega = 0.05$.

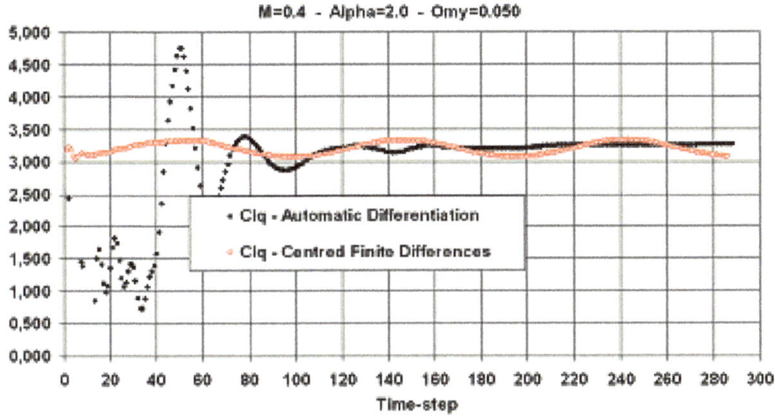

Fig. 10. Comparison between automatic differentiation and finite differences Clq derivatives $\left(M = 0.4, \alpha = 2°, \Delta\alpha = 1°, \omega_y = 0.050\, rad/s\right)$.

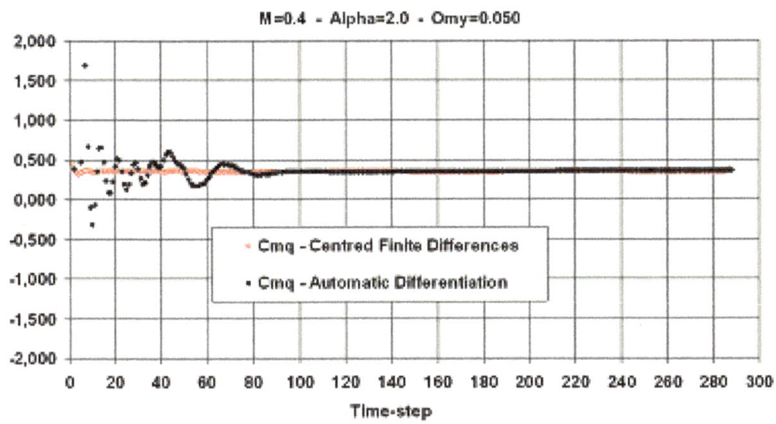

Fig. 11. Comparison between automatic differentiation and finite differences Cmq derivatives $\left(M = 0.4, \alpha = 2°, \Delta\alpha = 1°, \omega_y = 0.050\, rad/s\right)$

The overall progression of results is good, despite some local peaks. As a general behaviour the automatic differentiation provides results much more stable, because of its direct computation method based on a mathematical management of functions and not on a latter data handling with FD. Anyway some initial fluctuations have to be expected also in automatic differentiation, since anyway the relevant derivatives are managed by the augmented solver as normal functions which need some time to converge.

Finally, figures Fig. 12 and 13 show the evolution of C_{lq}, C_{dq}, C_{mq} as functions of frequencies.

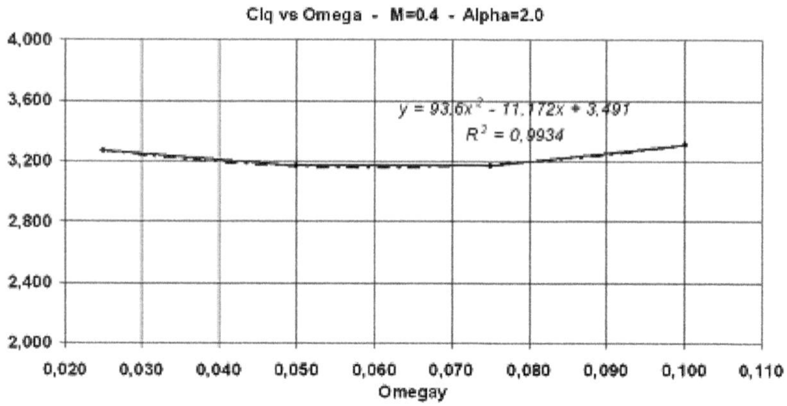

Fig. 12. Automatic differentiation Clq vs frequency progression $(M = 0.4, \alpha = 2°, \Delta\alpha = 1°)$

The relevant interpolating second order curves have been reported, as well as the least squares values computed as follows:

$$R = \sum_{i=1}^{n} [y_{i-AD} - f(x_i)] \tag{9}$$

The value of y_{i-AD} value is computed by using the AD, while $f(x_i)$ is computed by using the tendency function.

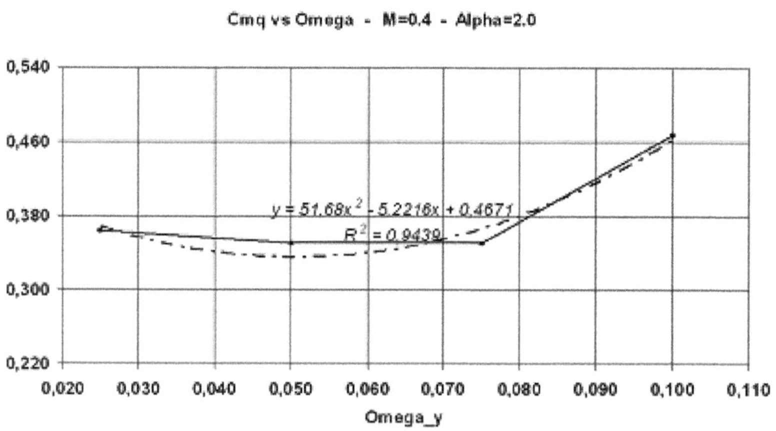

Fig. 13. AD Cmq vs frequency progression $(M = 0.4, \alpha = 2°, \Delta\alpha = 1°)$.

It is shown that all curves can be reduced to polynomials; however the physical evolution is well modelled by polynomials just in some cases, e.g. for Clq and Cmq, while in other cases (Cdq) such a reduction cannot be really accepted since curves are not smoothed and a polynomial reduction is only an algebraic manipulation

4.1.2 Variable incidence

The same approach described for studying the variable frequency has been applied to variable angles of attack. The most challenging dynamic behaviour is for $\omega = 0.1$, so some curves (Fig. 14-17) and tables are here discussed for such a frequency. The angle of attack ranges from $\alpha = 5°$ to $\alpha = 8°$.

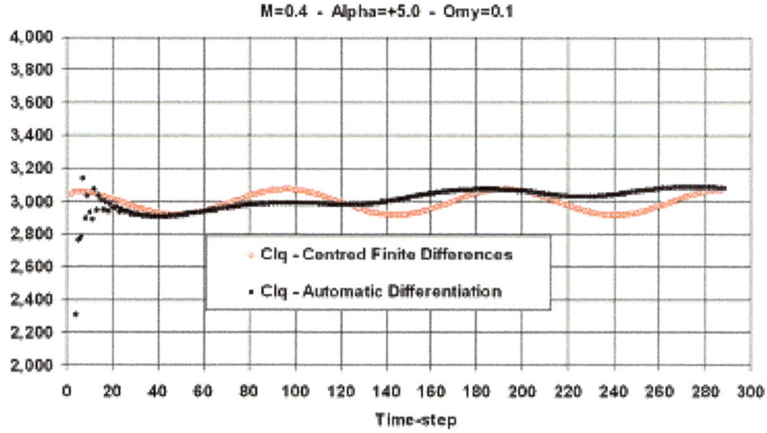

Fig. 14. Comparison between automatic differentiation and finite differences Clq derivatives $(M = 0.4, \alpha = 5°, \Delta\alpha = 1°, \omega_y = 0.100\, rad/s)$

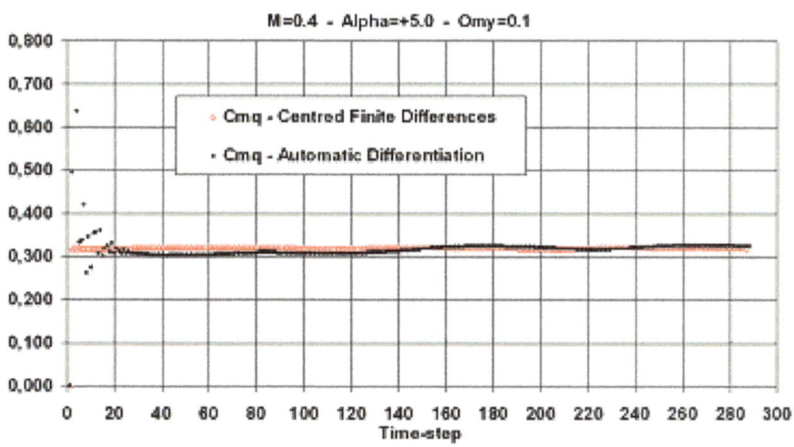

Fig. 15. Comparison between automatic differentiation and finite differences Cmq derivatives $(M = 0.4, \alpha = 5°, \Delta\alpha = 1°, \omega_y = 0.100\, rad/s)$.

For $\alpha = 5°$, the values of percentage errors for C_{l_q} show a strong variability, due to the high frequency. However, when the solution is stabilised, the percentage error between AD and

FD derivatives is not higher than 4.16 % (time-step 240) and in some points it is close to 0%. For C_{m_q}, the error reached after stabilisation is 2.03% (time step 260) so showing to be even more stable than C_{l_q}. Repeating the same analysis for $\alpha = 8°$, both C_{l_q} and C_{m_q} show a similar behaviour and anyway the percentage error between the two methods (AD and FD) is acceptable.

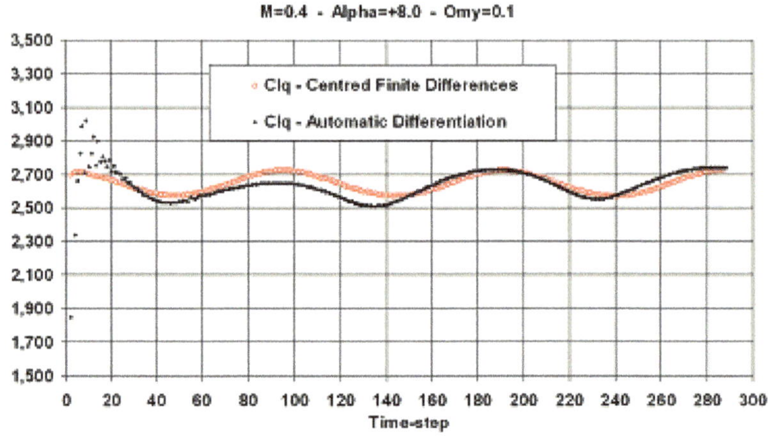

Fig. 16. Comparison between automatic differentiation and finite differences Clq derivatives $\left(M = 0.4, \alpha = 8°, \Delta\alpha = 1°, \omega_y = 0.100\, rad/s\right)$

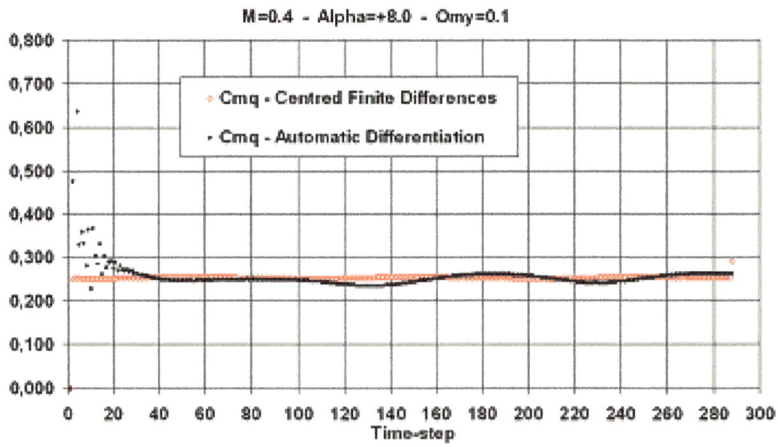

Fig. 17. Comparison between automatic differentiation and finite differences Cmq derivatives $\left(M = 0.4, \alpha = 8°, \Delta\alpha = 1°, \omega_y = 0.100\, rad/s\right)$.

Finally, Fig. 18 to Fig. 20 show the evolutions of C_{lq}, C_{dq}, C_{mq} as functions of the angle of attach, ranging in the interval $\alpha \in [-2°;+8°]$. The evolutions are well smoothened and their regularity is evident.

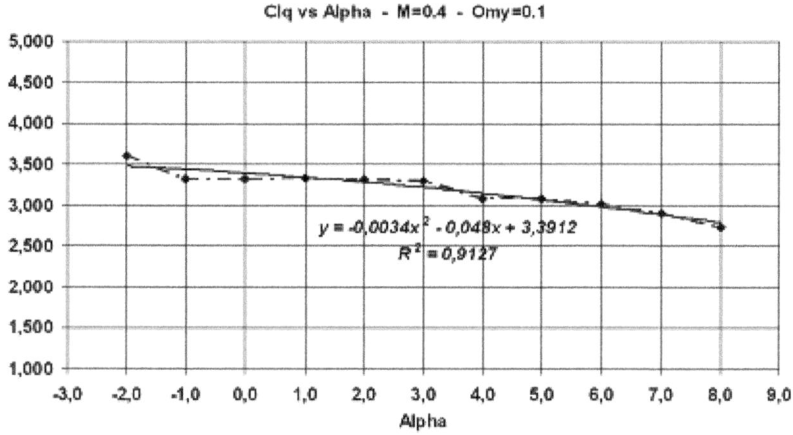

Fig. 18. Automatic differentiation Clq vs Alpha $\left(M = 0.4, \Delta\alpha = 1°, \omega_y = 0.100\, rad/s\right)$

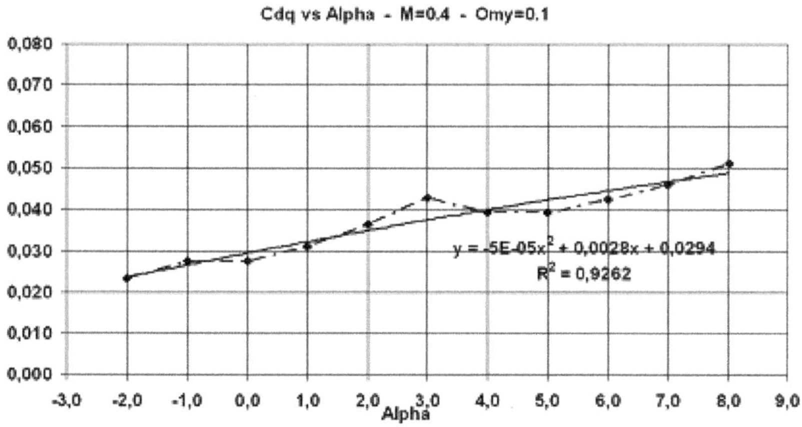

Fig. 19. Automatic differentiation Cdq vs Alpha $\left(M = 0.4, \Delta\alpha = 1°, \omega_y = 0.100\, rad/s\right)$.

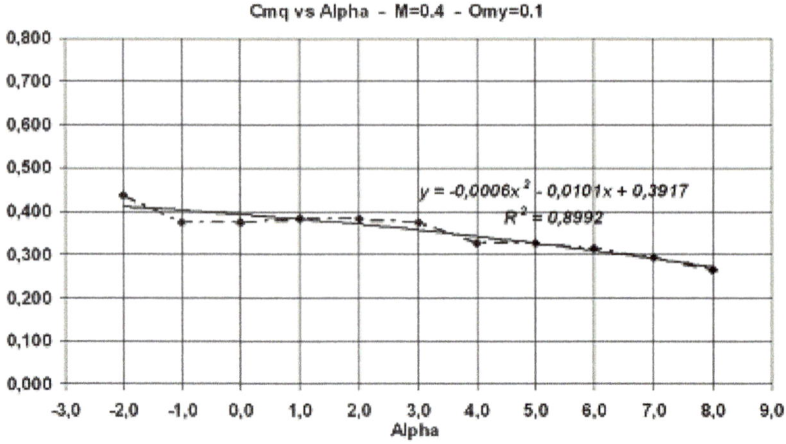

Fig. 20. Automatic differentiation Cmq vs Alpha $\left(M = 0.4, \Delta\alpha = 1°, \omega_y = 0.100\, rad/s\right)$.

4.2 F12

The second sequence of tests has been carried out on the DLR-F12 configuration, working in inviscid condition at $M = 0.202 \quad (u = 70 m/s)$. As already said, F12 has been one of the test benches used in SimSAC; within such a context, DLR and Cerfacs/ONERA generated two sets of meshes, unstructured and structured respectively, to be used by Partners in order to have a common computing basis. The Euler grid used by Alenia was the unstructured full configuration, 3802374 nodes, and 25263927 segments. The dimensions of the grid are the following:

$$-4.951e + 02 < \Delta x < 7.048e + 02$$

$$-5.998e + 02 < \Delta y < 5.998e + 02$$

$$-6.034e + 02 < \Delta z < 5.969e + 02$$

Figure 21 show two views of the overall mesh.

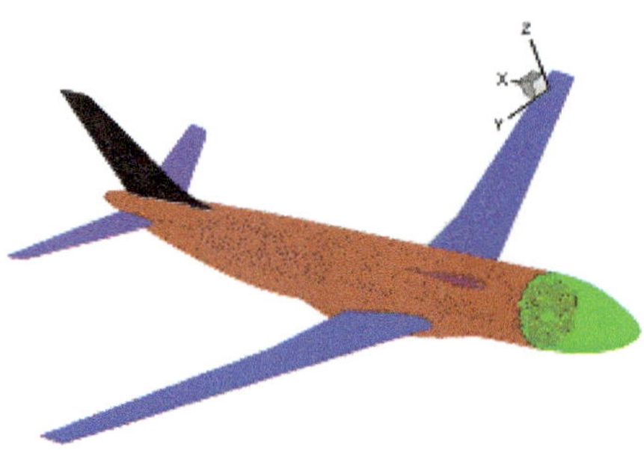

Fig. 21. F12 overall mesh.

An extensive numerical testing has been carried out by running 15000 iterations to reach the primitive steady solution PS1. After this, the unsteady solution has been computed by engaging the AD solver and running it for 15000 more iterations.
Computation has been run considering normal environmental conditions; in particular, the solver has been set to simulate a flight at $p = 1Atm, T = 25°C, u = 70m/s$. Numerical data have been compared to those acquired during the wind tunnel testing; the German-Dutch Wind Tunnels premise DNW-NWB, sited in Braunschweig, has been used for the acquisition of experimental data. Static testing included α- and β-sweeps. Figures 22 to 24 show the evolutions of computed lift, pitch moment and drag coefficients as functions of the angle of attack, compared to the experimental values. It is evident a large data spread, from different Partners, for angles of attack up to 8°, especially for pitch moment and drag. This is mainly due to the different numerical approaches used by different Partners. Fig. 25 to Fig. 27 show roll, pitch and yaw moments as functions of sideslip angle; in these cases, the data spread is evident just for pitch, while for roll and yaw moments data are much more similar with each other despite different numerical approaches. Figure 28 shows the sensitivity to the angle of attack; it is evident that, for an angle of attack 0°, the numerical result exactly overlaps the experimental one. This is the last result, computed in static condition, showed here. Figures 29 and 30 report dynamic sensitivities with respect to the pitch rate in the longitudinal plane, respectively for lift and pitch moment. Even in this case different numerical approaches lead to a wide range of results for the same configuration. Dynamic derivatives with respect to roll and yaw rates are shown in figures 31 to 36. Numerical data are in the latest cases much closer to experimental values and their spread is reduced to a minimum.

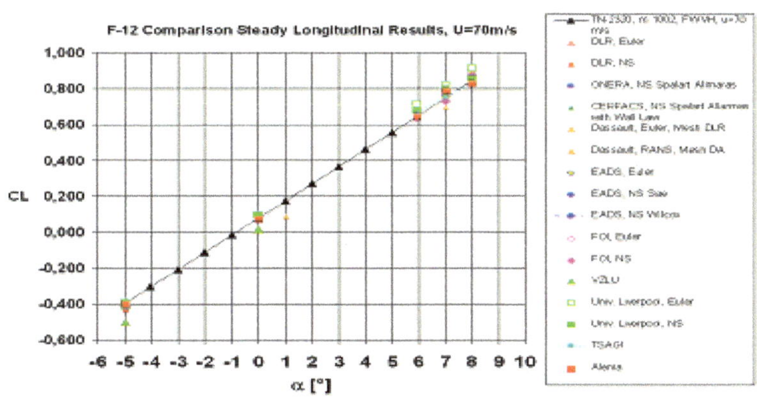

Fig. 22. CL Comparison - Numerical vs. Experimental Results.

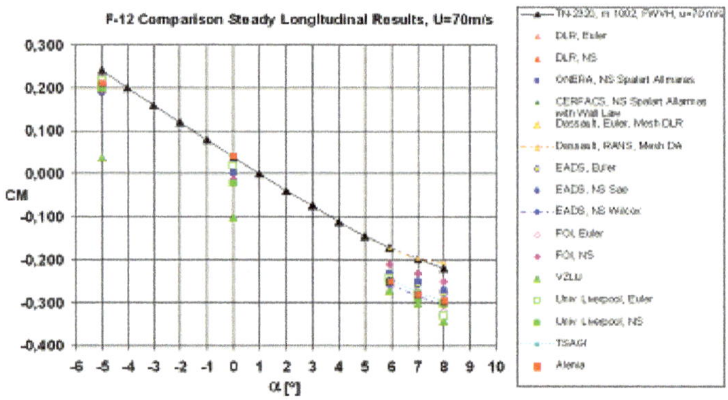

Fig. 23. CM Comparison – Numerical vs. Experimental Results.

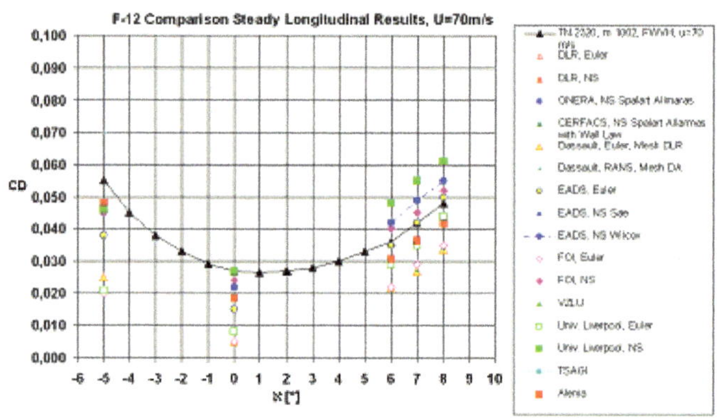

Fig. 24. CD Comparison – Numerical vs. Experimental Results.

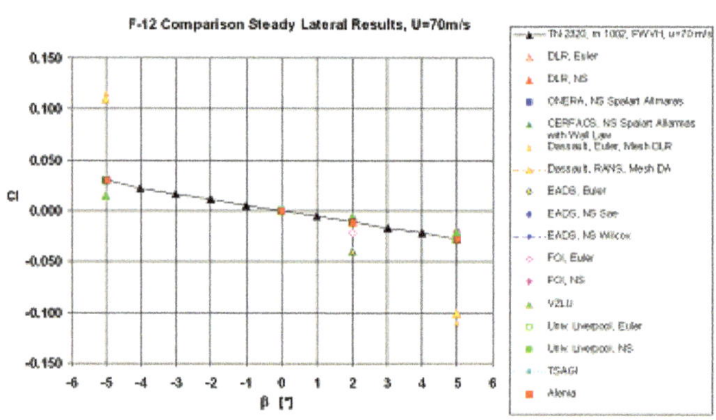

Fig. 25. Cl Comparison – Numerical vs. Experimental Results.

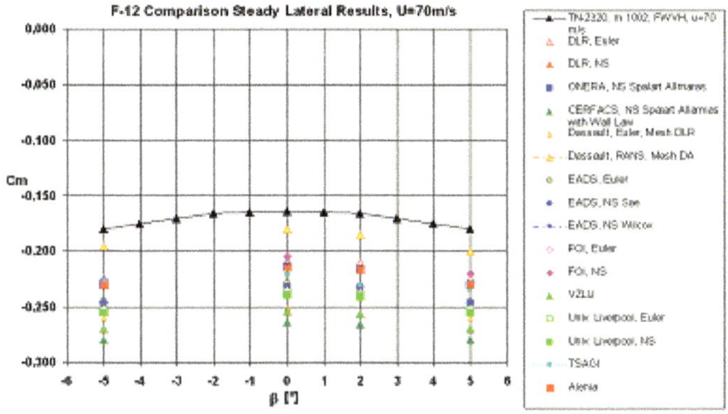

Fig. 26. Cm Comparison – Numerical vs. Experimental Results.

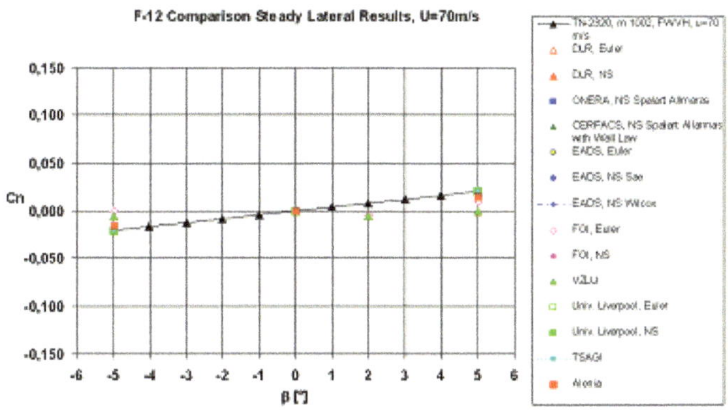

Fig. 27. Cn Comparison – Numerical vs. Experimental Results.

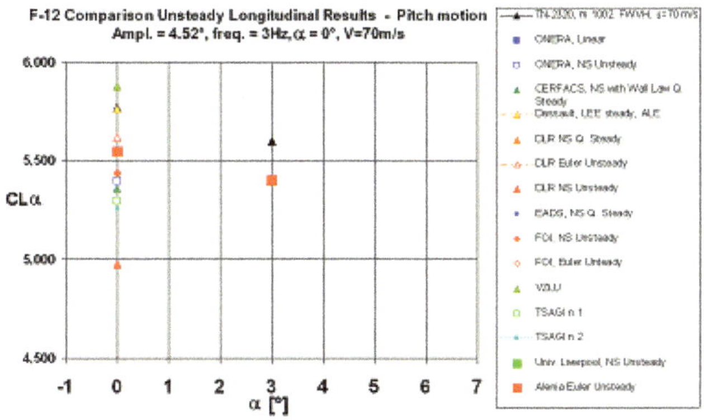

Fig. 28. CLα Comparison – Numerical vs. Experimental Results.

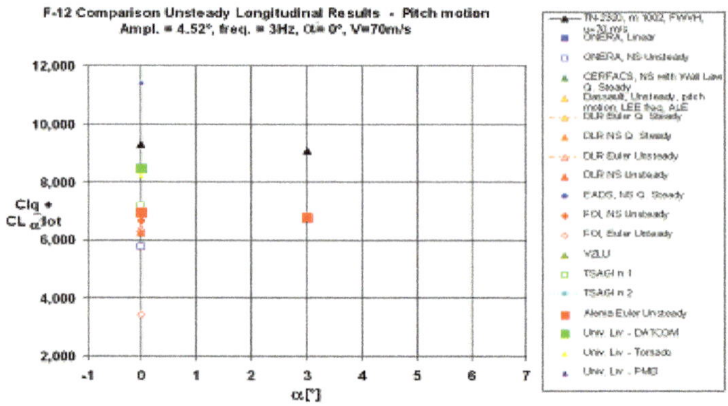

Fig. 29. CLq+CLαdot Comparison – Numerical vs. Experimental Results.

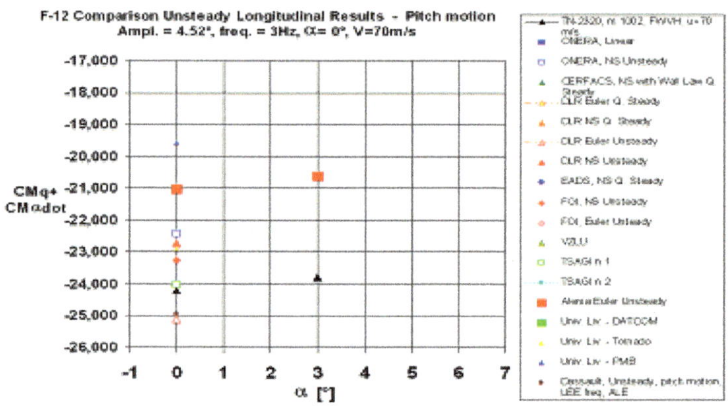

Fig. 30. CMq+CMαdot Comparison – Numerical vs. Experimental Results.

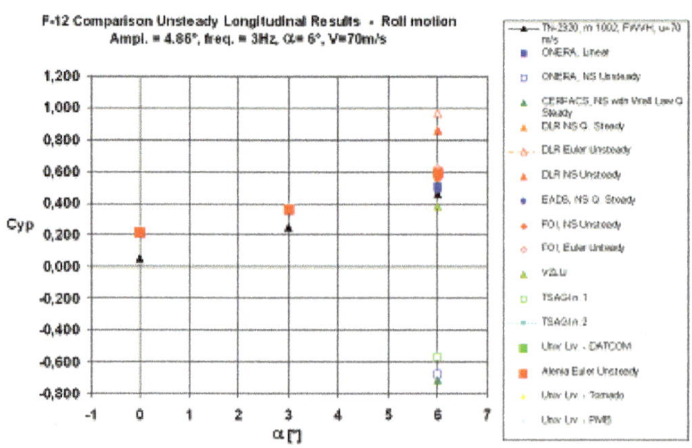

Fig. 31. Cyp Comparison – Numerical vs. Experimental Results.

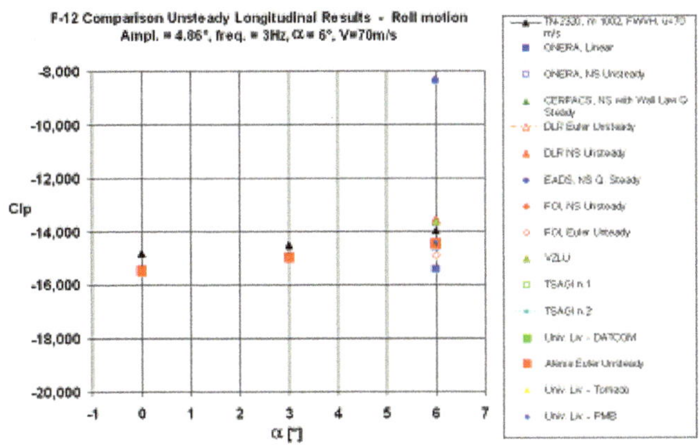

Fig. 32. Clp Comparison – Numerical vs. Experimental Results.

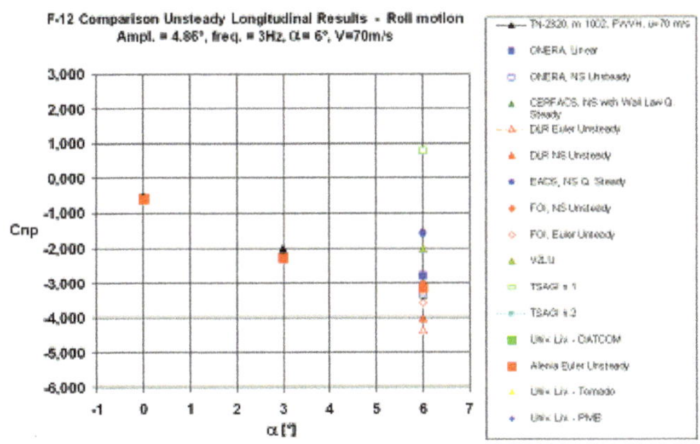

Fig. 33. Cnp Comparison – Numerical vs. Experimental Results.

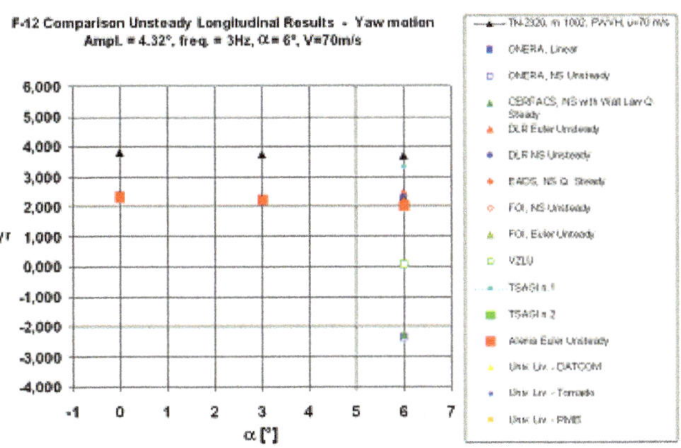

Fig. 34. Cyr Comparison – Numerical vs. Experimental Results.

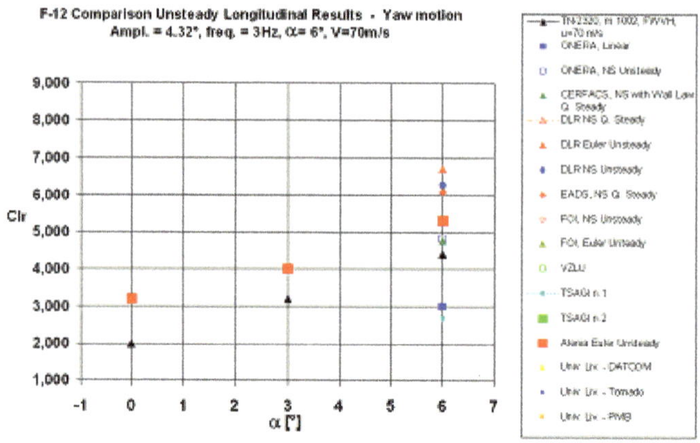

Fig. 35. Clr Comparison – Numerical vs. Experimental Results.

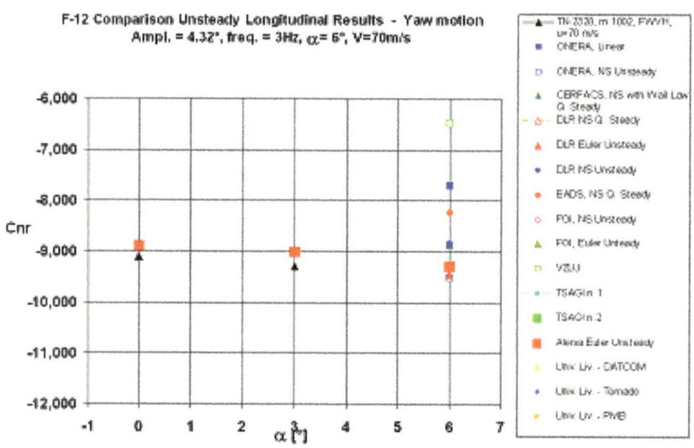

Fig. 36. Cnr Comparison – Numerical vs. Experimental Results.

5. Machine in use

All the runs have been carried out on a Quadrics multi-processor platform. The CPU is the AuthenticAMD working at 2594MHz. The system is made up of 7 computational nodes +1 node dedicated to the file system access. Each node has 4 processors. The physical memory of each node is 3943 MB and the virtual one is 8189 MB.

6. Conclusion

Automatic differentiation in static and dynamic condition has shown to be reliable for industrial application. Even if a comparison with the FD technique may be enough to qualify the AD approach, the final and definitive confirmation comes from the comparison with experimental data. According to what has been achieved, one can say that:

- AD is reliable;
- Alenia Aeronautica approach is reliable;
- AD provides a good alternative to the methods used by other Companies or Research Centres.

Following another work of the Authors (Ref. [12]) it is confirmed that a software reengineering activity is necessary after having generated the augmented code. This implies a cost in terms of time and deployed effort, for optimizing the augmented code lines, memory allocations and splitting the code in several components. Time saving achieved with AD is indeed remarkable if compared to other classical means to evaluate derivatives, and it provides an evaluation of the exact derivatives avoiding problems related to mesh refinement. Static and dynamic differencing procedures are clear enough to allow a daily use of AD features in the daily industrial activities. An extensive testing is now ongoing at Alenia premises in Turin in order to investigate:

- a faster procedure to obtain dynamic derivatives, avoiding three different computing phases;
- the application of AD for computing derivatives of higher order;
- the extension of AD to turbulence.

Further experiments related to the shape optimization, mesh adaptation and consequent resizing are currently in progress and are very promising.

7. Acknowledgments

The present work has been performed through a close cooperation between Politecnico di Torino, SimSAC project under funding from the Sixth Framework program of the European Union and Alenia Aeronautica S.p.A and through the use of Confidential Information and Data, property of Alenia Aeronautica S.p.A. which remains the sole owner of all such relevant IP rights.
The results of the present work shall be, therefore, property of Alenia Aeronautica S.p.A.

8. References

Almosnino, D. Aerodynamic Calculations of the Standard Dynamic Model in Pitch and Roll Oscillations, 32nd Aerospace Sciences Meeting & Exhibit, AIAA Paper 94-0287, January 1994.

Altun, M. & Iyigun, I. Dynamic Stability Derivatives of a Manoeuvring Combat Aircraft Model, Journal of Aeronautics and Space Technology, January 2004, Vol. I, No. 3, pp. 19-27.

Anderson, W. K., Newmann, J. C., Whitfield, D. L. & Nielsen, E. J. Sensitivity Analysis for the Navier-Stokes Equations on Unstructured Meshes Using Complex Variables, AIAA Paper 99-3294, June 1999.

Anderson, D. A., Tannehill, J. C. & Pletcher R. H., Computational Fluid Mechanics and Heat Transfer, 1st edition, McGraw-Hill, New York, 1984, pp. 39-46.

Guglieri, G. & Quagliotti, F. B. Dynamic Stability Derivatives Evaluation in a Low Speed Wind Tunnel, Journal of Aircraft, Vol. 30, No. 3, pp. 421-423, 1993.

Guglieri, G., Quagliotti & F. B., Scarabelli, P. L. Static and Oscillatory Experiments on the SDM at Politecnico di Torino, Nota Scientifica e Tecnica n. 74/93, 1993, Forced

Oscillation Technique-Reference Documentation, Vol. 3, Politecnico di Torino, DIASP.

Murphy, P. C. & Klein, V. Estimation of Aircraft Unsteady Aerodynamic Parameters from Dynamic Wind Tunnel Testing, AIAA Paper 2001-4016, 2001.

Newmann, J. C., Anderson, W. K. & Whitfield, D. L. Multidisciplinary Sensitivity Derivatives Using Complex Variables, MSSU-COE-ERC-98-09, Engineering Research Centre Report, Mississippi State University, July 1998.

Green, L., Spence, A. & Murphy, P. Computational Methods for Dynamics Stability and Control Derivatives, AIAA Paper 2004-0015, January 2004.

Hascoet, L. Analyses Statiques et Transformation de Programmes: de la parallélisation à la différentiation, Inria Database [online database], 2005, pp. 163-169, http://www-sop.inria.fr/members/Laurent.Hascoet/papers/ hdrHascoet5.html

Hascoet, L. & Pascual, L. Tapenade 2.1 User Guide, Hal-Inria Database [online database], September 2004, pp. 8-13, http://hal.inria.fr/docs/00/06/98/80/PDF/RT-0300.pdf [retrieved 19 May 2006].

Necci, C., Ceresola, N., Guglieri & G., Quagliotti, F. Industrial Computational Fluid Dynamics Tools for the Evaluation of Aerodynamic Coefficients, Journal of Aircraft Vol. 46, November-December 2009, pp. 1973-1983.

Selmin, V. Application of Automatic Differentiation to Aerodynamic Shape Optimization, ECCOMAS 2004, [online database],
http://www.imamod.ru/~serge/arc/conf/ECCOMAS-2004/ECCOMAS_V2/proceedings/pdf/611.pdf, [retrieved 28 July 2004]

Permissions

The contributors of this book come from diverse backgrounds, making this book a truly international effort. This book will bring forth new frontiers with its revolutionizing research information and detailed analysis of the nascent developments around the world.

We would like to thank Dr. Hyoung Woo Oh, for lending his expertise to make the book truly unique. He has played a crucial role in the development of this book. Without his invaluable contribution this book wouldn't have been possible. He has made vital efforts to compile up to date information on the varied aspects of this subject to make this book a valuable addition to the collection of many professionals and students.

This book was conceptualized with the vision of imparting up-to-date information and advanced data in this field. To ensure the same, a matchless editorial board was set up. Every individual on the board went through rigorous rounds of assessment to prove their worth. After which they invested a large part of their time researching and compiling the most relevant data for our readers. Conferences and sessions were held from time to time between the editorial board and the contributing authors to present the data in the most comprehensible form. The editorial team has worked tirelessly to provide valuable and valid information to help people across the globe.

Every chapter published in this book has been scrutinized by our experts. Their significance has been extensively debated. The topics covered herein carry significant findings which will fuel the growth of the discipline. They may even be implemented as practical applications or may be referred to as a beginning point for another development. Chapters in this book were first published by InTech; hereby published with permission under the Creative Commons Attribution License or equivalent.

The editorial board has been involved in producing this book since its inception. They have spent rigorous hours researching and exploring the diverse topics which have resulted in the successful publishing of this book. They have passed on their knowledge of decades through this book. To expedite this challenging task, the publisher supported the team at every step. A small team of assistant editors was also appointed to further simplify the editing procedure and attain best results for the readers.

Our editorial team has been hand-picked from every corner of the world. Their multi-ethnicity adds dynamic inputs to the discussions which result in innovative outcomes. These outcomes are then further discussed with the researchers and contributors who give their valuable feedback and opinion regarding the same. The feedback is then collaborated with the researches and they are edited in a comprehensive manner to aid the understanding of the subject.

Apart from the editorial board, the designing team has also invested a significant amount of their time in understanding the subject and creating the most relevant covers. They scrutinized every image to scout for the most suitable representation of the subject and create an appropriate cover for the book.

The publishing team has been involved in this book since its early stages. They were actively engaged in every process, be it collecting the data, connecting with the contributors or procuring relevant information. The team has been an ardent support to the editorial, designing and production team. Their endless efforts to recruit the best for this project, has resulted in the accomplishment of this book. They are a veteran in the field of academics and their pool of knowledge is as vast as their experience in printing. Their expertise and guidance has proved useful at every step. Their uncompromising quality standards have made this book an exceptional effort. Their encouragement from time to time has been an inspiration for everyone.

The publisher and the editorial board hope that this book will prove to be a valuable piece of knowledge for researchers, students, practitioners and scholars across the globe.

List of Contributors

Victor Udoewa
George Washington University, USAID Development Engineer, AAAS Science & Technology Policy Fellow, AAAS, 2009-2011, USA

Vinod Kumar
Mechanical Engineering, University of Texas at El Paso, USA

Ali Hadi Ghawi
Department of Civil Engineering, Faculty of Civil Engineering, AL-Qadisyia University, Iraq

Jozef Kriš
Department of Sanitary and Environmental Engineering, Faculty of Civil Engineering, Slovak University of Technology, Bratislava, Slovakia

Antonio Viviani
Seconda Università degli Studi di Napoli, Aversa, Italy

Giuseppe Pezzella
Centro Italiano Ricerche Aerospaziali, Capua, Italy

Jorge Flores-Velázquez, Irineo L. López-Cruz and Abraham Rojano-Aguilar
Department of Agricultural Engineering, University of Chapingo, Chapingo, Mexico

Guillermo De la Torre-Gea and Enrique Rico-García
Department of Biosystems, School of Engineering, Mexico Queretaro State University Querétaro, México

Xiana Romani Fernandez, Lars Egmont Spelter and Hermann Nirschl
Karlsruhe Institute of Technology, Campus South, Institute for Mechanical Process, Engineering and Mechanics, Karlsruhe, Germany

M. Muhasilovic
IPSA-Institute Sarajevo, Sarajevo, Bosnia and Herzegovina The CIM-Collaboration Program of the GiZ, German International Technology Agency, Germany

A. Mededovic and E. Gacanin
IPSA-Institute Sarajevo, Sarajevo, Bosnia and Herzegovina

K. Ciahotny and V. Koza
The VSCHT - Institute for Chemical Technology, Prague, Czech Republic

Samuel Santos Borges and Cassiano Antunes Cezario
Research and Technological Innovation Department, WEG, Brazil

Giuseppe Pezzella
Italian Aerospace Research Centre - CIRA, via Maiorise, Capua, Italy

Masroor Mohajerani, Mehrab Mehrvar and Farhad Ein-Mozaffari
Department of Chemical Engineering, Ryerson University, Toronto, Ontario, Canada

Zied Driss, Sarhan Karray, Hedi Kchaou and Mohamed Salah Abid
University of Sfax (US), Tunisia
National Engineering School of Sfax (ENIS), Tunisia Laboratory of
Electro-Mechanic Systems (LASEM), Tunisia

Avinash Khopkar
Dow Chemical International Pvt. Ltd., Pune, India

Michael D. Cloeter and Quan Yuan
The Dow Chemical Company, Freeport, TX, USA

Massimo Milani, Luca Montorsi and Fabrizio Paltrinieri
DISMI - University of Modena and Reggio Emilia, Italy

Rodrigo Béttega
Federal University of Uberlândia, School of Chemical Engineering, Brazil

Ronaldo Guimarães Corrêa and José Teixeira Freire
Federal University of São Carlos, Department of Chemical Engineering, Brazil

Wu-Chieh Wu
Institute of Mechanical and Electrical Engineering, National Taipei University of Technology, Taiwan, Republic of China

Nguyen Anh Tuan and K-David Huang
Network Operations Laboratory, Chunghwa Telecommunication Laboratories, Taiwan, Republic of China

Carlo Necci and Nicola Ceresola
Alenia Aeronautica SpA, Italy